SANDY HOOK

SANDY HOOK

AN AMERICAN TRAGEDY AND THE BATTLE FOR TRUTH

ELIZABETH WILLIAMSON

DUTTON

DUTTON
An imprint of Penguin Random House LLC
penguinrandomhouse.com

LIBRARY OF CONGRESS CATALOGING-IN-PUBLICATION DATA
has been applied for.

ISBN 9781524746575 (hardcover)
ISBN 9781524746599 (ebook)

Printed in the United States of America
1 3 5 7 9 10 8 6 4 2

Book Design by Ellen Cipriano

People should try to treasure each other more. Because life eventually disposes of everyone, and you don't know when it's going to happen . . . Sometimes I'm so tired, I have to force myself to do that. To do that extra thing. But then I'm like, "Remember."

—Veronique De La Rosa, mother of Noah Pozner

John J. Crowder
1963–2005
Rest in peace

CONTENTS

AUTHOR'S NOTE

THIS BOOK DOCUMENTS a battle by victims' families against deluded people and profiteers who denied the December 14, 2012, shooting at Sandy Hook Elementary School in Newtown, Connecticut, that killed twenty first graders and six educators.

The book traces a nearly ten-year effort pioneered by Leonard Pozner, whose six-year-old son, Noah, died at Sandy Hook, to sound the alarm about the growing threat posed by viral lies and false conspiracy narratives, a cultural phenomenon that eventually brought a mob to the United States Capitol on January 6, 2021.

These families' saga, and its societal implications and potential solutions, rests on more than four hundred interviews, including with Sandy Hook survivors, first responders, government officials, lawyers, researchers, political scientists, psychologists, journalists, conspiracists, and others, conducted over three years. My reporting on the exploitation of the shooting by profiteers like Alex Jones of Infowars and others is based on some ten thousand pages of court testimony; business, financial, police, medical, and court records; internal emails surfaced during legal proceedings; online exchanges, videos, and recordings of personal conversations, interviews, meetings, and courtroom proceedings I observed. I have traveled to Newtown more than a dozen times to interview participants and visit places described in the book.

I did not write about the Sandy Hook shooting at the time of the tragedy. I began researching this book in April 2018, when Noah Pozner's parents, Lenny Pozner and Veronique De La Rosa, and Neil Heslin, the father of Jesse Lewis, who also perished in the shooting, sued Alex Jones for defamation in Texas. My first *New York Times* story about the shooting, "Truth in a Post-Truth Era: Sandy Hook Families Sue Alex Jones, Conspiracy Theorist," appeared on May 23, 2018, on the day a separate group of victims' families and an FBI agent targeted by Sandy Hook conspiracists sued Jones in Connecticut. I have covered the lawsuits and the Sandy Hook conspiracy phenomenon for the *Times* since then.

The Sandy Hook family members' activities on the day of the tragedy and during the years following are based on my interviews and correspondence with them; the family members' own writings in books, in articles, or on social media; their media appearances and coverage; interviews with individuals who interacted with them at the time and thereafter; court testimony; and the work of academic researchers who met with and/or surveyed family members. The endnotes provide a guide. In referencing the Sandy Hook families as a collective, it is not my intention to attribute any action, emotion, opinion, or perception to the entire group. Their experiences and interpretation of the events described are highly individual and to be respected.

As I embarked on the book, and again closer to publication, I communicated with the broad group of Sandy Hook families, via an email sent through their designated family liaison, to explain the project and offer to answer their questions about it.

This book is not a treatise on gun policy in America. That is a distinct issue properly addressed by experts, advocates and lawmakers, many of whom have written books on the subject. The gun debate and the firearms industry's response to mass shootings appear only as context for the spread of false narratives. The Sandy Hook family members presented in this book hold a range of views on guns in America. This book does not explore them except as they relate to conspiracists' claims about and attacks on them. Mass shooting conspiracy theorists often use survivors' gun policy views—actual, assumed, or fabricated—to bolster false narratives. My choice to leave that debate to others is intended in part to deprive them of that opportunity. I know from years spent talking with people who believe conspiracy theories that this likely will not work, but that is my intent.

I reviewed sections of the book pertaining to the experiences, recollections, and emotions of the Sandy Hook families and others at the center of the book with them, in an effort to ensure that I portrayed their stories as accurately as possible. In reporting this book I was guided by recommendations from the Columbia University Graduate School of Journalism's Dart Center for Journalism and Trauma, and professionals in post-trauma counseling. In referencing the perpetrators of the Sandy Hook and other mass shootings mentioned in the book, I was guided by principles established by No Notoriety, founded by Caren and Tom Teves, whose twenty-four-year-old son, Alex Teves, died in the 2012 Aurora theater shooting in Colorado while shielding his girlfriend from gunfire. No Notoriety calls on the media to limit gratuitous use of the killers' names to discourage copycat attacks and to shift the focus from perpetrators to victims, survivors, and helpers.

I denote my personal observations, opinions, and views in the text by describing them in the first person. All errors are mine alone.

No culture can rest on a crooked relationship to truth.

—*Robert Musil,* The Man Without Qualities
(1930–1943)

PROLOGUE

TWO DAYS AFTER THE SHOOTING, before the funerals began, President Barack Obama arrived to plead with the families of the dead and their neighbors: "Do not lose heart."

The president drew on Newtown's history, its traditions, and the stories of heroism during the shooting to fortify them.

"As a community you've inspired us, Newtown. In the face of indescribable violence, in the face of unconscionable evil, you've looked out for each other. You've cared for one another. And you've loved one another. This is how Newtown will be remembered, and with time and God's grace, that love will see you through."[1]

In the darkened auditorium of Newtown High School that night, the president's words to Newtown sowed comfort, acknowledged anger, and brought the dawning of a desolate fact. More than three hundred years after Newtown's founding, technology with the ubiquitous, vaguely whimsical name Google would forever link their town with the murder of children. Unimaginably, in time the words "Sandy Hook" would come to symbolize a sinister development in America's cultural and technological history, transforming a massacre at school into a battle for truth. In an age when facts have never been more knowable and accessible, some Americans would insist that nobody died at Sandy Hook Elementary School and that everyone involved, from the children and educators to those who tried to save them, were actors in a government-led hoax. Driven by ideology or profit, or for no sound reason at all, conspiracy theorists would use technology created to unite the world to hunt and attack vulnerable people.

It has happened many times since, but Sandy Hook was the first mass tragedy to spawn an online circle of people impermeable and hostile to reality and its messengers, whether the mainstream media, law enforcement, or the families of the dead, for whom the torment by deniers added immeasurably to their pain.

From a decade's distance, Sandy Hook stands as a portent: a warning of the power of unquenchable viral lies to leap the firewalls of decency and tradition, to engulf accepted fact and established science, and to lap at the foundations of our democratic institutions.

This book tells the truth of how that happened.

AT NINE-THIRTY on the frigid morning of December 14, 2012, as twenty-year-old Adam Lanza drove down forest-fringed Dickinson Drive toward his former elementary school, Sandy Hook, Veronique Pozner bustled through hallways forty miles away, well into a busy Friday at Grove Hill Medical Center in New Britain, Connecticut. An oncology nurse, Veronique tended to patients in a sedate, lounge-style chemotherapy infusion room, its warm air dry and snapping with static electricity. A TV mounted to the wall chirped morning-show pap: holiday gift ideas, a cooking demo, yet another feature on PSY, the South Korean rapper whose "Gangnam Style" video Veronique's six-year-old son, Noah, played nonstop at home.

Noah's mother called him a "one-man tornado," a chatty, rambunctious little man with big eyes fringed by inky lashes, and a cherub's full lips. His classmates gave him a hand-decorated T-shirt for his sixth birthday that November 20, drawing hearts and stars in rainbow marker around their names: Joey, Daniel, Charlotte, Emilie, Grace, James, Sammy, Ana, Caroline, Chase, Ben, Maddie, Jack, Catherine, Jessica. A week later Noah lost his first tooth and posed for a photo, mouth open to show the tiny red gap in the bottom row. He loved karate, chocolate fondue, the beach, superheroes, and scaring people with the lifelike plastic insects his parents bought for him at the Rainforest Café in Orlando. A photo captures him on Halloween as Batman, grinning behind his blue plastic mask, his child's belly puffed proudly over a cloth utility belt.

At Noah's school, the high-pitched din of 350 children, kindergarten through fourth grade, excited about the coming holidays, filtered through the hallways and into the parking lot as Lanza parked his mother's black Honda Civic precisely, then walked the few paces to the door. He wore sunglasses and a black bucket-style brimmed hat, a black polo-style shirt with a black T-shirt underneath. He also wore an olive-drab fishing-type vest, whose pockets he'd stuffed with ammunition. He carried a Glock 10 mm handgun in the pocket of his black cargo pants, cinched tightly with a web belt to hold them up on his skinny frame. He had strapped another gun, a Sig Sauer 9 mm, to his leg.[2] In his hands, sheathed in fingerless gloves, he carried a Bushmaster AR-15–style rifle, similar to the one he'd used an hour earlier to murder his mother, Nancy Lanza, shooting her four times while she slept. Nancy had bought the rifle her son used to kill her, along with the Bushmaster and the two handguns. Six feet tall, Lanza weighed only 112 pounds. In his mind, starvation equaled self-control and power. He wore earplugs.[3]

Classes having started, the school's front door was locked. Lanza didn't touch the door handle; germs terrified him. Instead he shot out the wall-sized window to the right of the front door and stepped through, his black leather Nunn Bush sneakers crunching on glittering cubes of safety glass as he passed through the airy front lobby.

AT ABOUT 9:45 A.M., an emergency text message popped up on Veronique's cell phone. False alarm, she thought, but then the television began flashing sketchy details of a shooter inside Sandy Hook Elementary. Veronique's three youngest children went there: seven-year-old Sophia, and her twin babies, Noah and Arielle. Somebody—a patient, maybe a doctor—urged her *go*, as if she needed to be told. Filled with a primal terror, "like being in a trap," she tore from the building into the frigid air, wearing just her short-sleeved scrubs. She sped southwest on I-84. The engine light flashed on, and she prayed the car wouldn't stall out.

When she reached Sandy Hook thirty-five minutes later, she found the school cordoned off, a bedlam of police and crowds and cameras swirling around the placid village she'd left four hours earlier. She ditched the car and ran for the school. Intercepted by the police, she muscled her way through crowds to the firehouse, where her husband, Lenny, and dozens of parents had gathered. Lenny had found Sophia and Arielle in the parking lot, lined up with their classes. They stood with him, silent and withdrawn. Together the Pozners numbly watched the reunions, frantic parents grabbing children who wept and clutched at them or who stood with arms at their sides, shocked and staring. Noah did not appear.

SERGEANT BILL CARIO had run into the school with the first wave of police. In the lobby the Connecticut state trooper passed the bodies of the school principal, Dawn Lafferty Hochsprung, who had rushed into the hallway after hearing the shots in the lobby and confronted Lanza, and the school psychologist, Mary Sherlach, who had followed her. In a conference room on his right, Cario found Natalie Hammond, shot in the leg and hand, who'd played dead while Lanza stepped over her, then scrambled back inside the conference room where she, Dawn, and Mary had been having a meeting. Natalie had been holding the door closed while the shooting raged. Cario told her to stay put, and he would be back for her. He ran on, following a Newtown police officer into classroom 10, where he found Lanza, lying in a fetal position, so thin the trooper at first thought he was a child. His hat lay nearby, blown off his head by the shot from the Glock he'd used to end his life.

"His injury was not consistent with life and I did not check him for vitals or remove the weapons from him," Cario wrote in his report. The first victim he saw in classroom 10 was Vicki Soto, lying beyond the gunman and closer to the windows on the gray-carpeted floor. Nearby lay three of her students, including Jesse Lewis, a stocky little guy with a big voice who in his last moments yelled, "Run!" Cario dropped to his knees, checking, but none had a pulse. Not far from Jesse lay Dylan Hockley. He had autism and seldom spoke but had a treasured connection with his behavioral therapist, Anne Marie Murphy. She died with her arms around Dylan, shielding him. Moving more slowly now, Cario entered classroom 8. At first it looked empty, a relief. Then he saw the bodies of two teachers on the floor. Opening the door to what he thought was a closet, he saw a damp heap of cloth that in his shock he mistook for some kind of art project.

Describing what happened next in his report, Cario's detached official language turned ragged and anguished. "As I stared in disbelief, I recognized the face of a little boy . . . I then began to realize that there were other children around the little boy, and that this was actually a pile of dead children.

"The face of the little boy is the only specific image I have in that room."

IT HAD TAKEN THE GUNMAN less than ten minutes to murder twenty first graders and six educators, destruction that would radiate, like fallout, from those moments through years.

That night police erected klieg lights, like the kind used on nighttime construction projects, in the Sandy Hook school parking lot. Their glare illuminated the long, pale-brown tent where H. Wayne Carver, the Connecticut chief medical examiner, waited with his staff. Carver knew it would be easier if the families identified their loved ones from photographs, his team laboring to keep the worst of their wounds outside the frame.

A slow procession of stretchers and gurneys passed through the bullet-pocked front lobby, past a toppled houseplant, chairs sprinkled with broken glass, and a decorated bin labeled "Bags, Bears, Books & Basics," collecting holiday donations for Newtown's needy children. They moved down the terrazzo hallway, its glazed-brick walls festooned with paper snowdrifts and sparkling candy canes. Inside classroom 8, a bulletin board titled HOPES AND DREAMS displayed the first graders' Crayola images of their future selves.

Noah Pozner lay faceup, Batman's shield on his sweatshirt no protection against the bullet that passed through his back. Fourteen of his classmates died with him inside their classroom's bathroom, a 4'7"-by-3'6" space into which Lanza fired more than eighty rounds, while shouting and laughing. Only one of the children who signed Noah's happy birthday T-shirt survived, shielded by the bodies of her friends.

"You feel something snap," Veronique said, about a time that forever changed her family, the school and its families, the town and then the country. "And you know that everything that came before is going to have no impact on what's going to come after. Because you're living a totally different life now. And there's no going back."

1

NEWTOWN, CONNECTICUT
JANUARY 2019

"DO YOU SEE THE STARS?"

It was nearly dusk, the end of our car tour of Newtown, when Neil Heslin directed me to stop on a hilltop side street off Riverside Road. Below and across Riverside stands the volunteer firehouse, and down a long drive beyond it, Sandy Hook Elementary School.

It had been seven years since the shooting. Neil's question seemed at first like an effort to lighten the mood. Our conversation had been grim to that point, and I had noticed my mind edging away, like the lazy V of geese I saw retreating toward the silhouette tree line. I smiled and craned my head to peer through the windshield, then saw that Neil was pointing across the road to the sloping roof of the firehouse. Arranged in waves suggesting a constellation, twenty-six copper barnstars gleamed dully in the fading light. One for each of the twenty children and for the six educators who died trying to protect them. The five-pointed stars were so large—seven feet across for the educators', five feet for the children's, that I was surprised not to have noticed them right away. A local carpenter named Greg Gnandt, whose cousin was a firefighter for the Sandy Hook Volunteer Fire & Rescue, made them in his shop, coating them with lacquer to protect their shine. About two weeks after the shooting,[1] Gnandt and a crew of volunteers mounted them to the roof of the firehouse, the building where the governor had told twenty-six waiting families that no one else had survived.

One of the stars honors Jesse Lewis, Neil Heslin and Scarlett Lewis's six-year-old boy. He was Scarlett's second-born, the little brother of J.T., six years older. Jesse was Neil's only child and, as he has often said, his only family.

Jesse's star is positioned "front and center, just like he would be," Scarlett once told me. Scarlett moved from Fayetteville, Arkansas, to Sandy Hook in 1998. She grew up in Darien, Connecticut. An avid horsewoman and artist, she sought a quieter life in Sandy Hook with more room for both. She bought a picturesque spread on Great Ring Road, with a creamy yellow eighteenth-century clapboard house and a red barn for her horses, and named it Wild Rose Farm. Scarlett worked then for a technology firm in New York that had created a financial trading system, a job that involved visiting the financial houses that had flocked into Fairfield County. In 2000, two years after she moved to Sandy Hook, her son J.T. was born, but her relationship with his father frayed a few years later. Scarlett met Neil in the Blue Heron, an antique shop in Milford owned by her mother, Maureen, who had moved to Newtown to be near Scarlett and J.T. Neil had grown up in nearby Shelton and had a small construction business. They struck up a conversation about a tree that had come down on her farm, and Neil offered to remove it for her.

"Neil was really fun," Scarlett told me, a problem solver and good-natured partner in part-time farming. She wanted sheep and bought a "spinner's flock," one in every color. Thanks to inexperience and bad timing, they wound up raising lambs born in January inside the farmhouse, drying them and keeping them warm with a hair dryer. "We had lots of farm adventures," Scarlett said, laughing at the memory.

Their son, Jesse, was born in 2006. Neil and Scarlett split up when Jesse was a baby, but they shared parenting.

Jesse had deep brown, almond-shaped eyes, a broad smile, and a deep voice for a kid that age, bellowing "Heeeere's Jesse!" when he entered a room. He liked to play "army guys," wearing a flopping plastic camouflage helmet and rubber boots on missions around the farm, where he kept a shaggy burro named Turquoise, his pony named Chocolate, and his and J.T.'s drooling mastiff pup, Remington.

Jesse had stayed with Neil the day before the shooting, in the house Neil inherited from his parents in Shelton, about seven miles southeast of Scarlett's place, a modest ranch with a deep, wooded front yard. About 7:30 in the morning on Wednesday, December 12, Neil drove up Scarlett's gravel drive, past the red barn, to collect Jesse. While his parents chatted, Jesse scratched "I love you" into the light frost coating Scarlett's car door, surrounding the message with hearts.[2] Scarlett told him to stay put and ran into the house for her camera. Her last photo of Jesse is of him squinting in the morning sun, showing off his handiwork. His ski jacket was open, revealing an untucked rugby jersey with blue and black stripes, the shirt he died in.

NEIL IS TALL with a rangy build, partial to jeans, square-toed Western boots, and work shirts, more comfortable outdoors than in. His words emerge in slow eddies, his deferential demeanor accentuated by a shrug, a tic that's more noticeable when visiting uncomfortable topics. When angry, he speaks in flinty bursts punctuated by profanity.

Neil carried his grief like hard water in a metal container, blunting its corrosive power by staying in motion, starting projects, fixing things.

When Scarlett's livestock gate swung open one night while she was traveling and her horses strayed into the road, he went over to find and pen them. When Heart, her favorite old mare, died, he brought his backhoe and buried the horse on her land. One recent, difficult Mother's Day, he turned up at the farm to help her put in a kitchen garden. On this day he served as my guide to Newtown, pointing out the landmarks of his loss.

We talked best together in the car. Neil gave directions and thought aloud while I drove, his recollections flowing out while he gazed through the window, his face angled so hard against the glass sometimes that it fogged. He's one of the survivors who refer to the shooting as "the tragedy." Others call it "12/14," a term that, like "9/11," denotes a reverberating catastrophe. Any reference to the shooting by less sheltering terms sounds to Newtown ears like an obscenity, particularly when uttered by outsiders who did not live through it.

Newtown's story began in 1705, when settlers from Connecticut and New York "purchased" from its Native inhabitants an eight-mile-long, roughly six-mile-wide parcel on the northeast edge of Fairfield County. The site hugged the Housatonic and Pootatuck Rivers, the latter named for the Native people who traded their farming and fishing grounds for a small haul of goods, including four guns.

Newtown looks like what you might imagine when someone says "picturesque New England town." Curving narrow roads lined with low rock walls; grassy, misty culverts radiating from its downtown; a colonial hilltop idyll, distinguished even from afar by a steepled white clapboard church and the hundred-foot flagpole, a beloved local hazard, that in a fit of patriotic fervor during the 1876 American centennial was planted smack in the center of Main Street. Newtown also has its share of fast-food places and boxy tract houses, testimony to the New York exurban sprawl that nearly quadrupled its population between 1950 and 2010.

In 1956, on a vacant, scrubby parcel at the intersection of its colonial past and suburban future, Newtown built a new school for the children of the baby boom. Sandy Hook Elementary was designed to shield its young students from mid-century parents' chief terror: fire. Just two years prior, a blaze in a wooden annex at Cleveland Hill Elementary School in Cheektowaga, New York, had killed fifteen children,[3] prompting urgent calls[4] for new building standards. Sandy Hook school was brick, built low and squat, with big windows, walls of pastel-glazed brick, and terrazzo floors speckled like birds' eggs. In a last, proud touch, the builders marked the wooded drive leading to its entrance with a white wooden shingle hanging from iron brackets. SANDY HOOK SCHOOL 1956, it announced. VISITORS WELCOME.

Before December 14, 2012, taxes and unbridled growth, not crime or violence, remained Newtown's principal preoccupations. In a book marking Newtown's tercentennial in 2005, old-timers lamented the loss of farms and forests, portraying "more homes, lawns and roads" as the chief threat to Newtown's treasured ordinariness. If only.

NEIL IS NOT GIVEN to dark theorizing or magical thinking. But after Jesse's death, his search for meaning took him to places that his logical mind knew held no answers. Why, of all the mayhem-plagued hellscapes in America, had the shooting happened in this serene, prosperous place, which Pat Llodra, the town's first selectman, once called "close to big-city culture, without any of the big-city problems," and Neil called "Mayberry"? In Newtown the right way to live was literally mapped out on a poster that hung in the front lobby of Sandy Hook Elementary School on the day Lanza blew out the window. It depicted a tree and was titled "Cultivating Character: Newtown's Core Character Attributes." At the tree's roots, "Perseverance." Along its trunk, "Citizenship." Springing from that sturdy stem, "Caring," "Respect," "Trustworthiness," and "Responsibility."

What if General Electric had not moved from Manhattan to Fairfield County in the early 1970s, bringing scores of executives to Newtown? Maybe then Peter Lanza, his wife, Nancy, and their troubled younger son, Adam, would never have moved into a roomy subdivision whose children attended Sandy Hook.

We drove past the rolling grounds and empty red-brick colonial buildings of the former Fairfield Hills mental hospital, their white-painted colonnades peeling, broken windows gaping like sightless eyes. Fairfield State Hospital accepted its first residents in 1933. At its peak it housed four thousand patients on its eight-hundred-acre campus and employed hundreds of locals. It closed in 1995, an anachronism in an age of outpatient treatment. Newtown bought a chunk of the campus to house municipal operations and other potential projects.

Fairfield Hills' brooding emptiness lured urban explorers and ghost hunters, who prowled through its network of underground tunnels. Videos of the complex's deserted morgue appeared on blogs and later, on the nascent video-sharing site YouTube, narrators hyperventilating about phantom psychotic screams and sightings. Newtown responded to its first brush with internet-borne delusion by arresting trespassers and tearing down the morgue.[5] The *Newtown Bee* newspaper, since its founding in 1877 a civic guardian against "gossip, grudges, hate and scandal," posted a video of the demolition to YouTube.[6]

Neil marveled at the irony. His son was murdered in the exact town where two decades earlier a disturbed young man might have been hospitalized and supervised. If Fairfield Hills had remained open, maybe his son's killer would have been doing chores on Fairfield Hills' dairy farm, instead of holing up in the basement of a house with an arsenal, searching up mass shootings on Google and plugging them into a spreadsheet.[7]

Of all his what-if questions, the most torturous swirled around

Neil's last day with Jesse. His son had seemed uncharacteristically quiet, even sad. Did Jesse somehow sense his fate? Could Neil have done something differently and saved his life?

NEIL LET JESSE SLEEP in a little longer on Friday morning. When Jesse awoke, "he wasn't really himself, it didn't seem like," Neil recalled. "He just seemed a little more withdrawn—I don't know how you would describe it." Neil also felt a little down. His mother had died five years earlier on that day, December 14. Neil glanced at a framed photograph on the living room end table of his mother and an infant Jesse at Christmas. He had recently found Jesse gazing at it too.

"I don't remember Grandma," Jesse told him.

"'Well, you were young,'" Neil replied. Then "Jesse went back and looked at the photo and studied it, really studied it. And he said, 'I'll know Grandma when I see her.'

"It was strange. Ah, it was just strange."

Jesse dawdled that morning. Neil told him that if they hurried, they could stop at the Misty Vale Deli for breakfast on the way to school. They hustled to get dressed, and Jesse grabbed his backpack, printed with Pixar characters from the Disney movie *Cars*. That day it was stuffed with crumpled papers and a couple of library books, and his snack, a plastic bag of orange sections. After breakfast, driving down winding roads in brilliant sunshine, Jesse still seemed off-kilter. Neil pointed out all the things they had to look forward to, starting at two o'clock that afternoon, when he and Scarlett would join the first graders in decorating gingerbread houses.

Again, "I asked if he was all right. And I thought, 'I wonder if I should just keep him home?'" Neil told me.

Neil dropped Jesse off at school. "He gave me a big hug, which I can still feel, and told me, 'Everything's gonna be okay,'" Neil recalled.

Taken aback by that strangely adult reassurance, Neil reminded Jesse that he'd be back in just a couple of hours for the gingerbread-house party. But Jesse shook his head, telling his father, "'That's not gonna happen. We're not doing that today.'

"That was that. I walked him in, the bell went off at 9:04, he hugged me again," Neil said. Looking his father in the face, his hands on his shoulders, "he said, 'I love you. I love Mom too,' and kind of darted around the corner," Neil recalled.

"If we weren't making those damned gingerbreads, I probably would have kept him home," Neil said, adding, as he often does, "It is what it is."

Neil was running errands around Newtown shortly after ten that morning when he got the first emergency text message from Sandy Hook Elementary, the same type it sends for a snow day. "It said the school's on lockdown, and a couple of minutes later it said the school was on lockdown because there was a shooting in town," he recalled. Not overly alarming, except for the shooting part. After 9/11, schools tended to lock down even for distant threats.

Another message arrived: the shooting involved a school. Neil called Scarlett, or maybe Scarlett called him, after they both heard vague reports that the school was Sandy Hook, something about someone being shot in the foot.

"'Well, maybe it was a teacher, a disgruntled parent, a husband and wife, an isolated thing,' I don't know," Neil recalled wondering at the time. He set off for Sandy Hook.

Scarlett thought it was probably another false alarm. She figured she may as well go to the school anyway, in case Jesse was worried about whatever had happened and wanted his mom.[8]

Neil got there first. He found a scene "like a combat zone," Riverside Road and Dickinson Drive, the intersection nearest the school, jammed with cars and emergency vehicles. Inside and around the firehouse on Riverside he saw kids and teachers carrying signs with classroom numbers written on them. But he didn't see Jesse's room number, 10, and he didn't see Jesse. "Nothing was real clear," he told me. "All you knew was something happened, and they evacuated the kids. There wasn't a lot of information being given out there. I think the media"—whose camera crews and satellite trucks had already engulfed the firehouse parking lot and the hilltop overlooking it—"actually had more, but all I knew was, Jesse wasn't anywhere."

As a local tradesman, Neil knew a fair number of the cops in town. He approached officers milling around the parking lot, and talking in hushed voices amid the yellow reels of hose stacked behind the firehouse. No news, they told him, looking away.

Neil bypassed the officers gathering at the mouth of Dickinson to block access to the school, about three hundred yards beyond the firehouse. "I must have blended in well, I guess," he recalled of his circumventing them. He followed the wooded lane that was Dickinson and "walked up toward the school, where the glass was shot out." A police officer prevented Neil from entering, in what would prove an immense mercy.

Scarlett arrived to escalating chaos. She parked a couple of blocks away, the only place she could find, and ran with a throng to the firehouse. Parents and children, police, and emergency medical technicians swarmed. The firefighters had moved trucks out of the firehouse's seven bays, and terrible sounds echoed inside the yawning space. Evacuated kids, dazed or weeping, lined up with their teachers or wandered about. Parents found their children—many of whom had fled without their coats—draped them in their own jackets, and carried them away. Those who did not find their children began dreadful rounds of searching. Scarlett began beseeching anyone wearing a uniform for news. Police channeled her to a small spare room in the rear corner of the firehouse, where parents with missing children milled about, frantically working their cell phones.

In the late morning, unable to stay put, Scarlett went out to the parking lot. She was asking a police officer she knew for help when a woman she knew from the PTA came up. The woman couldn't be sure, she told Scarlett, but she thought Jesse was with a few kids who'd run past not long before. She directed Scarlett to a small cedar-shingled house, painted yellow with green shutters, a short distance from the firehouse on Riverside.

A few minutes later Gene Rosen, a sixty-nine-year-old psychologist retired from the Fairfield Hills hospital, opened his front door to find Scarlett panting with exertion and fear, her long blond hair tousled. "Her face looked frozen in terror. I've never seen a face look like that," he recalled to AP.[9] She gave him Jesse's name. Yes, six children had come, he told her, and also a bus driver. But that was some time ago. They had found the parents of several of the children, who had already collected them. But maybe, Rosen told her, some kids had gone to the daycare down the road?

He closed the door without telling Scarlett the terrible things the children had said. He'd come around from the backyard to find the six of them sitting in a circle on his front lawn, as if playing a game. As he approached, he saw their distress. A school bus driver hovered over them, reassuring them in a ragged voice. There had been an incident at the school, she told Rosen. He led them all into his living room, gave them some juice and an armful of stuffed toys from his grandson's toy chest.

One of the boys looked up, wild eyed, and said, "'We can't go back to that school. Our teacher is dead,'" he told Rosen. "I could not take that in. I could not accept that," Rosen told Erin Burnett of CNN.[10] "And then I just kept listening to them, and they talked more, and the boy said, 'Oh no, it was a big gun and a small gun'—and then I knew."

One of the boys said they saw blood come from their teacher's mouth. And that she fell. "Then they said her name," Rosen said, weeping. "It was that very pretty twenty-seven-year-old teacher."

Victoria "Vicki" Soto, age twenty-seven, was Jesse's teacher.

Neil called, and Scarlett told him to check the daycare. She hurried toward Dickinson Drive, hoping she'd find Jesse amid the pandemonium at the barriers now across it, where family members confronted law enforcement officers. Men in army fatigues with rifles and flash grenades on their belts drove past in jeeps, and Scarlett thought how impressed Jesse would be to see so many soldiers in real life.

The police directed Scarlett back to the firehouse. She obeyed. She could not think of anywhere else to look, but she still held out hope that Jesse was hiding in the woods, as some of the police had suggested. She crossed the firehouse floor, the place of reunions, to the room at the back, the place of fading hope. On a sheet of notebook paper laid out for the purpose, she added Jesse's name to a list of the missing.[11] A state trooper asked Scarlett for a recent photo of Jesse, and she chose one from among hundreds of him on her phone.

J.T., a seventh grader at Newtown Middle School, texted his mom. He was on lockdown in Spanish class but wanted to join her. Scarlett's mother, Maureen, and Maureen's partner, Bob, picked him up.

Scarlett and J.T. fled the firehouse when a parent broke down and screamed, making J.T. recoil. They returned to the parking lot and gathered with Maureen, Bob, two of Scarlett's three brothers and their spouses. Somebody brought lawn chairs, and the group arranged them in a tight circle, where they huddled as if around a campfire on a trip from hell. They kept moving their chairs, trying to escape the sounds of grief from despairing families. Scarlett cupped J.T.'s face in her palms, their knees touching. "Even if the worst has happened, and we've lost Jesse, we know exactly where he is"—in heaven and fine. J.T. shook his head violently, refusing to accept her words.

After another hour, maybe two, a middle-aged man who introduced himself as a psychologist came and knelt beside Scarlett. "There's no easy way to tell you this," he said. "Your son is dead." She thought that over the preceding hours she had prepared herself for these words, but Scarlett couldn't absorb them. She stared at the man, unable to move or think, and then he abruptly stood up and walked away. Scarlett remained motionless until J.T. erupted in tears, propelling her forward. She enveloped her son, kissing and reassuring him. Her family crowded in, encircling them both. They clung together like that for a long time, standing on a yellow-striped asphalt lot transformed into a wasteland of human suffering.

Neil, standing twenty feet or so across the parking lot, saw Scarlett's stricken face and her family's reaction. Neil intercepted the man as he walked away, to ask what he had said. Confused, the man told Neil that he had just informed Jesse's parents that he was dead. "*I'm* Jesse's father!" Neil roared. He lunged forward, and a state trooper stepped between the two men.

Scarlett never learned where the man had gotten the information about Jesse, but as a police officer told her that day, no one had authorized him to deliver it. Social workers, counselors, and clergy had poured onto the scene since the morning, summoned by family members and first responders. But some arrived unbidden, proselytizing, handing out business cards, inserting themselves. The man who notified Scarlett of her son's death was among them, Scarlett told me. "I can forgive Adam Lanza, but it's hard for me to forgive someone who would intentionally try to take advantage of our misery," she said.

In the mayhem of that day, there was no way to vet such people or to prevent them from sharing the terrible information circulating among first responders but not yet communicated to the families.

Police who responded and had seen the bodies inside the school were restrained from sharing what they knew. State protocol requires that victims be officially identified before notifying their next of kin. But that policy was unworkable in this situation. It would involve hours, maybe days of waiting, which would leave twenty-six families relying on secondhand reports and hearsay.

"It was like knowing someone was about to be hit by a bus, and you couldn't warn them," one trooper who was there told me. "We would be destroying their lives."

That role fell to the Connecticut governor.

When Governor Dannel Malloy heard the first reports in his Hartford office that morning, he thought, like many of the families did, that the event was an isolated domestic incident. But within an hour he was in a car speeding south on I-84 to Newtown.

I visited Malloy in Orono, Maine, where in 2019 he had taken a job as chancellor of the University of Maine System, leaving politics. It had been eight years since the shooting when we spoke, but Malloy slipped into the present tense, the scenes playing out in his mind.

As a four-term mayor of Stamford, Malloy had stood with families whose loved ones had commuted by train to Wall Street on September 11, 2001, and never come home. He had attended the funerals of warfighters killed in Iraq and Afghanistan. In those cases, protocol afforded safety and comfort. But this was different. These people had waited agonizing hours for news that was already leaking and in danger of being badly mishandled.

Arriving at the firehouse, Malloy conferred with First Selectman Pat Llodra, respecting boundaries, making sure she wanted the state's help.

She did. The governor, hands jammed in the pockets of his black overcoat, met with Colonel Danny Stebbins, the Connecticut State Police commander. Speaking in low tones inside a trapezoid of yellow police tape in the firehouse parking lot, Malloy posed a series of questions to the police commander. Was anyone unaccounted for, perhaps hiding in the woods or elsewhere in the school? How many were injured?

Stebbins told him that, of the three people taken to the hospital, one had survived. Two, both children, had died. Everyone on the list of missing, six female educators and twenty children, lay dead inside the school.

"The policy is we never tell someone that their loved one has expired until we can identify the body," Malloy told me. "So, I start to think about, 'Well, how are we gonna do that?'"

The teachers who knew the children best were also dead. "It's related to me that some of the injuries are pretty horrific," Malloy recalled, too horrific to ask other teachers to identify them. Janet Robinson, the school superintendent, asked[12] whether the children's new school pictures had arrived.

"They'd just got there the day before or two days before," Malloy recalled, and were somewhere in the principal's office. They summoned the school secretary, who had been out sick. Police guided her to the office, shielding her from the carnage, and she found the photographs.

At about noon Malloy entered the chaotic firehouse, introduced himself to the distraught families, and told them he would brief them regularly. Outside, he instructed his staff and the police to set up some chairs, while knowing what a futile gesture that was—imposing order while the families' lives crumbled.

He pressed Stebbins: "When are we gonna know?"

Stebbins told him he thought they had to transport the bodies up to the state morgue in Farmington, nearly an hour away, where medical examiner H. Wayne Carver and his team would make the identification before notifying family members. They didn't have enough vehicles to transport them all, so the whole process could take until 2:00 a.m.

Malloy knew then, "I'm going to have to tell these folks."

They thought about bringing each family, individually, to a small social service center several hundred feet from the firehouse. "I thought that would be far more humane," Malloy told me. But the walk to the building led through a gantlet of media, and once the first family knew their loved one's fate, all of them would.

Malloy paused. His chest heaved as a sob escaped him. "So I make the decision. The next time we get together with them, I'm going to find a way."

He entered the firehouse. At first he spoke so haltingly that few could hear. "Just tell us!" a woman, her voice jagged, demanded. Others in the room shushed her, but Neil joined her in demanding, "Give an answer! Is anyone in that school still alive?"

The governor searched for the least scarring words. He couldn't bring himself to use the word "dead."

"Two children were brought to Danbury Hospital, and expired," he said as a guttural wailing lifted and reverberated.

"The parents just were hysterical," Donna Soto, the mother of Vicki, Jesse's teacher, recalled later. "They were on the floor."[13]

Malloy paused, then plunged. "Nobody else was taken to a hospital," he said.

"I told the folks in the best way that I could. Obviously, I said, 'I'm sorry'—I don't know exactly what I said," Malloy told me. "What I was trying to say is, 'If you haven't been'—I used those words—'If you haven't been reunited with your loved one by now, you're not going to be united with your loved one.' And they knew."

Despair turned to fury. "What are you telling us?" a parent shouted. "They're all *dead*?"

Yes, Malloy replied. There were no more survivors in the school. He repeated the words more than once, to be heard over howls of grief, and to pierce the disbelief of those who stood staring at him.

Some families found him cold and clinical, a criticism that still sears him.

"I didn't do it callously, I just—" he told me, losing his composure. "I did it out of desperation. Because there was no other way to do it."

SCARLETT FOUND NEIL in the parking lot, tears streaking his face. They hugged tightly, then parted.

Scarlett could not return to her farmhouse, with its reminders of Jesse. She and J.T. retreated to her mother's home in Newtown.

Neil returned to the firehouse, where he remained after the other parents had gone. He also couldn't go home yet. Not to an empty house where the Christmas tree he and Jesse had erected stood in the living room, still awaiting its decorations.

He needed to see his son to believe he was dead.

It was about one in the morning when Neil quietly picked his way past the barricades and down Dickinson Drive. Police wheeled gurneys from the school into its klieg-lit parking lot, where inside the tent Dr. Carver's team and medical examiners from New York attended to their work.

Neil has never revealed who helped him to see Jesse that night, nor exactly where he was when given that opportunity. Neil cradled his boy in his arms. Jesse's rugby jersey was untucked as always. His carpenter pants were too short; he'd had a growth spurt that year. His face was composed and clean. It looked almost unharmed, except for the dime-sized gunshot wound near his hairline that had ended his life.

"When he was born, I was the first one to see him, and I was the last one to hold him and hug him when he went out," Neil said. "It meant a lot to be able to see him. It helps with anyone when you lose them, to put a closure to the death when they're gone, an understanding."

Scarlett and Neil soon learned that as the gunman began shooting in classroom 10, Jesse had shouted "Run!" during a pause in the carnage, when the gunman's rifle jammed or he changed magazines. Nine children obeyed him, and lived.

AFTER PRESIDENT BARACK OBAMA visited for the prayer vigil at Newtown High School on December 16, Scarlett returned to the farm to fetch Jesse's burial clothes. "I wanted him to be warm," she told me. She chose "flannel-lined jeans, a soft turtleneck, a flannel shirt, a polo sweater. Thick warm socks and his rubber boots he wore everywhere, even in the summer."

She stopped in the bathroom and scooped up his rubber ducks and plastic army men, still lined up along the sides of the bathtub. She took Jesse's favorite plastic camo helmet, part of a soldier playset a friend had given him.

Neil did not have anything to wear to Jesse's funeral. He was in the construction business and had no need for suits, and because business was bad, he didn't have much cash. He doesn't recall telling anyone about this or how it made him feel, but two days before Jesse's wake, an FBI agent pulled up at his house in Shelton. He drove Neil to Men's Wearhouse in Danbury and stood silently near the front of the big store while Neil chose a dark tweed suit and white shirt, a green-striped tie, shoes, and a belt. He refused Neil's offer to pay the bill and drove him home.

On December 19, Scarlett's mother and siblings accompanied her to Honan Funeral Home on Main Street. Jesse lay in a white casket, wearing his rugged warm clothes. Scarlett sat down, reached in, and held Jesse's hands in hers, trying to warm them. He still had dirt from the farm under his fingernails. He was a tough and sturdy boy. He weighed eleven pounds when he was born, and more than seventy when he died. He rode Scarlett's biggest horses. But alone with her, he liked her to carry him and to snuggle with him in bed, still her baby.

Jesse was one of five children buried in Newtown on December 20, 2012. Word of Jesse's selfless act had spread, and police from several states arrived to give him a fallen hero's funeral. Thousands of people attended his wake that clear, sunlit day, waiting in a line that stretched down Main Street and around the corner.

The viewing lasted several hours. When it was over, Scarlett and Neil approached Jesse's bier. They placed his rubber ducks, army men, and camo helmet inside the casket beside him. Scarlett took up a thick woolen Pendleton blanket with a colorful Native American design, a gift to Jesse from her mother. Scarlett had draped it around herself all that day, so it would carry her scent. She covered him with it, tucking the edges beneath him, bundling him up to go. Police officers approached Jesse's casket in pairs and saluted before they closed the lid.

At Jesse's funeral service at Beacon Hill Church on Old Zoar Road in Monroe, Jesse's uncle recalled his rule for living: "Have a lot of fun." J.T. spoke about the battles and secrets they shared. Scarlett came forward.

"People have been asking me what they can do to help," she said. "Do something that will help all of us, by turning an angry thought into a loving one. This whole tragedy began with an angry thought, and that thought could have been turned into a loving one. If it had been, none of us would be here today to bury a child we all loved so much.

"If you want to help, then please choose love."

Solemn police officers on motorcycles guided the hearse to Zoar Cemetery. First responders lined the route, saluting or placing hands on their hearts as the cortege passed. By the time they reached the cemetery it was nearly dusk. Mounted police officers from Bridgeport awaited them at the gates.

Neil remained in the background during the funeral, by nature not one to take center stage. His parents had died years before, and he lacked the warm bubble of family and friends that buoyed Scarlett. After everyone had gone, Neil remained behind. He watched the crew cover Jesse's grave. Then he sat down on the mound of raw earth, keeping company with his son in the dark.

These days Neil does not like to visit Jesse's grave alone, so he seemed almost relieved for the company on the windy and gray January day in 2019 when he guided me down aging roads to Zoar Cemetery. Town elders established Zoar, one of Newtown's original burial grounds, in 1767 on what had been a hilltop farm. Most of the children are buried in Newtown Village Cemetery and at St. Rose of Lima. Neil had found the clusters of small graves unsettling.

"I wonder if one hundred years from now, people will look at the gravestones for all these children who died on the same day and wonder—" Neil had said to me once, but then broke off, realizing that no one would ever wonder.

Jesse lies beneath a tree in a corner plot near the wooded back border of the cemetery, a short walk from Scarlett's farm. Some days, the braying of Jesse's burro, Turquoise, floats up on the wind. Neil pointed through bare trees to the Misty Vale Deli, where they'd eaten breakfast that day. He'd told me that Jesse ordered hot chocolate and a breakfast sandwich with sausage. "He used to order bacon," Neil recalled, "but he choked on it once, and it scared him."

Jesse's grave is flanked by a stone bench and a bath for the birds he loved. I laid a white rose among teddy bears, rubber ducks, and a platoon of his plastic army men scattered by the raw gusts.

Neil paced, grousing that the roots of the tree might be growing too close to the grave, that visitors' feet had chewed the earth into frozen ridges of mud. "These stuffed animals," he snapped, "they look like shit."

I lifted a matted bear, his frozen smile flecked with grime, and propped him gently against the cold granite.

"Let's go," Neil said. His eyes rested on the stone's oval portrait of Jesse, smiling in his first-grade photo, taken two months before his death. Neil kissed his first two fingers and touched them to the porcelain image. His boy, afraid of bacon, who confronted a gunman.

We continued on our driving tour of Newtown, from the old to a newer, upscale section of town. Neil directed me onto Yogananda Street, named for a peace-loving yogi, now a locus of Newtown's pain. In 1998 the Lanza family moved into a new house at 36 Yogananda, buying a four-bedroom colonial with green shutters on a sweeping two-acre lot. Peter and Nancy Lanza found Newtown a comforting reminder of rural Massachusetts and New Hampshire, where both of them had grown up. Peter was an executive in the tax department at General Electric, and Nancy a stay-at-home mom since their two sons were born, Ryan in 1988 and Adam in 1992.

Nancy told friends in Newtown how fortunate she was to be able to stay home, because clearly Adam needed her. He bounced from school to school in Newtown, spending time at Sandy Hook Elementary, St. Rose of Lima Catholic school, and Newtown High School. His mother withdrew him from school entirely for a while, a lapse that went without discernible scrutiny or intervention by the school district, even though staff at Danbury Hospital, where Lanza was taken during a crisis, recommended mental health treatment, according to a 2014 report by Connecticut's Office of the Child Advocate.[14] The advocate's investigation uncovered repeated failures to address Lanza's deteriorating mental health by most of the adults in his life. At age eleven, he washed his hands until they were raw and found loud noises, shifts in routine, and being touched intolerable. At thirteen, the Danbury Hospital staff noted, he weighed ninety-eight pounds, though he was about five feet eight. He was prescribed hand lotion for his irritated hands and a laxative for constipation likely associated with anorexia, diagnosed only after his death. Lanza was preoccupied with violence, yet his middle school writings portraying the murder of children again drew little substantive intervention.

The Lanzas had separated in 2002, and Adam lived mostly with his mother. In late 2006, when he was in ninth grade, he was evaluated at the Yale Child Study Center at Yale University. In an initial report, a Yale psychiatrist described Lanza as a "pale, gaunt, and awkward young adolescent standing rigidly with downcast gaze and declining to shake hands." He used to look at people, Nancy Lanza, hovering, hypervigilant, hastened to explain. The psychiatrist observed the teenager's mother laboring to comply as her son issued directives, telling her at one point to stand without leaning on anything. Asked by the doctor what his "three magic wishes" would be, Lanza said that "whatever was granting the wishes would not exist."

The Yale team's assessment was urgent. Lanza needed intensive treatment, possibly a special therapeutic school, for debilitating anxiety and obsessive-compulsive disorder, impairments whose severity they feared his parents did not fully grasp. "We are very concerned about [Lanza's] increasingly constricted social and educational world," the psychiatrist wrote. Trying to convince Lanza to take medication, another Yale clinician reported, "I told him he's living in a box right now, and the box will only get smaller over time if [he] doesn't get some treatment."

After a promising start Lanza shut down, and his parents acquiesced.

By the time Lanza's parents divorced in 2009, his mother had stopped seeking mental health treatment for him. In 2010 Adam Lanza broke off contact with his father, who lived in Stamford. He left school and refused to leave the house. He taped dark plastic bags over the windows of his downstairs bedroom, where he played violent computer games and compiled data on mass shootings dating to the 1970s. He communicated with his mother mostly by email. The Lanzas had gone to shooting ranges as a family in the past, and in the years leading up to the shooting, Nancy Lanza, desperate to bond with him, bought him guns and took him for target practice.

"It cannot be overlooked that as his mental health deteriorated and his isolation from the world increased dramatically, his access to guns did not diminish," the Office of the Child Advocate's report said.

Summarizing findings from Lanza's high school years: "If there is a single document that is most prescient regarding [Lanza's] deterioration, it is the October 2006 report from the Yale Child Study Center—an evaluation that so dramatically states the high stakes presented by [Lanza's] disabilities and the need for meaningful and immediate intervention," the report read. "The lack of sustained, expert-driven and well-coordinated mental health treatment, and medical and educational planning ultimately enabled his progressive deterioration. Though again, no direct line of causation can be drawn from these failures to his commission of mass murder."

In a news conference[15] after the report's release, one of its authors, Dr. Julian Ford, director of the University of Connecticut's Center for Trauma Recovery and Juvenile Justice, described Lanza as so profoundly isolated at the end of his life that "he was losing a sense of other people as human beings."

Nancy Lanza spent her last years in a misguided effort to shield him from a lengthening list of intolerable irritants: grease, dust, dirt, cooking smells, doorknobs, phone calls, sunlight, medication, therapy, and other people, ultimately including her.

SOME OF THOSE WHO perished at Sandy Hook were the Lanzas' neighbors; one child's family lived across the street. As we drove up Yogananda, we passed a neighbor on his lawn who looked up, unsurprised by now to see another voyeur with out-of-state plates drive down his secluded street.

As with the school and the firehouse, the Lanza house grew into a magnet for conspiracy theorists. A year after the shooting, the family gave the house and lot to Newtown. Determined that nothing inside find its way onto the murder-memorabilia market, the bank that handled the transaction burned the home's entire contents—its pricey furniture, bedding, and carpet, and the gunman's belongings that the police hadn't already removed.[16] The 3,100-square-foot home was razed in late March 2015, a local contractor obliterating even the driveway. The town intends for the site to remain vacant, reclaimed in time by the surrounding woods.

Neil and I traveled back through downtown and around the flagpole marking Newtown's heart. Facing it on one side of Main looms the elegant white clapboard Newtown Meeting House, and on the other, the tall stone carillon and red door of Trinity Episcopal Church. It was here in early January 2013 that Neil went to meet Adam Lanza's father.

IMMEDIATELY AFTER THE SHOOTING, the state police assigned a trooper to each victim's family to protect them, help them navigate its aftermath, and serve as a liaison with investigators. Two weeks after Jesse's death, Neil told a state trooper that he wanted to meet Peter Lanza. The officer passed word, and a few days later Neil received a text message in the wee hours of the morning from Lanza, offering to meet him at Trinity Episcopal at eight in the evening, in four days' time. Only two victims' families are known to have taken Peter Lanza up on his offer to answer their questions, and Neil was one. Though Scarlett was more forgiving of the gunman than Neil was, she did not want to meet his father. So Neil went alone.

He parked down Main Street, a block or so away from the church, and waited in the January chill. He wondered whether he would recognize Lanza when he saw him. He did. The father had darker hair than the son, his face an older, fuller version of the emaciated, staring visage in the media. Neil stepped forward. He doesn't remember whether they shook hands. They followed a woman who greeted them to a meeting room with stackable chairs and arts and crafts paraphernalia lying about.

They remained facing each other until after midnight, two fathers whose sons died in the same room. Lanza seemed nervous, even fearful, "like he didn't know what to expect of me, whether I was gonna knock him out or something," Neil told me. These were weeks when he "didn't have a clear mind," Neil said, so he has trouble remembering chunks of the conversation. He does recall Lanza saying he was sorry. A year later, Lanza would tell the *New Yorker*'s Andrew Solomon he wished his son had never been born.[17]

Lanza didn't defend his son, Neil said. "But his explanations were meant not to blame himself. And you can't blame him. Not one hundred percent. It was poor parenting, poor judgment, poor dynamics across the board.

"The whole thing was just a disaster looking for a place to explode. And it happened. I don't have a vendetta with him personally."

Lanza, then a vice president at General Electric, "is not by nature given to self-examination," Solomon wrote. Neil agreed. That night, as the father of Jesse's murderer rambled, talking almost nonstop, growing anger writhed behind Neil's quietly listening face.

When Lanza paused, Neil asked why it had taken him nearly two weeks to claim his son's corpse.

"He said he didn't want to put any more emotional strain on the families. And I thought, 'The fuck—your kid just killed twenty of our kids, and you're worried about putting emotional strain on us?' I'd been tempted to claim that body and throw it on Lanza's front porch."

Neil knew people who said the killer's remains were cremated and scattered in the Atlantic. Lanza did not confirm this, and he has said he never will reveal what he decided.

Walking out of the church into the night, Neil paused and turned. "Look at my face and my hands," he said he told Lanza. "If your son hadn't died, I would have killed him myself."

I had been driving silently, watching the road as Neil unspooled this story. But at that point I looked at him. "I said it out of anger and self-pity," he said. "But I would have." He thought Peter Lanza believed him, judging by the color draining from his face.

WE ENDED OUR TOUR that day at the new Sandy Hook Elementary School, which opened in 2016. The school day had ended, and we drove through open security gates into the parking lot.

Nestled among trees behind a slalom course of barriers, the school's warm timber facade resembles undulating waves. The windows are anti-ballistic, the surrounding rain garden a moat.[18] Mary Ann Jacob, the Sandy Hook librarian who had sheltered a group of fourth graders during the massacre, told me that the school's computers blocked all external information about Sandy Hook school, so the students "wouldn't see God knows what."

The best security money can buy arrived too late to save his son, Neil noted, adding that Lanza, who "weighed one hundred and ten pounds soaking wet," still could have slipped through its gates and bollards like a letter through a slot.

"That's the murder site," Neil said as we passed a gentle grassy berm topped with saplings, to the right of the entrance. Or is it the left? The berms match and are purposely unmarked, to discourage those without a legitimate reason to know where the bodies once lay. Neil and Scarlett had been angry about the decision not to mark the site. The state gave Newtown $50 million to build this school. Local politicians made sure their names were associated with it, they said. But what about the people who died here?

We drove back out Dickinson Drive and parked on the hilltop overlooking the firehouse. We ended our day as the last streaks of light left the sky over the placid building, the memorial stars on its rooftop the only sign that for twenty-six families, normal life ended there, in a crowded garage where the ladies' auxiliary brought water and food no one would eat and played cheerful cartoon videos for the children who had survived, to distract them from the keening of parents whose children had not.

2

NINETY SECONDS AFTER GOVERNOR MALLOY confirmed that their daughter Emilie was dead, Alissa and Robbie Parker surged toward a door in the rear corner of the firehouse, desperate to escape the plangent wails. Alissa was gasping for air. During six hours of waiting, she had nearly collapsed; paramedics had given her oxygen. Bursting outdoors, they found themselves in an empty space behind the firehouse. They held each other in silence. "I want to go home," Alissa said at last, and Robbie took her hand. Rounding the corner of the firehouse to the parking lot, they walked into the maw of the media. As they staggered in shocked confusion, engulfed by shouted questions and the whine of camera motor drives, a firefighter noticed them and barreled into the pack of reporters, ordering the journalists back.[1]

Jessica Hill of the Associated Press had already retreated to an unobtrusive position overlooking the firehouse—near the ridge where Neil and I would pause years later to view the rooftop memorial of copper stars. Hill's zoom lens swept the parking lot, finding Robbie and Alissa as they broke free of the mob.

In the AP photograph the young couple look almost like college students. Clad in jeans, gilded by the winter's slanting afternoon sun, they walk through blurred space that, except for the fluorescent-clad firefighter in the background, could be anywhere. He's thirty and she the same, their birthdays five months apart. They had known each other since they were thirteen. Robbie wears only a plain blue T-shirt against the cold, staring at the ground, lips parted, his face open with disbelief. His left hand is jammed into his front pocket and his right grips Alissa's shoulder, knuckles white as he presses her tightly into his ribs, as if a ruthless centrifugal force threatens to fling them apart.

Alissa wears a black parka, a strip of polka-dotted shirt visible beneath it. A long rose-colored scarf hangs in a hasty half hitch around her neck. Over her right shoulder she carries a black handbag with extra pockets on its outside, the kind a mom of three little girls under seven would choose to hold a jumble of last-minute needs: packets of half-atomized Goldfish crackers and fruit snacks, sunscreen and wipes.

Alissa is still attempting to evade the cameras. The slender, elegant fingers of her right hand cover her right eye and press her forehead in confusion, the tissue in her grip half obscuring her mouth, her lips pulled tight by a sob. Her left arm encircles Robbie's waist, hanging on as he propels them forward.

The photo appeared in news accounts around the world. Their families in Utah saw it in the *Deseret News*. The image brought home the devastation of the worst elementary school shooting in American history. For the Parkers it crystallized the moment they were thrust from peaceful, anonymous family life into public renown as the parents of a murdered child.

Five years after the shooting, Joanne Cacciatore, an Arizona State University professor who counsels people affected by traumatic death, interviewed fifteen Sandy Hook family members: fourteen parents and one person whose parent died. While many researchers have documented a mass tragedy's effect on a community or social and political movements, only a few have explored the impact on the families whose loved ones were killed. In late 2020, Cacciatore and Sarah F. Kurker, who teaches medical social work at Arizona State, published their findings in a paper titled "Primary Victims of the Sandy Hook Murders: 'I Usually Cry When I Say 26.'"[2] In it, the family members, quoted anonymously, describe how the public response to the shooting intensified and prolonged their suffering.

Hundreds of journalists descended on Newtown, feeding names and details, photographs and footage and commentary, to a world straining to digest the unfathomable. Legitimate inquiry or voyeurism, honoring their loved ones' stories or exploiting them—it soon all seemed the same to the families who felt, one said, like prey.

Within hours after their loved ones' deaths, the victims' families were struggling to retain control of their life stories. They worried that others speaking for them or the media would make mistakes. They feared that individuals, activists, and businesses would leverage the public outcry over the tragedy to promote themselves, push an agenda, or make money. Their fears were soon borne out.

"Our daughter is not just our daughter now. She's one of the children who was murdered at Sandy Hook," the mother of a girl who was killed told the researchers. She added, "It's not that I want that to be forgotten. But that is not her story. That is not her life story, that is only how she died."

In Cacciatore's interviews with the families, a majority said they had "a contentious, or at least conflicted, relationship with the media who 'camped' out at their homes, accosted them in public, and demonstrated disrespect for their personal tragedy."

Said one: "We go out of our back door and we walk around to the front, and we just get hit with this wall of photographers . . . We are trying to move past them and it was just so invading."

"My experience of living as a free person in America is gone," said another. "I do not live as a free person in America. I am a tragic public figure."

Reporters remained outside the families' homes for weeks. Photographers took positions in their trees. Some outlets offered to pay neighbors for photos of the children who died, a grotesque practice prohibited by the *New York Times* and most news outlets. As the families' phones and doorbells rang, business cards and handwritten interview requests piled up in their mailboxes, they agonized over whether to speak or remain silent.

"There are many comparisons made about the families, about who might be grieving well or not, or who is doing it right," a parent told the Arizona State researchers. "We were very private and we felt very alone and we felt that we were judged as something less. We were not as important. Our child wasn't as important because we wouldn't open up as easily. We felt forgotten."

Most of the family members, friends, and neighbors who appeared in news accounts of those early days had seldom if ever spoken to a reporter before the worst day of their lives, when they faced banks of cameras.

Breaking news from Newtown reflected early confusion among the multiple town, state, and federal law enforcement agencies that responded. Police found the gunman's older brother's ID card on the floor in the school, leading to early reports, later corrected, implicating him. Police detained a man outside the school, leading to speculation about a second shooter, but he turned out to be the father of a first grader who survived.

"First reports of this tragedy have turned out to be inaccurate,"

60 Minutes' Scott Pelley said at the opening of the CBS program on December 16. "We were told that the gunman's mother was a teacher at the school. That he was allowed in, because he was recognized, and that he targeted his mother's classroom with two handguns. Tonight, we know that all of that is wrong."[3]

Police investigators expect discrepancies in people's initial versions of destructive, confusing, emotional events. In the intelligence world, crystal-clear narratives and seamless explanations can garner greater suspicion than foggy recollections and scrambled timelines, for good reason.

In the conspiracy world these rules don't apply. Every error in the official record, no matter how swiftly corrected, is proof of lying, a dropping of the veil obscuring from the public what *really* happened. At the same time, the conspiracists portray officialdom—the police and FBI, government officials and mainstream journalists—as flawless actors in a sweeping, intricate plot. The conspiracists acknowledge neither this contradiction nor their own errors, papering over debunked claims with implausible new ones, or ignoring inconvenient facts altogether. As first responders, investigators, and media clamored to understand the senseless deaths of twenty-six people, the conspiracists used their mistakes against them, building a seamless explanation of why it didn't happen.

Errors in reporting were perhaps inevitable. But errors in good journalistic practice among some fueled a lasting reluctance in Newtown to engage with national media. Staff at the *Newtown Bee*, the local paper that functions more as a repository of local history and restorative narratives than a check on official power, stood appalled at the habits of their colleagues from away. John Voket, the *Bee*'s government reporter, was so familiar to officials that they let him stay inside the firehouse on December 14 while other media were banned. He talked with waiting family members and first responders, never publishing a word of what they told him.

Voket wrote an open letter to the New England Newspaper & Press Association, urging the media to get out of the survivors' front yards and leave them alone.

To protect the families, Connecticut legislators worked under the radar to craft legislation withholding records identifying individual victims. Officials refused to release 911 calls routinely shared with the media until forced to by a court. Media use these records to hold emergency responders accountable and debunk bogus claims already bubbling up among conspiracy theorists. But for families stripped of their basic sense of safety, the records requests fueled terror. Did the public also have the right to photos of murdered first graders lying on their classroom floors?

THE PARKERS HAD MOVED to Newtown just eight months before the shooting. They had met in middle school in Ogden, Utah, where they had grown up and where Emilie was born. Robbie and Alissa married in 2003, soon after Robbie's return from two years of Mormon missionary work in Brazil after high school. They embarked on a peripatetic life. Robbie works as a physician's assistant in neonatal intensive care, and his schooling, rotations, and job assignments took them from Utah to Oregon, Washington, Montana, and New Mexico before Robbie landed a position in the neonatal unit at Danbury Hospital, some ten miles west of Newtown. By that time the Parkers' boisterous household numbered five, including Madeline, a year younger than Emilie, and Samantha, born a year after Madeline. They seized on the job opportunity as a family adventure. One big plus: instead of working nights, Robbie would work twenty-four-hour shifts, allowing him a couple of days at home with Alissa and the girls in between.

After touring Newtown with a real estate agent, Alissa crossed all other options off the list. She had never lived in the East, and this seemed her ideal image of it: rolling colonial vistas, a folksy downtown, well-rated public schools. Sandy Hook Elementary reminded Robbie of his own grade school in Texas, where he spent his early childhood. It seemed a good fit for Emilie, their only school-age child, an expressive, empathetic girl with a gifted child's intuition. Blue-eyed and slight, with long, pale blond hair and a flair for the dramatic, Emilie favored dresses, tiaras, and the color pink. She went through a phase when she instructed everyone, including a church elder, to call her "Sleeping Beauty." As she wrote on an elaborately calligraphed royal blue poster: "I love being fancy." She also loved to paint and draw, developing calluses on her fingers while toiling into the night on her creations. She carried paper and a bag of markers and crayons around with her, drawing instant greeting cards for people she met who seemed sad or under the weather. If she misbehaved, she would present the wounded party, usually her parents or sisters, with a colorful picture captioned "I'm sorry."

By early spring the Parkers had found a fixer-upper in a wooded hollow on Country Squire Lane, a quiet Newtown cul-de-sac. They spent their savings on the four-bedroom gray house and its DIY renovation. Alissa's father flew out from Utah, and together they tackled a rewiring project, built new closets, and, downstairs, a bedroom and an adjoining art room for Emilie. They painted the walls pale gray to set off the bright pink furniture, pink-striped chenille bedspread, and frilly, homemade pink curtains. Alissa and Emilie stenciled luxuriant pink, blue, and black flowers on the wall over Emilie's four-poster bed.

Emilie entered Sandy Hook for first grade in fall 2012, again a newcomer. Chatty and outgoing, she made friends easily.

Emilie was a Sandy Hook student for less than a year. After her death, her bed remained unmade and her artwork half-finished on the table nearby, as she had left it on her last morning.

Alissa and Robbie had no extended family and relatively few friends in Newtown. They tended to keep to themselves, weathering what had proved a cruel season even before Emilie's death. On September 29 of that year, Alissa's father died after a crash during a long-distance bicycle race. Alissa, reserved and private like her mother, had shared a deep bond with her gregarious, exuberantly loving dad, whose death at only sixty-two left a deep void. Less than three months later, Emilie was killed.

After they left the firehouse and made it home that afternoon, Robbie and Alissa spent hours helping Madeline and Samantha understand that "Emilie has died and she won't be coming home again." Leaders from their church arrived to offer help with logistics and to pray with them. It was after midnight when police arrived at the house with official notification of Emilie's murder. Soon after, Robbie collapsed on the bed, still wearing the blue T-shirt and jeans he'd worn at the firehouse.

"The events from that day kept playing in my head like a scratched record," he wrote years later. "Suddenly a pain I'd never felt before stabbed me and burrowed deep into my chest. I laid motionless, letting tears stream down my face while the pain nestled into a space somewhere between my body and soul.

"'Welcome to real grief,' I thought. As the realization of her actual death took hold in me I was formally introduced to a new constant companion in my life."

Then the phone started ringing.

"Immediately after the shooting we started getting contacted by the media, and obviously a ton of family and friends in Utah, saying, 'Hey, local news has called me and they want to interview me,'" Robbie told me. "I didn't want to engage with anybody. I just wanted to focus on my family."

The state trooper assigned to the Parkers ordered reporters off their property. Their neighbors enforced an ad hoc security perimeter by parking their cars across the Parkers' driveway. But it soon grew clear, Robbie said, that "the media was going to put pressure on until somebody talked. And I didn't want anyone to say anything about my daughter that didn't come from us.

"I thought, 'If I make a statement, maybe the media will back off and leave me and leave my friends alone.'"

So on December 15, the day after the shooting, Robbie called his friend in Utah. He told him to tell the reporter who had contacted him to send a Connecticut-based colleague to the LDS church in Newtown at five that evening. He would meet the reporter there and provide a statement.

That day had been a disorienting nightmare. Emilie's body lay in the medical examiner's office in Farmington as the family worked with their church to plan her funeral in Utah, where she would be buried next to Alissa's father. Robbie's parents had just flown into town. Madeline and Samantha struggled to understand where their sister had gone. Alissa had withdrawn into a shell of anguish, refusing to eat or to see anyone outside a small group of family and friends. Robbie needed to speak for them both.

He put on a black suit, white shirt, and deep red tie. He composed a few simple remarks and favorite memories of Emilie in his head and wrote them on letter-sized paper. Alissa hugged him, and accompanied by his little brother, James, and sister-in-law, Natalie, he drove to the church, a single-story building edged with shrubbery and bordered by a parking lot. He and Alissa had avoided watching or reading the news, shrinking from the wall-to-wall coverage of Sandy Hook. So when he arrived at the church, Robbie had no idea he was the first family member to speak publicly.

"I showed up there thinking there was going to be one person and a camera," he told me. "It was twelve cameras, and the parking lot was full of people. I became really, really nervous.

"I didn't want to be there, to have any part of it. I was expecting someone to tell me what to do," he said. But the only ones there besides the reporters were their bishop and family members. He had worked on some notes, laboring to capture in a brief speech their Emilie: a chatty girl whose first word was "happy," painter of mammoth butterflies and tree-sized flowers, picky eater, ringleader big sister, Sleeping Beauty.

At 5:22 p.m., Robbie and the small knot of family and clergy, all dressed as if for services, filed outside. Someone had dragged a heavy wooden lectern from inside the church to the parking lot, a few feet from the church door.

Robbie stood suddenly alone in a narrow strip of space between the wall and the waiting lectern, bristling with microphones bearing media logos. He glanced apprehensively outward as a woman in a parka surged forward to place another recorder on the lectern, then at his family, who hung back, deferring to him.

Robbie turned as if to say something to one of them. Overwhelmed by the surreal nature of it, he smiled incongruously, tilting his head back toward the others in a gesture that in any other circumstance would suggest a cocky aside. "I gave this kind of chuckle, like 'What's going on?'" he recalled. "It was more like me taking a deep breath."

"Start?" he asked no one in particular as everyone watched him in silence. "Okay." Clutching his notes in his left hand, he moved from the sheltering yellow glow of the church entrance as the camera strobes came on, exposing him with a daylight glare. Still grinning, he laid his speech down and his smile died when he looked at the words, as if finally absorbing what they meant. His breath emerged slowly and his jacket opened as his arms moved back, hands searching briefly for his pockets. Lips pressed tight, he closed his eyes and opened them to the waiting reporters, blinking as if emerging from underwater. He breathed sharply in and out to settle himself, the forceful rush of air the only sound. Then he began.

"So, my name's Robbie Parker. My family's one of the families that lost a child yesterday in the Sandy"—he paused, then rushed out the sentence, getting rid of it—"Hook Elementary School shootings here in Connecticut." He cut off a sob.

"I would really like to offer our deepest condolences to all the families who are directly affected by this shooting . . . This includes the family of the shooter. I can't imagine how hard this experience must be for you.

"My daughter Emilie would be one of the first ones to be standing and giving her love and support to all those victims, because that's the type of person that she is. Not because of any parenting that my wife and I could have done, but because those were the gifts that were given to her by her heavenly father," he said.

"As the deep pain begins to settle into our hearts, we find comfort reflecting on the incredible person that Emilie was, and how many lives she was able to touch in her short time here on earth.

"Emilie was a mentor to her two little sisters, delighting in teaching them to read, dance, and find the simple joys in life.

"Emilie's laughter was infectious, and all those who had the pleasure to meet her would agree that this world is a better place because she has been in it," he said.

The reporters asked about Emilie and her sisters, where Robbie was when he learned about the shooting, what his last conversation was with Emilie. Their last words together were in Portuguese, which he was teaching her: a brief "good morning" and "I love you." Then he left for work at the hospital, which a few hours later went on lockdown after news of the shooting that killed her.

"She is an incredible person, and I'm so blessed to be her dad," he said, giving way to raw grief.

Mormon church leaders in Newtown shouldered the planning for Emilie's funeral in Ogden, Utah. US Airways offered to fly Emilie's body there. The route went from Hartford to Charlotte, North Carolina, to Phoenix and finally Salt Lake City. At each stop airline employees signed messages of sympathy and loaded the underbelly of the plane with toys and flowers. When Emilie's body arrived in Utah, the plane's passengers remained on board while her white casket was carried off. Airport employees stood in tribute, lining its slow route from the belly of the plane to a hearse waiting on the tarmac.

Alissa and Robbie arrived on a separate flight. Police escorted them on the drive from Salt Lake to Ogden. Overpasses, fences, and posts along their route were festooned with thousands of pink ribbons and posters honoring their eldest daughter.

THE KILLING OF SUCH young children and their teachers just before the winter holidays prompted a global spasm of heartbreak and generosity. Thousands of letters, cards, and murals decorated and signed by schoolchildren flowed in. "Our hearts are being sent to you all the way from Châteauguay Quebec!" "A Sympathy Card from the Children of Liberia to the Children of Sandy Hook Elementary School in Connecticut, U.S.A." "I wish I could make your heart happy." "Thanks for all of y'all helpin everyone out."

Coworkers, including Robbie's, donated their vacation time to the grieving. Songs, symphonies, and plays were written. Alongside the road near one of the many ad hoc memorials, an elderly man sat in the open hatchback of his car for hours, playing soothing music on his violin. On an easel sign next to him he had written, "Our tears are on your shoulders, and our hands are in yours."

Elderly people sent antique dolls and toys they had treasured as children. Children mailed their bedtime teddy bears and blankets. Classrooms across the country knitted and crocheted winter scarves in green and white, Sandy Hook's school colors.

Such gestures moved the families deeply. "There was so much love, it was beyond imagining," Scarlett told me. She recalled the anonymous donor in North Carolina who, watching scenes of the firehouse on the news, noticed the volunteer firefighters' Christmas tree lot alongside, stacked with pines for its traditional fundraiser. The donor bought twenty-six trees, and volunteers decorated them on the spot, personalizing them for each victim. "I visited Jesse's tree a lot," Scarlett told me, each time finding new decorations and adding some of her own.

Americans tried to heal Newtown's pain and their own under a tsunami of cash and goods.[4] Tens of millions of dollars poured into Newtown. Corporate swag arrived by the queasy-making truckload. Overwhelmed Newtown leaders pleaded with donors to stop and redirect their largesse to their own communities. But it kept coming, sowing resentment and division, ensnaring well-meaning town leaders in bitter controversy that among some persists to this day.

Monsignor Robert Weiss, pastor of St. Rose of Lima Roman Catholic Church, was one of those leaders for whom the gifts would prove a burden. He recalled those months for me over lunch in spring 2019, at one of his favorite haunts, a restaurant in a small shopping center a few minutes from the rectory.

On the day of the shooting the monsignor awoke expecting to spend a rare quiet morning at a hectic time of year. Founded in the mid-nineteenth century to minister to Irish immigrants working in Sandy Hook's mills, St. Rose of Lima had swelled into one of the region's biggest Catholic parishes. "Father Bob" was settling down in the rectory to wrap Christmas presents when the police called. Lock down the school, they told him; there's been a shooting.

The priest strode across a parking lot from the rectory to the church.

Interrupting Friday-morning mass, he told the priest and the students gathered there to shelter in place. He summoned two associate priests and took his BMW sedan to the Sandy Hook grade school, less than five minutes away.

They were the first clergy to arrive. Father Bob knew most of the Newtown police officers, and a couple of them quietly asked him to perform last rites for the uncounted dead still inside the school. He did not know it then, but nine were his parishioners.

Father Bob crossed the school threshold and paused in the lobby, standing on the broken glass. He was afraid, but not for himself exactly. The police had warned him about the horror around the corner from where he stood. He worried that going farther inside would render him useless to his congregation. He closed his eyes and prayed, then walked slowly out.

After the governor delivered the searing news to the families, Father Bob stayed behind in the firehouse, talking with one victim's husband and another's father, both deeply shaken. Then he drove home.

"The church was already packed. This was like four-thirty or so. The mass was at seven. Then I heard that the two U.S. senators are coming, the governor was coming.

"It was just amazing, the church that night. There were as many people outside as there were inside, you know? All the windows were left open, the doors left open, so they could hear.

"We left the church open for days," he recalled.

Father Bob wept often during our conversation, a percussive sound of grief bursting forth, like a cough.

During the two weeks after the shooting, Father Bob officiated eight victims' funeral masses. Two were scheduled so closely together one morning that mourners followed a child's white casket from the church, then reentered for another child's rites.

Some children's wakes took place inside the church as well. In the wee hours of Tuesday, December 18, Father Bob, sleepless, walked over to the church to check on a girl whose funeral was scheduled for later that morning. The priest knew her well. Mourners would wear pink hair ribbons, her trademark, at her service, and Father Bob would pin one to his vestments.

Inside the dimly lit nave, he said, "I saw this police officer sitting next to the casket, reading to the child."

Father Bob recalled the trooper looking up and telling him, "'If I were home, this is what I'd be doing.'"

"Every family was assigned a state police officer, you know, and this one, he stood at attention. For hours. And I got worried about him, because I was afraid of what was happening to him," Father Bob said, pausing to apologize about the grief surging again.

The trooper, a former Marine, assured the priest: "'I was in trouble a lot, so I'm used to standing on duty.'"

Funeral services and activity jammed St. Rose's campus in those weeks. Then came Christmas, a normally joyful time turned wrenching. A few days afterward the post office called to say its trucks hadn't been able to get through, and could someone come and pick up the church's mail?

"So I go down to the post office. I could not believe it. I mean there were just racks and racks and racks, just for us," Father Bob recalled.

People had learned of the parish's losses, and sent cash, about $1.6 million in all. They had heard stories like that of Caroline Previdi, who died in classroom 8. She attended preschool at St. Rose, and at Christmas one year presented Father Bob with a plastic sandwich bag filled with coins, to buy toys for children who hadn't any.

"Everybody brought toys," the priest recalled. The parish rented tents to store them all.

"We called a lot of daycare programs, like from Danbury, Hartford, New Haven," all thrilled to take the toys because it was Christmas.

"Somehow I lost my watch in the mix of confusion at Sandy Hook school. And I said to this friend of mine, 'I cannot find my watch. There was so much hugging going on, it must have come loose, and it just dropped.'

"Within two weeks, I got a watch from Burberry, I got a watch from Cartier," he said. "To this day I can't figure it out. This is the Burberry watch," he said, pushing up the sleeve of his black suit. He received about nine watches, and donated them to a raffle to benefit children with special needs.

But soon the stuff overwhelmed them.

"We could hardly walk in our rectory. I mean, we got everything from hams to Christmas cookies to all kinds of religious things, ornaments. I mean it was hundreds and hundreds of thousands."

Church volunteers worked in rotation to sort the goods. Father Bob limited them to two-hour shifts because the work was so emotionally taxing.

The priest invited homeless shelters, food pantries, and relief organizations to come take what they needed, but they reached the saturation point. He placed full vats in the rear of the church and outside along Church Hill Road, but they sat untouched.

"And that was just St. Rose," he said. Every church, school, and service organization received a flood of items, he said, shrugging and raising his arms, as if snowed under.

Unvetted businesses jammed switchboards with offers of public relations advice, pet therapy, estate planning, and "free facials/nails/makeup/waxing for lifetime for mothers of children who were killed." In the two weeks following the shooting, people from miles around deposited sixty-eight thousand plush toys in Newtown, plus hundreds more in surrounding villages. The toys rolled into muddy streets, clogged intersections, and lay in mounds at the foot of the flagpole on Main Street, sodden, painful reminders of babies in warm beds.

At the end of December, First Selectman Pat Llodra invited the families to pay a final visit to the many outdoor memorials that had sprung up and to take whatever they wanted of the messages, toys, candles, and flowers. Then, late at night the public works department dismantled and carted them away. Voket, the *Newtown Bee* reporter, came along with his camera, trusted by Llodra to record the scene. The town composted the mounds of flowers and burned the rest. Llodra called the ashes "sacred soil" and promised to use them in a memorial to the victims one day.

The post office opened a new substation to process several thousand pieces of mail a day. Crates of toys, books, school supplies, and personal care products arrived. Cast-off clothing and used household items arrived from donors who seemed to confuse a violent crime in a prosperous town with a hurricane that leveled the place.

In a cavernous warehouse Newtown rented downtown, volunteers deposited sacks of mail and parcels into boxes larger than those that hold watermelons in summer, one aisle per victim's family. Alissa and Robbie arrived at the center to find their section overflowing. They filled and refilled their minivan, stuffing everything into their basement.

"Twenty-six of everything was getting delivered," Robbie told me. "We have pictures of our entire basement full of stuff."

"I feel so strongly about honoring the intent in which it was given. But honestly, what am I gonna do with it? What do we do? That becomes a huge burden, another distraction and another source of energy that you have to use."

Newtown residents received invitations to plays, symphonies, art exhibitions, and parades, lifetime memberships to museums, aquariums, and zoos. The Boston Bruins hockey team and the U.S. men's national soccer team provided clinics for Newtown kids. The New York Giants chartered nine buses to transport four hundred residents to a game. Disney premiered *Frozen*, its children's blockbuster, in Newtown and gave away trips to its parks. But not everyone could attend these events, or felt like going. Judgment crept in, and disputes among neighbors.

More bicycles, skateboards, baseball bats, soccer balls, and footballs arrived than Newtown has children. Robbie recalled the parents of another child who died describing residents scrambling to score a pair of green-and-white Nikes with "Sandy Hook" emblazoned on the heel. A Sandy Hook child had requested the sneakers for his classmates as a kindness, and Nike did not overtly publicize the gift. But nobody asked the families how it would feel to see a reminder of their loved ones' murder on their neighbors' feet.

"The shooter wanted to do something, and he used our children to accomplish it. So you feel very, very sensitive to that sensation of being used," Robbie told me.

"So yeah, when a truck rolls in and says, 'Hey, everybody, here's this stuff,' you do kind of question: 'What are you using us for? What did you do that for?'"

Looking at his neighbors, Robbie said he wanted to tell them, "Do you realize why you get to enjoy this?"

MOST DIVISIVE WAS THE MONEY.

Within a month nearly eighty different fundraising groups mushroomed online, some of them ill-equipped or qualified to manage what they collected, and a few of them fraudulent. Nouel Alba, thirty-seven, of the Bronx, set up a scam appeal on Facebook, bilking sympathetic donors to a "funeral fund" for Noah Pozner. Alba falsely claimed that she was Noah's aunt and that she had entered the school to identify his body. She received an eight-month jail term for wire fraud and lying to federal agents.

Americans lead the world in charitable giving, no more so than in the wake of tragedy. In Newtown the donations ignited a bitter public struggle between the families of the victims and the local United Way, which collected millions in the days following the shooting.

Even before the death toll was known, United Way of Western Connecticut had teamed up with the Newtown Savings Bank to establish the Sandy Hook School Support Fund. United Way, as an established nonprofit able to immediately receive tax-deductible contributions, possessed the reputation and infrastructure to collect and distribute the money. But it had no intention of giving it directly to the victims' families. Indeed, United Way's charter prohibits its giving money to individuals. Its Sandy Hook School Support Fund was unrestricted, meaning United Way could distribute the money as it saw fit. By a week after the shooting, the fund had taken in nearly two million dollars.

United Way collects money in keeping with its philosophy that the entire community needs support after a disaster. But its practices had come in for harsh criticism by survivors of past mass shootings, including Caren and Tom Teves, whose son Alex Teves, twenty-four, died in the 2012 Aurora theater shooting in Colorado, and Anita Busch, whose cousin Micayla Medek, twenty-three, was also murdered there. Caren had noticed Cristina Hassinger, Dawn Lafferty Hochsprung's daughter, raising questions about the United Way's fundraising in a local Connecticut newspaper. She called Anita, who gathered victims' families from previous mass shootings, to try to stop donations from being diverted away from the Newtown victims' families.

"Cristina was the first one to stand up and say 'wait, where are the donations going?'" Anita told me. "When you're so traumatized, it's really hard, but Cristina was able to compartmentalize. She was sounding the alarm bell and leading the charge."

Anita, the Teveses, and other survivors of high-profile mass shootings critical of the United Way held a conference call with about ten Sandy Hook family members, including Cristina and Robbie.

Robbie, a person of deep faith who after Emilie's murder had publicly consoled the family of the gunman, directed his full rage toward the United Way.

"You can get as graphic as you want about what our kids suffered. That's the only reason that this money exists," he told me.

"They were saying to us, 'You guys can also sign up for the services, just like everybody else.' You mean, I'm gonna have to submit to your form of questioning and try and prove to you that I'm deserving of services for my family, for my daughters?

"That was one of the things that really got me fired up."

United Way of Western Connecticut eventually raised $10.2 million in Sandy Hook–related donations. In response to the families' criticism, it created the Newtown-Sandy Hook Community Foundation, led by a five-member board composed of community leaders. They would decide how to distribute the money. The move politicized the distribution, several family members said.

One of the five board members was Father Bob.

He and the others scoured donor notes and emails, trying to determine the donors' intent. They decided that of the $10.2 million, $7.7 million would be divided among the families of the twenty-six victims, two wounded educators, and twelve families of surviving children who were in classrooms where children and educators were killed. The remaining $3 million and any future donations would stay with the foundation.

That did not satisfy all the families. They sought out Kenneth R. Feinberg and his colleague Camille Biros, who had worked on similar issues pro bono after 9/11 and other mass tragedies. They began their work with a series of town hall meetings to gauge public sentiment. The meetings in Newtown were among the most wrenching Feinberg had ever attended.

"It was very, very raw," Feinberg told me. Parents wept and fled the room. One mother said the dispute was "compounding the death of my child." Some residents were appalled at what they saw as a revictimization of the survivors by a trusted charity. Others leveled accusations against the community foundation board.

"Sitting on that committee, we were accused of stealing money, accused of taking money to go on trips and buy cars—it was outrageous," Father Bob told me.

"I don't want to paint anyone in a bad or negative picture. Nobody knows what they were feeling or experiencing. But on this side it was a challenge. We were just five local citizens trying to do the best we could. We were overwhelmed."

"There were no villains," Feinberg told me. "But how many people who watch on TV the death of these kids send a check to the United Way, as opposed to helping out these families?

"All the money should go to the families."

The Connecticut attorney general, whose office oversees nonprofits, found that the distribution plan was legal, but criticized the United Way and the Newtown bank for a lack of transparency.

When I wrote about the controversy for the *Times*, Kim Morgan, then United Way of Western Connecticut's chief executive officer, declined to comment for the record, except to say, "We continue to be concerned and have compassion for everyone affected by this terrible tragedy." After multiple requests and a phone conversation, Morgan declined to comment on the record for this book.

Speaking with me about the controversy years later, Robbie brought up *Man's Search for Meaning*, psychiatrist Viktor Frankl's 1946 message for humanity, written after he survived three years in Nazi death camps, including Auschwitz, where his pregnant wife, parents, and brother were murdered.

"My favorite quote in that book is 'An abnormal reaction to an abnormal situation is normal behavior,'" Robbie told me.

"So yeah, the town fought over money . . . You're hurting, and part of the process is you're gonna have anger. And if you throw money into that, then that is just like gasoline on a fire."

The battle prompted other places beset by tragedy to create funds directly benefiting victims. But for Newtown it was an unforeseen additional trauma, with others soon to follow.

Eventually the flood of goods slackened. The town archived examples of the gifts, stored some, and invited the families to drop off anything they didn't want, to be rendered into more "sacred soil."

The families found that of all possible acts of remembrance, the quietest often meant the most.

Detective Rachael Van Ness has never spoken publicly nor to the media about her work with the Sandy Hook victims' families, a refusal to take credit that further cemented their gratitude. This account is drawn from public records and the recollections of people with firsthand knowledge of her work.

Van Ness was among the first state troopers to reach the school on December 14. A photograph of her from that day shows her evacuating a single-file line of children, a little girl in a pale blue top near the front walking with her mouth open in terror. Van Ness wears a navy blue state police jacket and had traded her trademark high heels for ballet flats. Her face intent, long brown hair swinging, she encourages the children forward.

During the weeks after the shooting, Van Ness traveled between Newtown and the state police barracks in Southbury multiple times a day, working sometimes until the wee hours of the morning. She guided families through the school, answering agonizing questions. On Christmas Eve she accompanied Scarlett on her visit to Jesse's classroom. Scarlett laid some greens from the farm on the place where Jesse died, and collected a glass fragment from the lobby of the school, which a friend would set into a pendant for her.

Lenny Pozner, too, met Detective Van Ness at the school. Together they walked through the makeshift plywood door to the hallway the gunman had strewn with bullets. Lenny entered classroom 8. Their desks and cubbies and the HOPES AND DREAMS bulletin board hung with the first graders' depictions of their hopes and dreams were gone. The bathroom where Noah died was stripped to the studs, barely recognizable. Yet standing there brought Lenny peace.

"There's this distance between you and your child when your child is dying without you. My child was getting killed, and I was oblivious to it. We all were," he said. "Learning about his last few seconds, I grew closer to him."

Van Ness and her team held evening meetings at the Edmond Town Hall, updating the families on the investigation. They returned the handbags, cell phones, earrings, and rings of the women who died. They brought a child's backpack to his home, then returned for the bullet fragment his parents found inside. Van Ness delivered a letter to a family that a postal worker had plucked from a processing line, recognizing the scrawled name in the return address as that of a child killed at Sandy Hook.

Some did not want to see the clothing their child, sister, mother, or wife wore on the day they died. But when relatives asked, Van Ness entered the police evidence room in Southbury and retrieved it.

Sometimes the clothes could be professionally cleaned. When that wasn't appropriate, she carried the parcel downstairs to the ladies' locker room, where she kept soap.

Van Ness worked in silence over the white basin, scrubbing T-shirts and jackets, scarves and ruffled tutus until the water ran clear. A silent ritual for the dead, performed in a police station basement.

Van Ness chose boxes and baskets to hold everything, decorating them with paint, decals, and ribbon. She knew Emilie Parker's favorite color from the clothes she wore that day: pink shirt, pink ruffled skirt, pink leggings.

On the sunny morning of March 23, 2013, Van Ness walked up the stone walkway to the Parkers' home, carrying a small white trunk decorated with pink polka dots, Alissa wrote in her book.[5]

Van Ness placed the trunk in Robbie's hands. Robbie removed a note tied with pink ribbon to the top of the little valise:

> "How lucky I am to have something that makes saying good-bye so hard." —A. A. Milne

They looked up, but Van Ness had gone.

The couple sat together on the sofa, sun from the windows warming the room.

Robbie lifted the trunk's lid.

Alissa wrote: "My eyes were immediately drawn to the item on top—a purple fleece scarf that my mother had made for Emilie just one week before the shooting. I had forgotten about it until I saw it in the trunk. My mind went back to that morning when I had stood in the living room with Emilie, wrapping the scarf around her neck and tying it in a knot before she headed for the bus stop."

Robbie lifted the scarf and unfolded it. Six frayed round holes marked the path of the bullet that entered Emilie's neck. Her parents wept.

LESS THAN A YEAR LATER the Parkers left Newtown for good. They moved back to the Pacific Northwest, to a miniature farm on three acres of land dotted with lush cedars and firs. Robbie found a new neonatology position, again with shifts giving him more time at home. He and Alissa had lived in this misty coastal region when the girls were babies. Starting out, without much money for vacations or expensive outings, Robbie and Alissa had jogged on its woodland trails, and Alissa had spent days with the girls on its beaches, where she and Emilie would creep up on seagulls before they burst into flight. It was "where my family used to be whole," Robbie told me. "That was comforting. I was going to be around people who don't live and breathe Sandy Hook and Newtown every single day."

Returning meant escape from the constant reminders and aching tributes, the reporters and battles and the ocean of stuff. They could reclaim their anonymity and breathe. Or so they thought.

3

A FEW DAYS BEFORE Noah's death, Veronique Pozner finished bath-and-story time in the house in Sandy Hook. It was a ritual Noah delighted in prolonging, coming up with last-minute requests and excuses so outlandish they made her laugh. After she finally tucked him into the toddler bed he had nearly outgrown, then kissed Sophia and Arielle in their bunk beds across the room, she went downstairs to her own bedroom. She was in bed with a book when she looked up to see Noah again, standing bare-chested in the December chill.

"What are you doing out of bed, and what are you doing without your pajama top?" she asked, exasperated.

"I just wanted to give you one more hug."

"Okay, but why is your pajama top off?"

"So I could feel your hug better," he said.

On Wednesday morning, December 12, Veronique had to leave early for work. She dropped all three children at Sandy Hook's pre-care program before school. Lenny would pick them up after school that day and keep them through the weekend. "Mommy loves you," Veronique said, squeezing them all quickly, trying not to be late. "See you on Sunday."

She remembers how it felt to hug Noah, the warm solidity of him in that suspended moment before he let go and went on his way.

"A hundred times I've said to myself, why didn't you hug him one more time?" Veronique told me.

"People should try to treasure each other more. Because life eventually disposes of everyone, and you don't know when it's going to happen. So make coffee for your partner, tuck your children into bed. If God forbid that was the last time, you would have that to remember.

"Sometimes I'm so tired, I have to force myself to do that. To do that extra thing. But then I'm like, 'Remember.'"

VERONIQUE HALLER AND LENNY POZNER met in 1987 while they were both still in college, introduced by a woman who had attended Vassar with Veronique. Lenny was studying at Brooklyn College, part of New York City's university system. He worked part-time at Globe Office Supply in Brooklyn as a tech consultant. Computers had been a passion since his childhood, when the industry was in its infancy.

Both foreign-born, Lenny and Veronique shared a sense of otherness that bound them. Lenny was born in Latvia, then part of the Soviet Union, and spent his early childhood in Lithuania. His father worked in aeronautics, and his mother as the assistant principal of a Vilnius music conservatory. The family lived in Israel and Italy before emigrating to America, where they settled in Brooklyn.

Veronique was born in Geneva. She lived in France until she was twelve, when her father took a position with a European bank based in Manhattan and relocated the family to the New York suburb of Scarsdale. Veronique had an offbeat and ironic sense of humor that appealed to Lenny, who shared it. She eventually left Vassar for the College of New Rochelle in New York, where she received a nursing degree.

The couple dated but split over the question of children—Veronique wanted them; Lenny, not right away. Lenny moved to Miami, a city whose international character he loves, and in 1991 founded an information technology consulting firm. He got into motorcycles and spent his vacations in the 1990s taking long road trips, riding to Wyoming and back, Colorado, Maine.

Veronique married a financier and had two children, Danielle and Michael, born a year apart in the mid-1990s, but her marriage ended shortly thereafter. In 2000, Lenny returned to New York and the couple reconnected. It took Lenny, a creature of New York City, six hours to find Veronique's house in Scarsdale.

The couple married in 2003. Looking for a cheaper, quieter place to live, they moved to Bethel, Connecticut. Within three years, Lenny went from a bachelor to a father of five: Danielle; Michael; Sophia, born in 2005; and the twins, born in late 2006. "I was ready for that," he told me.

In 2005 the family moved to a bigger house in Sandy Hook, a four-bedroom colonial on Kale Davis Road with space for an office for Lenny, most of whose clients were located in New York.

Soft-spoken but iconoclastic, with strongly held opinions, Lenny never quite fit into the stoic New England culture of Newtown. "In New York, when people say 'Fuck you,' they mean 'Hey, how you doin'?'" he said. "But in New England, when they say 'Hey, how are you?' it means 'Fuck you.'"

By the time Sophia entered kindergarten in 2009, Lenny and Veronique had begun clashing over the usual things that end marriages: money, stress, different visions for where to live and what to do when the kids grow up. They separated in 2011, and Lenny moved to an apartment in a building in Naugatuck he'd bought as an investment. They co-parented the kids. Lenny, with his more flexible schedule, did more of the school pickups, stopping off at a playground to let Noah run around with Jesse Lewis, whose dad, Neil Heslin, would stand on the sidelines with Lenny and chat.

Lenny and Veronique insist on presenting what Lenny calls "a 3-D perspective" when they talk about Noah. Even before the conspiracy theorists hijacked Noah's story, his parents resisted efforts by well-meaning people to valorize their son. They disliked superlative-laden valedictories imbuing the children with attributes that seemed only to make their parents miss them more.

"There was the salesman in him," Lenny said, chuckling at the memory of Noah sidling up to sweet-talk his way around the rules governing bedtime or treats.

"He was a character, very clever, very sneaky in his five-year-old world," Lenny told me. "He was empathic with others, but at the same time he would get in trouble, and notes would get sent home.

"During naptime he had to nap far away from everyone else," to keep him from chatting.

Veronique agreed. "It was a little hard to get him to sit down and do 'criss-cross applesauce' and pay attention," she said, smiling.

But during a parent-teacher conference when Noah was in kindergarten, his teacher, Lisa Dievert, shared a moment "that really lifted my spirit," Veronique recalled. Dievert told Veronique about Noah's bond with a disabled girl in his class. At snack time he took her under his wing, opening her apple juice and snack before he would sit down to eat his own.

"I thought, you know, that he had a good soul. I had no premonition of what was going to happen. I just remember that day being really uplifted that he was able to show kindness like that at a young age."

Did she tell him?

"I told him. And you know, he kind of brushed it off and was embarrassed, but I wanted him to know that it made me really proud of him, and of being his mother. That he was a good kid, and he was going to grow into a good man."

Kindergarten ended with singing that spring of 2012. Noah, in the second row, belted out "This Land Is Your Land," sweeping his arms wide, making motions to accompany the words. "Noah is a sweet, inquisitive boy and I feel very fortunate to have had him in my class this year," Dievert wrote in Noah's report card.

"I will miss him very much and wish him all the best in first grade."

When Veronique arrived at the firehouse, Lenny was waiting with Sophia and Arielle.

Sophia had been in her second-grade classroom on an adjacent hallway when the gunman shot through the front window and stepped through the hole into the school. She was nearly finished with an art project, a 2013 calendar colored in vivid holiday hues. Minutes later she escaped, evacuated in single file with her class. The route from Sophia's classroom to safety ran past the bodies of Dawn Lafferty Hochsprung, the principal, and Mary Sherlach, the psychologist, Veronique said. Sophia's teacher led the line, telling the second graders to place their hands on the shoulders of the child in front of them and to keep their eyes closed. But that proved difficult for panicked, stumbling seven-year-olds.

"The only thing Sophia ever told me is that she was told not to open her eyes—and she told me she did," Veronique said. "She was sorry for that and she wished she hadn't. Because what she saw was really bad."

Lenny also recalled Sophia telling him she had peeked and seen a woman on the ground, and blood. She developed an abiding fear of calendars, harbingers of a new year that to her marked the end of her normal life. "She sort of instinctively knew that that date would be burned into every year's calendar for all of us," Veronique told me. First came the twins' birthday, on November 20, then Thanksgiving, then December 14. "Sophia knew that this was going to be our fate."

Inseparable at home, Arielle and Noah shared the spiritual, wordless world of twins. "There was no rivalry, just a natural sharing of resources, whether it was food or toys or parental attention. If one was out of sight, the other would soon be looking for him or her," Veronique recalled.

The Sandy Hook administrators had recommended to Veronique that they be placed in different first-grade classrooms to gain a bit more independence. When they reunited at the end of the school day, the twins acted as if they'd been apart for weeks, crashing bellies in a hug, Arielle's long dark hair spilling over both their shoulders.

Arielle was in Kaitlin Roig's class, her classroom two doors down and across the hall from Noah's. Arielle and fourteen classmates were rummaging in their backpacks and through the shelves along the wall under the window that morning, finding favorite books to read aloud for a visiting reading specialist. They were waiting, excited to show off their skills, when Roig heard percussive sounds and cascading glass several paces away from her classroom door. Other teachers recall thinking it was construction noise, or that a heavy piece of furniture had come crashing down. In the library, Mary Ann Jacob heard noises over the school intercom and figured it was another of Dawn Lafferty Hochsprung's Friday-morning principal's office "dance parties" gone a little rowdy. She called down[1] there to tell them she and the fourth graders could overhear them. The school secretary answered from under her desk. "There's a shooter," she whispered.

But Roig knew right away.[2] She leapt from her desk chair, running to close the classroom door and switch off the lights. The door did not lock from the inside, and she had left the keys she needed across the room. The heavy wooden door had a round, porthole-style window; Roig had taped midnight blue construction paper over it for a lockdown drill months ago. That she'd forgotten to remove it may have saved all their lives.

With urgent whispers Roig guided fifteen students toward the room's tiny bathroom, so small the sink was mounted on a wall outside. She pressed, lifted, and stacked them inside, three atop the toilet, one standing on the paper dispenser. Then she wedged herself in, locking all of them in together.

Sweating, running low on air, they listened as the school's public address system, tripped on during the chaos, broadcast the deafening sounds of rapid gunfire, pleading, and moans. "I don't want to die before Christmas," one child whispered. A boy volunteered to lead the way out: "I know karate," he said. Roig told them they needed to stay silent and wait for the good guys to come. Forty-five minutes later they did.

Soon children and parents stopped rushing into the firehouse. After about an hour, Lenny took the girls for a walk, then drove them to stay with friends, and returned.

A police officer approached, somber, not sure where to look. What had Noah worn to school that day?

"Lenny and I talked about it, and between the two of us we were able to figure out everything. I knew his coat because he was wearing his favorite jacket," Veronique told me, a brown corduroy bomber jacket, its fleece lining matted from many washings. The jacket was a child's version of a tough guy's coat, and Noah wore it nearly every day.

Lenny remembered that Noah was wearing his red Batman hoodie.

"They also asked me about his backpack and his lunch box. They were trying to reconcile everything, and possessions, and um . . . ,"

Veronique said, and looked at the ceiling, remembering. "What a grim, horrible task it must have been for those people too."

At around three o'clock, Malloy, the Connecticut governor, entered the firehouse. Veronique was enraged by the hours the families spent with no information. It was she who tore an interruption into Malloy's halting opening lines.

"What are you telling us?" she shouted. "They're all *dead*?"

Her demand sounded so ragged that someone asked whether she needed an ambulance. Others glared at her, eyes filling. Malloy would remember her voice as an accompaniment to his own words, ending their waiting and stealing their hope.

"I just remember the wailing. People just falling to their knees," Veronique said. "It was like hell. There's no other way to describe it. I mean—it was just hell."

The Pozners' rabbi, Shaul Praver of Congregation Adath Israel in Newtown, stood with them,[3] begging Veronique, catatonic, to breathe, trying to reach her by telling her that no matter what his fate, as a being with a soul, Noah was not lost. He seemed to reach her by saying, "Death doesn't really exist—it's just a transformation. Because we all come from God, and everything in the world is from God."

NOAH'S FUNERAL TOOK PLACE on Monday, December 17, in Fairfield.

Veronique and Lenny had delayed the service by a day because President Obama came on Sunday the sixteenth for the prayer vigil at Newtown High School. The president hugged Veronique. She told him about dreams in which she sought her lost boy, and the president told her that surely, Noah was reaching out for her too.

Noah and his classmate Jack Pinto were the first children to be buried. Their services began at the same hour, 1:00 p.m., on that raw and rainy Monday. A crowd of reporters and more than two hundred mourners stood in line beneath umbrellas, waiting to enter the Abraham L. Green and Son Funeral Home, a white-painted Victorian house whose gingerbread facade seemed too cheerful for the gathering that day.

Veronique's work as a nurse put her in daily proximity with lives that end too soon. The euphemisms surrounding death irritate her. From the beginning she referred to Noah as "murdered," a word too brutal for some to utter in those early days.

"I kept hearing things like 'All these little angels went to sleep and went off to heaven.' *No.* They weren't angels and they didn't go to sleep. They were brutally murdered with a military-style weapon," she told me.

"Sometimes reality needs to be looked at right in the face," she said. She meant Noah's face, destroyed by gunfire. She wanted Governor Malloy, who was attending Noah's funeral, to see it.

NOAH'S WAS THE FIRST of many Sandy Hook services Malloy and Lieutenant Governor Nancy Wyman attended. Years later Malloy unspooled his recollection of that afternoon in the present tense.

"She meets us in a kind of anteroom of the funeral home in Fairfield. She makes it very clear that she wants to show me what they did to her son," Malloy said.

"'We're having an open casket,'" he recalled Veronique telling him. "And she grabs me by the hand and we kind of go."

It was three days after Veronique had confronted him in the firehouse. Malloy understood she had found his words lacking that day, and he worried he would disappoint her again. He knew from his years as a prosecutor that crime scenes and evidence of human brutality traumatized him. That was why, standing at the bullet-scarred entrance to Sandy Hook Elementary that Friday, he had turned back.

"I can still remember the steps between meeting her in this anteroom and then getting to where the body is displayed," in a small private room at the rear of the funeral home. "I'm thinking to myself, 'I'm going to pass out. She's going to show me open wounds and I'm not going to handle it very well.'"

But he could not fail this woman in a somber dress and dark knitted shawl, leading him to where her son lay, twenty minutes before they closed his casket.

Noah wore a dark suit and a tie gifted by a friend. Folds of a blue-and-white tallit gathered about his shoulders. A square of delicate white fabric hid his mouth, much of which had been destroyed by a bullet.

"Noah actually looked like he was sleeping. He looked very peaceful," Veronique told me. "I wouldn't have taken it to that level," she said, meaning asking Malloy to gaze on Noah's uncovered wounds.

"But he was still looking at a dead child. And that's what I wanted to show him. A child who practically the day before had been running around like a little locomotive, full of life."

After Noah's death there surfaced in Veronique a rare eloquence. She found unerring words to describe her youngest son, whose death, as she said, "mattered."

At the front of the rows of seated mourners inside the funeral home's large and airy main chamber, Sophia and Arielle draped themselves over Lenny's lap. Veronique rose and, carrying her notes, walked to the head of the small mahogany casket to eulogize her son.

"I will miss your forceful and purposeful little steps stomping through our house. I will miss your perpetual smile, the twinkle in your dark blue eyes, framed by eyelashes that would be the envy of any lady in this room," she said, as a ripple of laughter warmed the room, heavy with the scent of perfume and cold rain on wool.

"Most of all, I will miss your visions of your future. You wanted to be a doctor, a soldier, a taco factory manager. It was your favorite food, and no doubt you wanted to ensure that the world kept producing tacos.

"You were a little boy whose life force had all the gravitational pull of a celestial body. You were light and love, mischief and pranks. You adored your family with every fiber of your six-year-old being. We are all of us elevated in our humanity by having known you. A little maverick who didn't always want to do his schoolwork or clean up his toys, when practicing his ninja moves or *Super Mario* on the Wii seemed far more important.

"Noah, you will not pass through this way again. I can only believe that you were planted on Earth to bloom in heaven. Take flight, my boy. Soar. You now have the wings you always wanted. Go to that peaceful valley that we will all one day come to know. I will join you someday . . . Until then, your melody will linger in our hearts forever."

Noah's grave site sat nearly alone in a windswept new section, bordered by a low stone wall, at the rear of B'nai Israel Cemetery in Monroe, ten miles southeast of Newtown. His headstone of luminous blue granite came from an anonymous donor. Twin green canopies shielded the mourners from the weather.

After the traditional Jewish prayers, Lenny dropped the first shovelfuls of earth onto his son's coffin. The soil's heavy wetness amplified the hollow sound. Relatives and friends joined, some placing a comforting hand on Lenny's back as they stepped forward, then passed the shovel to the next. When Noah's grave was filled, Veronique, Lenny, and the girls let go two bouquets of balloons, one white, one blue. An AP film crew, positioned at a distance, captured the private service and posted it online, the audio silent except for the wind and the rattle of camera autowind. When Lenny saw it, he demanded they take it down.

Noah's name looked unnaturally solitary on the far-left edge of the vast, blank tombstone. Lenny remembered wondering why they chose such an imposing monument for a little boy. "Maybe because we wanted to just climb into the ground along with Noah," he said.

"WHAT DO YOU WANT people to know about Noah?" CNN's Anderson Cooper asked Veronique.

On the evening of December 21, a week after the shooting and five days after Noah's funeral, Veronique stood before the Edmond Town Hall in downtown Newtown, telling Cooper about her little boy. She wore black, with a deep-toned shawl wrapped around her shoulders, her black leather gloves slightly worn at the fingertips as she gestured, palms open, conveying her loss.

"He was a 6-year-old little boy. He loved running and playing with his siblings. And he loved bubble baths and fireflies. And he loved eating the inside of Oreo cookies," Veronique said.

"He was just a bundle of energy, like he was supposed to be."[4]

Behind her the windows of the red-brick, white-columned building, a hub for community and police meetings after the shooting, shed yellow light onto the granite steps. It was decorated for the holidays, evergreen wreaths gracing the arched transoms of the main entrance and windows. While Veronique spoke, a girl walked up to read the messages attached to bouquets of yellow, pink, and white roses mounded from the steps to the sidewalk, their cellophane wrappings quivering in a faint breeze.

"How are you holding up?" Cooper asked Veronique.

"Most of the time, I'm kind of numb," she replied. "You know, I think every mom out there can relate to the fact of how long it takes to create a baby, those nine months that you watch every ultrasound and every heartbeat. And it takes nine months to create a human being, and it takes seconds for an AR-15 to take that away from the surface of this

Earth. And it wasn't just my son. It was 25 other souls that left this Earth that day because that weapon fell into the hands of a tormented soul. And that haunts me."

She didn't have all the answers on gun policy, she told Cooper. She wanted Americans to remember Noah and all the victims, and understand that this could happen to them too.

"You know, if I asked everybody in this world who has ever loved someone, who has ever had a human being in their life who was essential to their well-being to raise their hand, I don't think there would be many hands down in this world," she said. "And every one of those hands is a reason why those weapons should not be out in the general public."

4

ALEX JONES IS A BARREL-CHESTED, vain man with a taste for muscle cars, Grey Goose vodka, and chunky Rolex watches. Though he seeks to portray himself as a feral, growling renegade, in fact he grew up a cosseted child in the affluent Dallas suburb of Rockwall, son of a dentist and a homemaker.

A community-college dropout, Jones had made his name and serious money pushing wild conspiracy theories, making up in desk-pounding passion what he lacked in facts. Most centered on his belief that a totalitarian world government, the American federal government, and powerful "globalist" business interests—either separately or in concert—aimed to subjugate freedom-loving people like him.

"Ladies and gentlemen, it is Friday. Thank you so much for joining us, the fourteenth day of December 2012," Alex Jones began his daily four-hour Infowars radio broadcast on the day of the shooting, his voice as usual sounding like twenty miles of rough road. He distorted his mouth as he spoke, speaking out of the left side while grinding his teeth, a tic of long duration.

Jones, then thirty-seven, wore a mud-brown collared shirt, his expansive brow moving up and down beneath the beginnings of a comb-over. He leaned into the chunky metal mic in front of him, making wagging, chopping gestures with hands resembling inflated rubber gloves, an unlit cigarette stub jammed between his right index and middle fingers. "There is a reported school shooting, ah, in Connecticut, one of the states that has draconian restrictions on gun ownership," he said. "The media will hype the living daylights out of this."

It was 11:00 a.m. Texas time in Alex Jones's studio, located behind security cameras and smoked windows in a sun-blasted industrial complex hunkered on the dusty outskirts of Austin. Jones liked to boast that Infowars' location is a secret, although an entry-level Google search reveals it. Years later, he would tell me that snipers positioned on the roof protected his enterprise, then uncharacteristically called me back to acknowledge he had lied.

Less than three hours after the gunman entered Sandy Hook Elementary, Jones, his brow furrowed in theatrical concern, began speculating, not incorrectly, that the shooting would reinvigorate the national gun control debate. He also was one of the very first Americans to speculate, falsely, that the event was staged,[1] a fact he would deny over subsequent years, insisting he was only echoing the claims of others.

JONES IS NOT AN IDEOLOGUE; he is a salesman and a diagnosed narcissist.[2] Chasing his mostly male, mostly Christian conservative audience, he has shifted over the years from Ron Paul–style hippie libertarian to evangelical to mouthpiece for Vladimir Putin's authoritarian rule in Russia. But Jones's roots are planted in the globalist paranoia of the far-right John Birch Society. Though exiled by Republican Party leaders a half century ago, the Birchers' brand of radical conservatism reemerged and engulfed the party by the mid-2010s, helping to elect the next president, who shares some personality traits with Jones.

Jones has always been staunchly opposed to limits on gun ownership, real or perceived. Guns fit into his perception of himself as an avatar of patriotic, Christian conservatism. By 2012, Jones had long been a celebrity among far-right "patriot" militias, including some of those behind armed standoffs against the federal government in the West. A race baiter who protests that he is not racist, Jones was an early adopter of the "birther" lie, claiming Obama is a foreign-born Muslim.

This twisted hierarchy of needs drove Jones's conspiratorial interest in Sandy Hook. For Infowars, his radio and online outlet, a horrific massacre of children meant engagement, attention, and a chance to spin a myth of official fraud and cover-up starring himself as crusading truth-teller.

FROM THE BEGINNING, Jones's viewers prodded him to say that Sandy Hook was a "false flag" operation, a term of art conspiracists borrowed from military and spy parlance. It originally described an archaic wartime maneuver in which a navy warship flew the flag of another country, a noncombatant or an enemy, to draw closer to an enemy target, raising its true colors just before engaging in battle. In the conspiracy world, the term describes an event staged by one actor, usually the federal government, aiming to pin it on others as a pretext for broader, tyrannical action. The year 2012 saw a sharp increase in mass shootings and, correspondingly, in the number of people claiming without any basis in fact that they were staged by the government as an excuse to confiscate Americans' firearms.

All familiar terrain for Jones, who five months earlier had speculated that the mass shooting at the theater in Aurora, Colorado, was a false flag operation and that the twelve people killed and seventy injured were somehow actors in the plot.

"We got some bad news," Jones told his listeners about an hour into his show. "They were first reporting that they'd shot the shooter. Now it looks like at least one child killed, some others wounded, and the shooter is a father of the student.

"I can just feel the bad vibes during the holidays. Twenty-seven dead? Okay, this is . . . oh my God." The number seemed to stagger him. But he barely missed a beat.

ALEX JONES, CONSPIRACIST, SALESMAN, conjurer of dark American impulses, comes from a long line of demagogues and mass media manipulators. Inspiring a cult-like following with passionate exhortations to prayer and armed defense of Christianity, Jones has been a onetime admirer of fellow Texan David Koresh, the leader of the Branch Davidian sect who died with scores of his followers in an FBI siege on his compound near Waco, Texas. Jones, a Koresh admirer, raised money and a team of volunteers to rebuild Koresh's church.[3]

JONES'S STIRRING OF EXISTENTIAL FEARS among some white Americans that they are a declining force brings to mind 1930s radio broadcaster and antisemite Father Charles E. Coughlin, or the Nazi Party propagandist Paul Joseph Goebbels. All three men have espoused "great replacement theory," an antisemitic lie that Jews and darker-skinned accomplices aimed to usurp white Christians' position at the top of society, by whatever means necessary. It's an ancient and durable falsehood, promulgated most recently by some members of Congress and Fox News' Tucker Carlson.

Jones grew up a bully and negative-attention-seeker, his fanciful theories about wrongdoing by authority figures scoffed at by his peers. His family's move from Rockwall to Austin while he was in high school was likely precipitated after Jones repeatedly slammed and kicked a classmate's head, fracturing his skull in multiple places, and then received a severe beating himself as retribution. Jones has called most of this story, first reported by Welsh author and filmmaker Jon Ronson, "horse crap." But the participants confirmed it.[4] Jones did not dispute that his father, David, paid the injured teenager's hospital and neurologist bills, bargaining unsuccessfully for his son to be left alone.

Jones got his start in the early 1990s, with simultaneous shows on Austin radio station KJFK and on Austin public access cable. He grew into a cult draw in the Texas capital, whose counterculture motto is "Keep Austin Weird."

In 1993, after the siege by federal law enforcement and resulting inferno at the Branch Davidian compound killed four law enforcement officers and about eighty Davidians, Jones maintained, evidence to the contrary, that the sect and its leader, David Koresh, were a peaceful religious community targeted by the government for murder. In an early sign of his influence and bankability, he raised $93,000 from Infowars listeners to rebuild the compound's church. In 1995, Jones pushed bogus claims that the government plotted the bombing of the Alfred P. Murrah Federal Building in Oklahoma City, whose true perpetrator, domestic terrorist Timothy McVeigh, expressed rage at the Waco compound's destruction.[5]

Jones met his first wife, Kelly Nichols, on a sweltering day in downtown Austin in 1998. Jones was jumping up and down on a busy city street, wearing a fat-rumped bumble bee costume as an advertising stunt for KJFK.

The future Kelly Jones had been a PETA activist, skilled at ginning up media attention.

Initially, Alex and Kelly made a good team. Together they founded Infowars around 1999, in an unused nursery in the couple's Austin house, with "choo-choo train" wallpaper, as Kelly likes to tell people. Jones used an ISDN line to broadcast his theories to radio stations across the country. The couple produced conspiracy-themed, feature-length videos. They sold them by mail or gave them away, urging people to pass them around and to spread the word.

Jones owes some core conspiracy themes and his product-marketing model to the late Gary Allen, a former speechwriter for Alabama governor George Wallace and reactionary author of the 1971 bestseller *None Dare Call It Conspiracy*. The book posits that a cabal of global bankers and power brokers, not elected officials, controls American policy making. Allen, one of the John Birch Society's most revered writers and thinkers, sold his books, filmstrips, and cassettes by mail order. (Allen, incidentally, is the father of Mike Allen, professionally accomplished and personally quirky Washington journalist now at the news outlet Axios.)

Jones found Allen's book on the shelves of his father, who had a string of successful dental businesses. David Jones bankrolled his son's ventures and, as he had when Jones was in high school, tried—often unsuccessfully—to keep him out of trouble.

In 2000, Kelly, nicknamed "Violet," accompanied Jones and his producer, Mike Hanson, on a trip to infiltrate Bohemian Grove, an annual summer glamping retreat for international business and political leaders, near Monte Rio, California. Author Jon Ronson accompanied Jones to Bohemian Grove, a trip that became a chapter in Ronson's book *Them*, about his travels with extremists. In the book Ronson documents how he and Jones and their associates sneaked into the redwood-enshrouded summer camp in Northern California for its "cremation of care," a pyrotechnic spectacle that symbolized the group's suspension of worldly cares for the weekend. Afterward, Jones ridiculously claimed that what Ronson judged to be a fratty, midsummer gathering for aging global power brokers was in fact "a pagan ceremony worshiping the earth and engaging in human sacrifice."

The trip yielded hours of paranoid Infowars chatter. Murky footage Jones and Hanson shot inside the grove became the VHS video *Dark Secrets: Inside Bohemian Grove.* Jones sold the full video by mail order, streaming footage from it on his website. The "first ever video from inside the northern Californian globalist retreat" helped turn Alex Jones into a fringe internet sensation. Not long after, a heavily armed man entered the grove and set a fire.[6] He had seen Jones's broadcasts and had come to stop the pedophilia and human sacrifice he believed took place there. Nobody knew it at the time, but the incident was an early warning of problems to come.

LENNY POZNER LISTENED TO Infowars broadcasts in his car in those early years, traveling between clients. A congenital skeptic, Lenny indulged in conspiracy theorizing himself in his youth. He enjoyed the intellectual exercise and imagination of rethinking historical narratives. "When I thought of conspiracy theories, I thought of *The Da Vinci Code*," he told me.

Lenny enjoyed listening to how pundits "on the edge of information" interpreted the news. "I didn't want to hear what everyone else was hearing," he told me. Through his company he "understood the Web and most of the sites and search engines," he said. "I'd seen it evolve through the 1990s and was aware of it as it was developing." So he found the Infowars business model intriguing.

"A lot of what he does is plant sneaky ideas to lead people to search on something only he's talking about," Lenny said. "When he would say 'look it up,' he was making it up. If he can control millions of searches on a particular question, he's shaping what's trending. If he says the shoe bomber had dinner with the Bushes the week before, and 'look it up,' he's shaping internet traffic."

Jones expanded his reach, his radio show airing on nearly one hundred stations across the country on the conservative Genesis Communications Network. His career high came in summer 2001, when he says he "predicted" the September 11, 2001, terrorist attacks. On his show in July of that year he begged Americans, "Call Congress. Tell them we *know* the government is planning terrorism." Jones mentioned the World Trade Center and speculated that the government would blame the attack on Osama bin Laden, "the boogeyman they need in this Orwellian, phony system." (Bin Laden and al-Qaeda's involvement in the 1993 bombing of the World Trade Center had been known for years.)

As the nation reeled on September 11, Jones trumpeted an "I told you so" on his show, claiming the attacks were a U.S. government plot. Nearly three-quarters of the stations carrying his show canceled him, he claims. He described this period to me as a test of his gumption, hyperbolizing about the rich film deals and TV commentator jobs he could have gotten had he renounced his conviction that the attack was an inside job.

He needn't have worried. His September 11 notoriety propelled Jones from Austin stoner recreation to a hero among conspiracy-minded 9/11 "truthers," the internet having expanded Infowars' audience exponentially. Jones harbored Hollywood ambitions too. In a bit of deft typecasting, director Richard Linklater, who also got his start on Austin public access TV, gave Jones a bit part as a bullhorn-shouting street prophet in his 2006 cult film *A Scanner Darkly*, starring Keanu Reeves and Winona Ryder. The film, Jones's second cameo with Linklater, is based on the Philip K. Dick novel, set seven years into a future dystopia, in which the government spies on a drug-addicted America.

By 2012, Jones had launched an ingenious new business model. On every Infowars broadcast, Jones interspersed his rants with ads for bizarre diet supplements, air- and water-filtration systems, dried food, untraceable gun components—everything needed to survive the end of times, available in the online Official Infowars Store. In a dark Infowars world where the feds poisoned food, water, minds, and individual freedoms, Jones sounded the warning and sold an answer:

"Sick of globalist, eugenicist control freaks adding poison to your water and laughing as you get sick and die? Start purifying your water with ProPur . . . ProPur is the best gravity-fed filter out there. It's what my family uses."

Some of the products and manufacturers, like the water and air filters, weren't out of the ordinary. SimpliSafe home security was another Infowars advertiser in those years. But others, like the libido-enhancing, brain-building supplements, were pure snake oil. Either way, it was Jones's pitch that was specific, adapting any product as an answer to the perceived need by his audience to gird themselves against the coming globalist takeover.

Jones proved a gifted salesman. His father, David, the dentist-entrepreneur who brought Infowars into the supplements business and took over the company's books, said in a court deposition that when Alex pushed a product online, sales spiked, moving a stream of merch out of the warehouse behind the studio.

Jones started living like a celebrity, hanging out with Dave Mustaine, frontman for thrash metal giants Megadeth; the filmmaker Linklater, who lived in Austin; and the Smashing Pumpkins' Billy Corgan. He partied in California with his fellow 9/11 truther Charlie Sheen, star of the hit CBS show *Two and a Half Men*, and at the time the highest-paid actor on television.

In 2011, the same year that Jones told *Rolling Stone*, "The globalists want us turned toward Hollywood and the TV so they can poison us," he invited Sheen onto his show. Sheen, in the throes of a self-destructive spiral, proceeded to blow up his career. He lobbed antisemitic insults at his show's cocreator Chuck Lorre and called Alcoholics Anonymous a cult that "brainwashed my friends and my family." Infowars had planned a joint media project with Sheen that year dubbed Planet Sheen. It never got off the ground for reasons that are unclear but likely related to the actor's meltdown.

In those years Kelly, Jones's right hand, was preoccupied with homeschooling the couple's three young children and tending to the demands of their new, moneyed lifestyle. The couple had a house on Lake Travis, but their base was a walled compound outside Austin, a heavy-beamed, stucco former ranch house with a safe full of firearms. The kids tumbled about with the family's four dogs: Kelly's Chihuahua, Bambi, a Labrador named Biggie, and two French bulldogs, Captain Fantastic and Sparky.

Kelly was working with contractors to build "this Disneyland pool that was Alex's idea" near the compound's gates, with multiple levels of basins, waterfalls, stone dining grottos, a restaurant-sized barbecue, and a pool house that doubled as a tap dance studio for Kelly.

The Joneses netted $5 million a year in 2012 and 2013, according to tax returns submitted in court proceedings.

"We were making more and more money. It was more than we needed. It was just this weird thing when you have so much money that you're like, 'What am I gonna buy next?'" Kelly told me. "It's just actually, like, cumbersome. And you have multiple people to manage, and your life isn't really private, and it becomes this, like, material monster, you know?"

In those years, "Alex wouldn't let us fly at all, because he wouldn't go through TSA," Kelly said. "He bought this horrible tour bus, and every family vacation was just like an Infowars broadcast," with the bus driver stopping in various towns so Jones could party with his fans. Alex loved it, she said, "Listening to heavy metal on this horrible diesel tour bus with three kids, and I'd be in the back with motion sickness."

Lenny, flipping the channel to Infowars on long rides from Connecticut to clients in New York, noticed a change in Jones's show. Jones offered fewer interpretations of the news and more paranoid rants.

Jones nurtured a bromance with the former Minnesota governor, actor, and World Wrestling Federation performer Jesse "the Body" Ventura, and they would discuss their shared paranoia about post-9/11 security measures. "The content lost interest," Lenny said. "He had Jesse Ventura on his show a lot, and they would talk about the airport scanners, and it didn't make sense to me. Why is he against airport scanners and why is he making it the object of his show?"

But Ventura's presence on the show, like Jones's hiring of former wrestler Dan Bidondi as a cameraman, was a tell, Lenny told me. He thinks Jones has crafted his broadcast around WWF, now known as World Wrestling Entertainment (WWE), the money-spinning franchise built on wild personalities like Ventura, "Macho Man" Randy Savage, and Andre the Giant, engaged in brutal, staged ring combat. "I'm convinced he liked that model, the showmanship of it," Lenny said. Like Donald Trump, who famously tangled in the ring with World Wrestling Entertainment chairman Vince McMahon, Jones saw WWE as a magnet for the same audience he sought. Jones brought WWE's style of fakery to Infowars, whose newscasts bore about as little resemblance to actual journalism as the Trump-McMahon "Battle of the Billionaires" did to Greco-Roman wrestling.

"It's what I see when I see Jones," Lenny told me. "He's kind of smirking when he does his schtick." In order to move products like My Patriot Supply storable food, solar generators, Supernatural Silver, Real Red Pill, and Tropic Mind Power supplements, Jones and his "reporters" mined for viral content that fit his agenda. He urged his fans to get out there, "crowd-source," and "investigate." These "citizen journalists" did most of the work, if you could call it that, combing official reports of major events for "anomalies" that to them suggested official lying and cover-up.

"I was kind of a governor on him when he went over the edge, saying like, 'No, don't do that,'" Kelly told me of her role when she was more involved in the business. But success had drawn them apart: "Our family was an adjunct to Alex, who was leading a completely separate life."

Immediately following Sandy Hook, Jones tread carefully, avoiding implicating the victims or their parents in his "gun grabber" narrative. But it wouldn't last. Outrage and sensation, traffic boosters of the internet, soon proved an irresistible lure for one of its most insatiable attention gluttons.

JONES WAS WORKED UP about Obama's reaction to Sandy Hook even before the president's remarks from the White House briefing room the afternoon of the shooting.

Obama was a favorite Infowars target, but professionally, Jones had mixed feelings about him. A Democrat in the White House elevated his listeners' fears of liberal-led plots and goosed Infowars' ratings. Jones estimated in court testimony that a Republican in the White House "will cause a thirty to forty percent, maybe even more, reduction in audience." In December 2012, Obama had just been reelected. During the campaign, Jones had seemed so dejected at the prospect of a Mitt Romney presidency that his staff speculated that Jones had voted for Obama, the man he called a "demon." He has not answered that question, as far as I can tell.

At 3:15 p.m. on December 14, Obama addressed the nation.

"We've endured too many of these tragedies in the past few years," he said, pushing tears from the corners of his eyes. "And each time I learn the news I react not as a president, but as anybody else would—as a parent. And that was especially true today.

"We're going to have to come together and take meaningful action to prevent more tragedies like this, regardless of the politics."[7]

On his show that day, Jones treated Obama's call to action on gun policy as a cataclysmic threat: "They're gonna come after our guns, folks. Get ready. Stay with us."

Jones canceled most of his guests that day and instead took calls from listeners. Most of them sought Jones's guidance. *What* was Sandy Hook?

John, a caller from Connecticut, got right to the point. Did Jones think the shooting was a staged act?

"People like these who are doing these shootings are being put in place by some agency or the government to promote a gun grab or keep the public in fear," John said. He noted that mass shootings seemed to be happening "like every weekend" that year. To him, that was evidence not of a societal problem but of a government plot.

"Well, we know Columbine was staged," Jones said, hedging by retreating to a well-worn theory. "They reported four different gunmen, there were one hundred and twenty-seven bombs in the building, half the school was empty, there was a stand-down, there were government mind control programs connected to both of the youth . . .

"Regardless, the White House is gonna use this to go after our guns."

Jones picked over initial reports on the shooting, using mainstream outlets' efforts to correct early reports of multiple shooters as evidence of a cover-up: "They've started scrubbing articles on this," he said. On it went through the afternoon broadcast, Jones's listeners offering scraps of theories and "anomalies" in the official reporting.

In their calls that day, Jones's followers urged him to be their oracle, to help them understand Sandy Hook as a phony event. Dan Friesen studies Jones's role in America's long history of political conspiracism for his podcast, *Knowledge Fight*. To understand how Jones came to be the foremost vector for the Sandy Hook conspiracy theory, Friesen listened to how Jones talked about the Sandy Hook shooting from the day of the tragedy forward, tracking his narrative arc.

On the day of the shooting, "Jones was trying to dance around and figure out how to make a false flag narrative out of this. He wanted to suggest it was fake immediately—but he also didn't seem to care that much," Friesen told me.

"It's only after he finds out that lots of kids have died that he moves into the territory of trying to talk himself into it being a false flag."

At that point Jones could not ignore the threat the massacre posed to his pro-gun agenda.

"Once there are kids that are dead, Alex can recognize that denial may be a useful tool. On some level he knows that if these events are real, it's a decent argument for gun control," Friesen said.

"There's no counterargument he can come up with other than 'it's fake. This guy didn't do this.'"

AFTER A BREAK to pitch Infowars supplements, Jones said solemnly, "I feel like we should say a prayer for all the children being killed around the world by the globalists."

And then he went for it.

"My gut is, with the timing and everything that happened, this is staged. And you know I've been saying the last few months, get ready for big mass shootings."

Later, "Why did Hitler blow up the Reichstag?" Jones bellowed. "To get control! Why do governments stage these things? To get our guns! Why don't people get that through their head?

"Heaven help us."

5

Sandy Hook was the first mass shooting to ignite viral, fantastical claims that it was a phony event staged by the federal government as a pretext for confiscating Americans' firearms. Since Sandy Hook, virtually every high-profile mass tragedy has generated similar online theories.

The worst elementary school shooting in American history, Sandy Hook capped a brutal year for mass shootings. It was viewed by Americans on both sides of the gun policy debate as a watershed event and by a segment of gun-owning conservatives as a signal threat.

The tragedy occurred amid surging growth in the use of social media, whose algorithms boosted sensational and outrage-producing content. Sandy Hook drew months of online furor, including among people who out of delusion or political expediency decided to cast it as a hoax.

The most lethal American school shooting before Sandy Hook was in 2007, when an undergraduate with a history of mental health problems killed thirty-two and injured seventeen at Virginia Tech, in Blacksburg.

I called Lori Haas, whose daughter Emily was injured at Virginia Tech. The shooting spurred a similar effort to change the nation's gun laws, which Haas joined, participating in protests across the country. But she doesn't recall any conspiracy theories or harassment.

"The reaction to Virginia Tech was much the same as the reaction to Sandy Hook. That this could happen was beyond comprehension, that someone could shoot your children down in cold blood," Haas told me. "We didn't have the disinformation campaigns and the fuel that social media platforms generally give them."

In 2007, the major social media platforms hosted only a fraction of the users they did just five years later, when Sandy Hook ignited millions of posts pushing false claims. Facebook, by far the largest social platform, had twenty million global users in 2007, compared with more than a billion in December 2012. YouTube averaged an explosive one hundred million videos viewed per day in late 2006, at the time of its sale to Google. But by 2012 a single video—"Gangnam Style," Noah Pozner's favorite—was the first YouTube video to exceed one billion views. Twitter was barely a year old in 2007, with five thousand tweets per day. By the end of 2012, users sent five thousand tweets every *second*, or four hundred million tweets every day.

Most consequentially, Sandy Hook occurred on the cusp of a profound shift in American politics, in which politically expedient "alternative facts" muscled out objective truth. From the dregs of the internet to the White House, the decision by a broad swath of Americans to replace our nation's self-evident truths with partisan lies helps explain how the murder of children and educators in a quiet New England town grew into a battle over truth itself.

SHOOTINGS CLAIMED MULTIPLE VICTIMS nearly every month in 2012.[1] Several of the attacks took place in schools. On February 27, 2012, in Chardon, Ohio, a seventeen-year-old boy entered the Chardon High School cafeteria and took aim at four students, killing three. On April 2, a forty-three-year-old dropout of Oikos University in Oakland, California, lined students and a school administrator against the wall of one of the university's nursing classrooms and began firing, killing seven of them. On May 30 in Seattle a forty-year-old man with a history of mental health problems began a shooting spree in a café, ultimately killing five people and, after a standoff with police, himself. A dozen people died and scores were injured in Aurora, Colorado, on July 20 when a gunman opened fire during a midnight showing of *The Dark Knight Rises*. Less than a month later, on August 5 in Oak Creek, Wisconsin, six people died when a white supremacist opened fire in a Sikh temple, then fatally shot himself. On September 27, a thirty-six-year-old man fired from his job at a Minneapolis sign company shot and killed six people there, including the company owner, two managers, employees, and a delivery driver before turning the gun on himself.

Most of the weapons involved were purchased legally, including the rifles and handgun used at Sandy Hook. The firearms industry's main lobbying group, the National Shooting Sports Foundation, happens to be based in Newtown, and a number of its employees live there. I visited its headquarters in 2019. A security guard stood behind concrete bollards off the parking lot and stopped me to ask what business I had there. Inside, Steve Sanetti, the foundation's president and chief executive, told me the foundation stayed silent for a few days after the massacre. Then it began promoting a program called Project ChildSafe, distributing free gun locks to firearms owners.

He told me he met with victims' family members, who asked the group to support legislation for universal background checks. Sanetti told them background checks would not have prevented this shooting, so "you're barking up the wrong tree."

"They didn't like that," Sanetti said.

Within an hour of the shooting, the National Rifle Association's chief lobbyist had relayed his list of the most likely gun control legislation to be proposed in its aftermath, including high-capacity magazine bans, expanded background checks, and legislation addressing sales of assault-style rifles, according to *Inside the NRA*, a 2020 book by Joshua L. Powell, a former adviser to the association's leadership.[2]

As it usually did after mass shootings, the NRA went silent for several days after Sandy Hook. A week after the massacre, Wayne LaPierre, its executive vice president and the paranoia-mongering face of the organization, held a news conference in Washington. In a belligerent, half-hour speech, LaPierre blamed the shooting on violent video games, movies, news media that "demonized gun owners"—and the lack of guns in schools.

"Does anybody really believe that the next Adam Lanza isn't planning his attack on a school he's already identified at this very moment?" LaPierre asked.

"The only thing that stops a bad guy with a gun is a good guy with a gun," he said, delivering a line that would become a cliché. The NRA's sole policy prescription was for armed guards inside every one of the nation's ninety-nine thousand schools.[3] The speech shocked many parents and drew derision from law enforcement officers, school administrators, and politicians across the political spectrum.

But the NRA dug in, catering to the most radical elements among a membership it claims is four million strong.[4] Donations surged. So, too, did gun sales.

A month after Sandy Hook, James Yeager, a forty-two-year-old former small-town mayor and firearms trainer in Tennessee, made a YouTube video in which he said, "I'm not going to let anyone take my guns. If it goes one inch further, I'm going to start killing people." Tennessee authorities suspended his handgun carry permit. Yeager later apologized.

Six months after Sandy Hook, an NRA vice president declared at its annual meeting that the NRA was waging a full-on "culture war."

The NRA's frenzied reaction to Sandy Hook helped cement its reputation as "the organization of no," Powell wrote. "We just became known for inciting riots on any effort that would impact the rights of gun owners. And as a result, we were seen as the problem. Even by gun owners." He cited a *Washington Post*–ABC News poll taken the month after Sandy Hook, in which only 36 percent of Americans had a favorable opinion of the NRA.

With the gun lobby under pressure, some of its members found common cause with mass shooting conspiracists. Renewed public interest in gun safety legislation fed a corresponding rise in false flag claims, including some that falsely cast Obama as the crimes' hidden instigator. Larry Pratt, who led Gun Owners of America, a radical competitor of the NRA, appeared on Infowars a week after the Aurora shooting, agreeing with Jones's false claim that it suggested a United Nations–led plot to impose a global gun ban. And at least one NRA executive would make a cynical attempt to exploit Sandy Hook conspiracism too.[5]

IT'S HARD TO KNOW whether Alex Jones truly believed that Sandy Hook was a false flag or that it didn't happen.

Jones did seem convinced that Sandy Hook posed an urgent threat to gun ownership. That, and his need to cater to his big audience, is why Infowars grew into a prime vector for Sandy Hook conspiracy lies. Jones steadfastly denies, including in a 2019 deposition, that he makes money from spreading bogus theories about Sandy Hook. Jones offers no proof of this. He has refused for years to produce substantial business records, even when ordered to by a court. What's clear is that the more people who visit Jones's Infowars store, whose products he promotes constantly on his broadcast, the more likely they are to load up on the diet supplements, air and water purifiers, survival gear, and other merchandise he sells there. Just as on social media, the more outrageous or infuriating the content Jones produces, the more people flock to his website.

"It's hard to make the case that it was all Sandy Hook–related, but around that time Jones's traffic was spiking like crazy," Dan Friesen, the Infowars researcher, told me.

On December 16, the Sunday after the shooting, Jones addressed the families of Newtown, as if any of them were listening.

"First off, I just want to extend my condolences to the families and the survivors from the tragic Connecticut shooting that happened Friday morning," he said.

Then he got to his main goal: stoking the gun policy fight.

Jones echoed the NRA's stance, calling for an end to gun-free zones around schools, which he called "free-fire zones." He had been predicting that the military would go "door-to-door" confiscating Americans' firearms. He accused the mainstream media of campaigning for gun control among Newtown residents. "They were chasing them around at grocery stores, in their yards, saying, 'Well, how do you feel about guns now that this happened?'" Jones told his audience. "The police are now begging reporters to leave grieving families alone. Well, they're not going to leave them alone, because they want to demonize the Second Amendment."

Jones brought on his friend Stewart Rhodes, a former Ron Paul staffer and founder of the Oath Keepers, an antigovernment militia group whose members, many of them veterans or former law enforcement officers, pledge never to obey a list of ten imaginary "orders," topped by surrendering their guns.[6]

Not many Americans had heard of the Oath Keepers in 2012; Rhodes had founded it only three years earlier. Since then the group has earned a reputation as one of the nation's most dangerous far-right extremist groups. Convinced that the presidential election was "stolen" from President Trump in 2020, the Oath Keepers participated in the Capitol insurrection on January 6, 2021.[7]

"Can you imagine these women that went and shielded the children from bullets, what they would have done if they had a gun?" Jones asked Rhodes.

"Yeah," Rhodes replied. "They would have killed the guy—they would have put a bullet in his brain."

ROUGHLY TWO MILLION PEOPLE listened to Jones's radio show that week. A month after the shooting, Alex Seitz-Wald wrote for *Salon*, a liberal news website, that Infowars online traffic had surged in the aftermath of the shooting,[8] according to figures from the analytics company Alexa, a subsidiary of Amazon. Jones's two websites, Infowars and PrisonPlanet, drew 11.5 million total visitors per month at the time, and more than 28 million page views, making Infowars.com the 390th most popular website in the United States.

Still, that meant 389 other websites were better-read than Infowars, including those of every major news media outlet.

"You often hear comparisons between this one piece of fake news that got X million views, and this other story that was true got one-tenth of that," Joe Uscinski, professor of political science at the University of Miami and coauthor of *American Conspiracy Theories*, said in an interview. "Because there's probably two thousand stories that have that official view in them, so people don't always have to go to just one, whereas there's only one Infowars.

"You probably had some people like, 'Wow, you know, now they're going to be calling for gun rights legislation, so it's probably a hoax to take away our guns.' You had some other people thinking, 'This just can't happen. There must be some other explanation; it's probably the government,'" Uscinski said.

But "Sandy Hook seems to be a watershed," he said. Ever since, "if a shooting gets a lot of coverage, you're going to have people saying, 'This is fake.'"

A 2013 poll by Fairleigh Dickinson University[9] suggested that a quarter of Americans thought that facts about the shootings at Sandy Hook Elementary the prior year were being hidden. An additional 11 percent weren't sure. More disturbing, 29 percent of Americans thought that "an armed revolution in order to protect liberties" might be necessary in the next few years.

Uscinski's research suggests that by 2020, one-fifth of Americans believed *every* school shooting was faked. And not just school shootings: virtually all high-profile mass shootings draw doubts about the official narrative.

Uscinski says he found that ideologically, people who believe mass shootings are fake "tilt slightly right. But once you account for people's conspiratorial worldviews, their partisanship isn't much of a predictor of their conspiratorial beliefs."

In other words: your psychology, not your politics, is a more consistent determinant of whether you routinely disbelieve official narratives and openly espouse what most of us would consider antisocial falsehoods, such as the government planning a murder of twenty-six people in a gun-control-policy gambit. People who expand their list of culprits from the government to the families of the victims are operating on a plane even further removed from pure politics. Lenny Pozner, who lived near Pulse nightclub in 2016, noticed false flag conspiracies spreading online before the gunman who killed forty-nine had even left the premises. In Las Vegas, survivors still in their hospital beds after the 2017 shooting there that killed fifty-eight got onto their phones or laptops to report that they were alive, and found vicious attacks from people calling them "crisis actors."

In a now-familiar pattern, Sandy Hook conspiracies surfaced in isolated message boards and groups, then caught fire across social media, their spread fueled by algorithms that select what users see based on their past preferences and choices.

John Kelly is a network sociologist and a pioneer in mapping online communities. A year after Sandy Hook, Kelly founded Graphika, a forensic digital investigations and digital marketing firm in Manhattan. He has written software that maps the flow of disinformation through online communities.

People used to define "community" geographically. Now they view it in online terms. On this invisible terrain, community is determined "by how we get filtered and sorted based on all these little micro-decisions we make about what to like and who to follow," Kelly said. "All of that activity conditions algorithms and creates direct relationships that form this kind of cyber human connectivity. That's how narratives flow, that's how information flows, that's how some people become influential and others follow them."

Kelly quoted the sociologist Max Weber's observation, that "man is an animal suspended in webs of meaning he himself has spun." Now those webs are spun collectively. When a big event occurs, members of an online community gather to decide on its meaning. Gathering strands of information and misinformation, they weave and embroider, then circulate the finished product. For a message to go viral, it must confirm the guiding principles or interests of one of these cyber communities, then jump and spark across many of them.

"So now you've got all these communities out there, that when something hits the news, a big event like this, shocking and horrible, they're quickly going to develop their own take on it. And they're going to do that in a way that is consistent with the fundamental values and interests of their community," Kelly said.

"Narratives get filtered into meta narratives, and those meta narratives don't change very easily," he said.

"The battle lines are drawn, and it all unfolds."

6

FIVE DAYS AFTER THE MASSACRE, Robbie and Alissa Parker flew to Utah for Emilie's funeral. On the day before her burial, Jones mocked Robbie on his broadcast, calling him "a soap opera actor."

Infowars.com posted an article about Robbie Parker's media appearance a day after Emilie's murder, in which he reminisced about her life.

> FATHER OF SANDY HOOK VICTIM ASKS 'READ THE CARD?' SECONDS BEFORE TEAR-JERKING PRESS CONFERENCE

Robbie didn't say any such thing. He had smiled and asked whether he should "start." He did not use cue cards or index cards. He carried 8½-by-11-inch paper, the type you would use in a computer printer, folded in half lengthwise in his left hand as he approached the lectern. The sheets contained notes he had written himself. All this is clearly apparent to anyone who views his news conference, which lives on YouTube to this day. Jones was instructing his followers to disbelieve their own eyes.

Jones and other conspiracists accompany their flimsiest claims with the dodge "Just asking questions," and many Infowars headlines are written that way. This one asked: IS THE ESTABLISHMENT MEDIA TRYING TO STEER THE VICTIMS' REACTIONS?

The article was accompanied by a video titled "Sandy Hook Shooting Exposed as a Fraud."

"It appears that members of the media or government have given him a card and are telling him what to say as they steer reaction to this event, so this needs to be looked into," Jones said in a statement in the article.

Jones spent most of his broadcast that day dourly predicting the end of the Second Amendment. Then he homed in on Robbie's press conference.

"We're going to play a clip from a CNN press conference with Robbie Parker," Jones said. Mischaracterizing Robbie's appearance as being somehow organized by CNN, a big Infowars target, was another way of undermining its credibility.

"I've seen this before, and he's laughing, looks really excited," Jones said, frowning. "And then he walks up like he's an actor, and then breaks down on camera."

Jones launched into a hammy, concocted rendition of Robbie facing the media, focused on Robbie's brief smile and side comment to his family as he stepped to the lectern that night.

"I mean, it's like, '*HAHAHA! Yeah, huh? Oh, I read this card? Okay. OOOH HOOOO,*'" Jones said, wailing in mock grief.

"*Whoa, WHOA,*" Jones said. "And maybe under the stress, that's just what happened. And then maybe in front of the camera, maybe it's—I mean . . .

"Folks, we gotta get private investigators up to Sandy Hook right now. Because I'm telling you, this—this stinks to highest heaven."

He told his staff to play the footage again.

"Unbelievable," Jones said. "I mean he does look like a soap opera actor. Handsome guy, he has a very suave walk. Perhaps that's how you handle it when you're really nervous, but that is a smile of absolute satisfaction."

Jones couldn't stop talking about those six seconds in an eighteen-minute appearance filled with grief.

"So he's really smiling, he looks happy, and all of a sudden, suddenly—it looks totally fake. He may be a good-looking guy but he's not a good actor.

"I don't know what's up, man . . . Maybe he should explain it to us."

Jones continued to toggle between describing Robbie's grief as real and condemning it as fake. "You know after you lose your daughter, they put you on some antidepressants or something, but I thought those take a month to kick in. I mean, it's like a look of absolute satisfaction, like he's about to accept an Oscar."

Jones bellowed more fake grieving noises.

"Let me ask you guys, does that look fake to you?"

One of Jones's employees chimed in off camera: "Yeah, that's an actor."

Jones ended his tirade with a non sequitur. "Well, our condolences go out to him," he said.

Emails and calls poured into Infowars, offering bogus theories and singling out individual victims and their families as frauds. One email referred to a photo the Parkers' surviving children took with Obama during his December 16 visit to Newtown for the prayer vigil. Emilie's little sister wore a red-and-black holiday dress. The deniers found a photograph of Emilie on a memorial page, wearing the same dress. They decided her sister, wearing Emilie's hand-me-down, was the dead child, smiling with the president.

"People tend to get a little aggravated when they are being lied to," the email read. "This is the pretext to accelerate the desire of the gun grabbers in the long march to disarm Americans."

The Parkers' high school friends received abuse and threats on the Facebook page they created to collect money for Emilie's funeral costs. The Parkers took down a Facebook memorial page for Emilie because deniers were posting links to Jones's broadcast with comments like "'How much money are you making by selling out to Satan?'" Robbie told me. He got a Facebook friend request from "Adam Lanza."

"When I did that statement, I had no idea I was the first person to come out and say anything," Robbie said. "I became an easy target."

He came to regret ever speaking publicly about Emilie. When CNN asked him to respond to the falsehoods on Anderson Cooper's show, he declined. "I didn't want to give them more opportunities to create false content," Robbie told me, referring to Jones and his followers. He sent CNN a statement instead:

"As a country we cannot let ourselves become derailed by the preposterous claims that are being made by a tiny number of people," it read. "We cannot let these false claims distract us from the things that matter most to all of us."

The Parkers' surviving daughters were three and four years old when Emilie died. "My approach to the conspiracists in the beginning was 'I'm just not going to engage.' I learned as a kid that you don't engage with a bully," Robbie told me.

"That was my way to protect myself and my family."

WHO WAS THE "PATIENT ZERO" of the Sandy Hook conspiracy, the very first deluded person to decide the shooting didn't happen? I tried to find out, and learned there is no single "Sandy Hook Conspiracist" archetype.

Some content has been taken down, accounts changed or closed, blurring the social media trail. But the records that do exist brought a disheartening realization: within hours of the shooting, a mass of people more or less simultaneously decided that the shooting was faked. They congregated on social media, a bottomless well of imagined evidence in support of any belief, digging for "anomalies" in the flood of news coverage, collaborating to surface new questions. Errors in mainstream coverage they took as evidence of a cover-up, while any reported details that chimed with their false claims were deemed proof from a credible source. They called this "research," and in a suggestion of the immense psychic income they derived from it, they dubbed themselves "the truth community."

The deniers soon coalesced around an illogical narrative of deception and cover-up, in which the families, their lost loved ones, first responders—everyone involved—were enablers helping the Obama administration to perpetrate a fraud.

The Sandy Hook hoax theory cast vulnerable people as powerful accomplices in a crime, deserving of exposure and punishment. For the internet's aggrieved and idle, it grew into an endless pursuit of painful details and personal data used to confront and torment people at the lowest point in their lives. If the victims' family members didn't cry or collapse, if they smiled or laughed, if they looked too old to be parents or too attractive to be grieving, they must be lying.

In a *Twilight Zone* version of what journalists do, the conspiracists competed to be first, surfacing ever more outrageous claims. Like the psychologist B. F. Skinner's pigeons,[1] they pecked away at the truth, rewarded by social media's ingenious reinforcement system with followers, shares, and likes.

A YouTube video titled "The Sandy Hook Shooting—Fully Exposed," which served up a buffet of bogus claims, drew 5.5 million views within a week after it was posted. Five days later the number had doubled.

"Me and my friends used to research 9/11 in high school," the video's anonymous creator explained to Max Read of *Gawker* in an email.[2] "That's what really got me started when it came to researching government cover-ups."

He had decided the Sandy Hook shooting was a false flag based on "feeling deep down that these people and the whole town had this artificial vibe about them."

He would have spent more time on the video, the auteur told Read, if he "had known it would explode."

Wolfgang Halbig, a retired school safety administrator with a scant record in law enforcement, launched a website, Sandy Hook Justice, and raised tens of thousands of dollars in donations to fund his "investigation."

Halbig said he started out believing the shooting really happened and emailed the Newtown School Board, offering to investigate. Halbig wasn't volunteering his services, however. He wanted to be paid.

Offended when the school board didn't respond, Halbig attacked the same Newtown officials he had pitched for business, accusing them of covering up a phony crime. A couple of weeks after the shooting, Halbig emailed Sally Cox, the school nurse, who had hidden in a closet during the massacre:

> Are you a registered nurse?
> Why in the closet for four hours?
> Why close your eyes when you have seen blood before you are a nurse?
> Wolfgang W halbig

Halbig began emailing and calling Newtown officials and the victims' families. He filed hundreds of public records requests and made some two dozen trips to Newtown from his home in Florida.

Halbig grew into one of Alex Jones's on-air Sandy Hook "experts." Infowars helped him raise money, and sent camera crews with him to Newtown.

JONATHAN LEE RICHES WAS likely the first conspiracist to travel to Newtown.

After police learned the gunman's identity on December 14, they evacuated his neighbors and broke into his mother's hilltop home. About seven-thirty that evening, Connecticut State Police detective Jeffrey Payette, covered in a white Tyvek hazmat suit, videotaped the scene inside the house. As he moved slowly up the carpeted stairs to Nancy Lanza's bedroom, the only sound besides the detective's steady breathing was the insistent ringing of her telephone.

Among the callers was Riches, a convicted online fraudster who pioneered computer-based identity theft through "phishing," in which the thief uses email to extract personal information from unwitting marks. After Riches secured the data, he would impersonate the victims in visits to banks, stealing hundreds of thousands of dollars from their accounts.

Riches, then thirty-five, was imprisoned in 2003 for fraud and conspiracy, and denied use of the internet. He achieved social media celebrity regardless, by filing some three thousand handwritten, frivolous lawsuits that he dreamed up in his jail cell, targeting celebrities, political leaders, and corporate titans, plus the Nazi Party and the "13 tribes of Israel."

Riches's lawsuits were all about attention-seeking, says Staci Zaretsky, an attorney with the legal news website *Above the Law*, who has covered Riches's exploits for years.

He clogged the courts with so many meandering briefs that at one point he was denied writing paper in his jail cell, Zaretsky told me. In 2010 he was enjoined from filing any more. He went on a hunger strike in protest.

Riches was paroled eight months before the Sandy Hook shooting, and his access to the internet restored. Within hours of the shooting he told his Facebook followers that he had called the Lanza house to "warn" Nancy Lanza that her son was "mind-controlled and manipulated to go in and allegedly shoot little kids."

Two days later Riches violated the terms of his probation by leaving his home in Pennsylvania to drive to Newtown. He made a video of himself driving to the gunman's house, which as of this writing still lives on YouTube. "We are truth-seekers," Riches said in his clipped, nasal accent. In the car's back seat, two cloth dolls had photocopied photographs of the gunman and his brother taped over their faces. Police turned Riches away.

Riches next drove to downtown Newtown, stopping at one of many impromptu memorials to the victims. He told mourners he was Lanza's "Uncle Jonathan," and that Adam had taken antipsychotic drugs. As news cameras whirred, Riches knelt among the flowers, candles, and homemade cards on the sidewalk, as if to pray.

Riches landed back in jail. By then his Newtown YouTube video had circled the world. Released three years later, he has scammed press attention by impersonating a "Muslim for Hillary," wearing a MAGA cap at a Trump rally, and protesting in support of both sides during Black Lives Matter demonstrations. In 2021 he traveled with groups mobilizing in support of Trump's election lies.

THE SELF-STYLED INVESTIGATORS toured Newtown with cameras, showing up outside the firehouse and the fenced-off, guarded, and empty school.

Their presence was a factor in the town's decision to demolish Sandy Hook Elementary. Contractors signed a nondisclosure agreement and agreed not to remove anything from the site without authorization, not even dirt. Trucks carted the rubble to undisclosed locations, where it was melted, crushed, and milled to dust.

To the conspiracy theorists, the destruction of the school, like that of the gunman's home, was further "proof" that Newtown had something to hide.

After school reconvened for the children who survived the Sandy Hook shooting, conspiracists spied on and telephoned the school's temporary location, convinced that it was a decoy and that no students actually went there.

Reports filed by the state troopers assigned to the families described people coming to their homes and digging through their trash.

Conspiracy theorists swore at family members on the street, looked into their windows, and vandalized their homes and the impromptu memorials to the twenty-six victims.[3] They sent them emails demanding, "Repent for your sins." One parent was barraged with phone calls and emails saying, "Your daughter is not dead. Your daughter is alive."

Gene Rosen, the retired psychologist who sheltered the children who ran from the school, then opened his door to Scarlett Lewis, had talked with several reporters about his experience. That made him a target too.

Deniers found a social media account for a different Gene Rosen, a member of the Screen Actors Guild, and wrongly insisted it was Rosen the Sandy Hook "crisis actor." White supremacist websites speculated about how much the "emotional Jewish guy" had been paid to lie. Rosen and his wife worried for their safety. But police could do nothing unless he received a direct threat.

Brendan Hunt, a twentysomething conspiracy enthusiast from New

York, turned up at Rosen's house asking to interview him for a liberal public radio station.[4] Rosen turned him away, but Hunt returned at night to film Rosen's house. He also visited Sandy Hook Elementary, approaching the school from the woods, filming the barbed-wire perimeter, and speculating about the role played by a unit of "tactical police officers." Hunt's visit seemed more bumbling than sinister then, akin to his online theorizing about the death of Nirvana frontman Kurt Cobain. But Hunt, the son of a retired family court judge in Queens, remained a dedicated conspiracist. He eventually fixated on the false notion that the 2020 presidential election was stolen. After the January 6, 2021, Capitol riot, Hunt posted a video online urging, "Get your guns, show up to D.C. put some bullets in their fucking heads." He was convicted of threatening to assault and murder members of the U.S. Congress.

THE SANDY HOOK CONSPIRACY drew a number of professors. In academia, Socratic questioning and thoughtful dissent is generally encouraged, hand in hand with proper peer review, and protected by rules enshrining academic freedom. But this crowd's theories emerged from hunches and paranoia, not research and debate. The only peer review they received was from one another. Some had spent decades concocting fanciful alternative narratives to explain major historical events. For others, Sandy Hook marked their maiden voyage down the rabbit hole.

One of Alex Jones's frequent guests was Steve Pieczenik, a Harvard-trained psychiatrist with a doctorate from MIT who served as a deputy assistant secretary of state under Henry Kissinger, Cyrus Vance, and James Baker. In the late 1970s, Pieczenik claimed to have been part of the negotiating team that reached the Camp David Accords between Egyptian president Anwar Sadat and Israeli prime minister Menachem Begin.

He consulted on Tom Clancy novels, including his Op-Center series, providing the inside-government view.

At some point Pieczenik apparently lost touch with reality. Two weeks after the shooting, Pieczenik emailed Infowars a racist, antisemitic screed, accusing then attorney general Eric Holder of funding the "NON-EXISTENT Newtown Massacre!"

"Welcome to the New Reichstag!!" Pieczenik wrote in an email released in court proceedings. "Is a black president above reproach for having lied continuously to the American people about Osama bin Laden; Benghazi; Newtown Massacre; and of course, folks, Obamacare?"

Pieczenik went on like that for three pages. He hammered the same theme on Infowars for several years.

JAMES FETZER, A BOMBASTIC, self-promoting philosophy professor retired from the University of Minnesota Duluth, had repelled his academic colleagues with his wacky theories since the early 1990s. He lurched from political left to right over time, claiming, for example, that both Gerald R. Ford and Lyndon B. Johnson helped plan John F. Kennedy's assassination. In 2006, Fetzer attended a 9/11 conspiracists' conference in Los Angeles. He appeared on a panel moderated by Alex Jones, making Jones look sedate by comparison.

Days after the shooting in Newtown, Fetzer blamed Mossad, the Israeli intelligence service, of orchestrating it. He soon revised that, saying the shooting didn't occur at all but was a Federal Emergency Management Agency drill.

Fetzer often boasted that his Sandy Hook "research group" included six PhDs, and he imagined himself their bard. He eventually compiled a collection of their articles into a book, *Nobody Died at Sandy Hook*. The free PDF version of the book has been downloaded more than ten million times.

Maria Hsia Chang, a retired China scholar at the University of Nevada, Reno, with a doctorate from the University of California, Berkeley, was a Fetzer collaborator.

Chang had a blog, *Fellowship of the Minds*, that she used to malign the Sandy Hook families in the guise of exposing "oddities" in official records, most of them the result of her own sloppy "research." Chang published her attacks on the families under the pseudonym "Eowyn," the name of a noblewoman in Tolkien's *The Lord of the Rings*.

Chang included the victims' birth dates in a blog article shared on scores of Facebook pages, touted in YouTube videos, and reprinted in the Fetzer book. For another blog post, she looked up the dead children's home addresses on real estate websites, and having failed to find the purchase prices, insinuated that their parents had been given their homes for free, in exchange for faking their children's deaths. Chang posted the families' addresses. Having doxxed the entire group, she acknowledged in the final sentence of her three-thousand-word article that she had little idea what she was talking about. "Your guess is as good as mine," she wrote. "I'd appreciate input from readers."

Articles like this were one reason why strangers appeared at the families' homes, followed them, and dug through their trash. Internet archives suggest the post circulated for at least six years.

Chang lives in Northern California with a collection of rescue cats and several birds. I heard one of them trilling in the background when I called to ask about her Sandy Hook articles.

"I do not want to be associated with this," she scolded. "I have received death threats!"

The Newtown families had also received death threats, I told her.

"I am aware of that, and all of that is deplorable," she said. "But I believe Americans still have the right of free speech."

How did a former professor at a well-regarded, taxpayer-funded university wind up spewing online poison?

"This is the first thing I've ever researched," Chang told me. "What provoked me is that there are all these oddities. I don't know what to make of it.

"I dig up the evidence, I present the evidence, and it is up to the reader. I would hope that *you*, as a reporter with America's premier newspaper, would have the integrity—and I do want to emphasize the word 'integrity'—to look at this," she scolded.

"It's not as if the government has never lied."

I conceded that last point. I asked whether she had children of her own, adding that several mothers had told me that at first it was difficult to believe the shooting took place. "I challenge the very premise," she interrupted. "I don't have human children of my own."

"Just the cats?" I asked. She refused to confirm the existence of her cats.

"I was shocked, and I was traumatized by the event when it first happened, like any human being," she said. "I have read newspaper accounts of the families claiming they have received death threats . . . I don't know for sure they have received death threats."

I told her I had heard recordings of telephone death threats leveled against a Sandy Hook family.

"The addresses are all in the public record," she said.

But Chang hadn't just published the information. She, a highly educated academic with teaching credentials, had implied repeatedly that they had committed a crime, and had published their addresses to expose them. What if an unstable person believed her and paid the families a visit?

"Who would do that?" she said, growing indignant. She insisted that her work and mine were the same: researching and reporting. I told her that if the *Times* put thirty families' home addresses online and insinuated without facts that they were criminals, we'd be sued.

"All I can say is with the benefit of hindsight, I would have been more careful. At the time this whole thing was just a mystery," she said, backpedaling. "Now that I am aware of the phenomenon of doxxing, I would be more careful," she added.

But doxxing was a weapon she and her conspiracy cohorts have used against the families from the start.

"I am not responsible in any way for what any reader does with that information," she said. Soon after, she hung up.

About a week later Chang announced online that she was shutting down her blog.

"In my professional life, I had specialized in the study of totalitarian Marxist-Leninist states wherein citizens were denied fundamental freedoms of speech, publication, assembly and religion," she wrote.

"I never thought or even dreamt that the United States of America could devolve into that."

JAMES TRACY WAS AMONG the first conspiracists to speculate that the family members were "crisis actors," a term used to attack survivors of every mass tragedy since.

Tracy was a forty-seven-year-old journalism professor at Florida Atlantic University (FAU) in Boca Raton. He posted his "questions" about Sandy Hook on his personal blog, called *Memory Hole*, a reference to George Orwell's *1984*. References to Tracy's false claims were reprinted in more than eight hundred news accounts. Though the vast majority did not give them credence, repeating them helped them spread. His status as a journalism professor granted him an air of credibility that drew adherents and imitators. Tracy reveled in the "national notoriety" (his words) he received and rebranded himself as "a media scholar, educator and political analyst."

Tracy grew up in Hornell, New York, a town of about ten thousand south of Rochester, on the old Erie railroad line, which he described as "a nice place to visit but you wouldn't want to live there." Hornell is closer to Cleveland than to New York City, and Tracy considers himself more an easygoing midwesterner. In conversation he seems quiet-natured and reflective. It is hard to believe he's the same person who sent a certified letter to Lenny and Veronique, demanding documents proving Noah was their son.

Tracy was an only child. His mother was a schoolteacher for a time, and his father had a succession of jobs, including as a representative for a firm helping to refurbish New York City subway cars. Tracy's paternal grandfather had been a well-regarded physician in Hornell.

"I think my mother was a pretty trusting person. My father kind of questioned official authority and things like that," he told me.

Tracy left Hornell to attend San Jose State University, California, then received a master's degree in media studies at the University of Arizona. He earned his PhD in journalism at the University of Iowa, which is where he met his wife, who studied library science. In 1996 he and his wife were given the rarest of opportunities: a double appointment at FAU, with the potential for tenure for both of them. Tracy taught in the communications department, and his wife took a position in the library. South Florida offered the growing family lower housing and living costs, and warmer weather than Iowa.

Tracy is an engaged father of five children, the eldest of whom was nearly the same age as the children killed at Sandy Hook in 2012. He is raising his children in the Catholic faith but emphasized that they are free to leave it once they reach adulthood. He mentioned the hours he devoted to one of his younger children, who has a life-threatening chromosome disorder.

Tracy said fatherhood made him more politically conservative than he had been as a "radical left" student.

"I began to question my colleagues and some of the history and ideas I was teaching," he told me. Teaching his students about censorship and propaganda during the Iraq War, he came to believe Americans were lied to about the existence of weapons of mass destruction in Iraq, the reason for the war. Though many Americans shared his belief, for Tracy the doubts were jarring.

Tracy began teaching a course called Culture of Conspiracy, in which students examined propaganda and official lies. He says it was popular, and he taught it four times.

"I think that the public fears questioning. They are inclined to respect and accept a narrative of a particular event out of emotional comfort," he said. "They might watch CNN for five minutes and Fox News for five minutes and think they have a balanced view of the world. Things are a lot more complex than that."

Beginning in the mid-2000s, he had noticed that more people online were questioning and reexamining their understanding of "everything," he told me. "What was the 'truth movement' a few years ago is becoming much more mainstream. I think that points to the lack of veracity that can be found in the news media."

Tracy said he at first believed that Sandy Hook had happened and, as a parent, deeply sympathized with the families. Then he started doubting the official story, as he often did, and shared his questions on his blog.

Not long after, an Infowars staffer wrote to Jones's team: "YOU HAVE TO GET THIS GUY ON THE AIR!!!"

On January 8, Infowars posted an article: COLLEGE PROFESSOR SAYS 'CRISIS ACTORS' MAY HAVE PLAYED PART IF SANDY HOOK WAS INDEED A HOAX.

It quoted Tracy's online speculation about the survivors. "After such a harrowing event why are select would-be family members and students lingering in the area and repeatedly offering themselves for interviews?" Tracy wrote. "A possible reason is that they are trained actors working under the direction of state and federal authorities and in coordination with cable and broadcast network talent to provide tailor-made crisis acting that realistically drive [*sic*] home the event's tragic features."

This of course was not true. No one selected those who were interviewed that day except reporters themselves, who approached people on their own and were often rejected.

CNN's Anderson Cooper, who had taken regular trips to Newtown, made Tracy and other conspiracists the subject of his evening broadcasts on January 11 and 15.

"Normally, we would not dignify these claims with airtime," Cooper said. "But it turns out one of the people who's peddling one version of this conspiracy theory is actually a tenured associate professor at Florida Atlantic University, a state university that gets taxpayers' money. His name is James Tracy.

"He's not convinced the parents whose children were killed are really who they say they are," Cooper said. "Tracy even cites a company called 'Crisis Actors' that provides actors to use in safety drills and the like. Apparently, that is supposed to bolster his case."[5]

Internet searches for "crisisactors.org" surged after Cooper's broadcast, aided by Infowars, which jumped on Tracy's reference to the website.

The term "crisis actors" first surfaced in 1977, used in a report by Michael Brecher, a professor at McGill University in Montreal, to describe nations in conflict.[6] It gained its current meaning after the Aurora theater shooting in 2012, when Visionbox, an actors' studio in Denver, set up crisisactors.org to promote its services as participants in active-shooter drills, to lend realism to the event and improve the emergency response.

The company gave Cooper a statement: "We're outraged by Tracy's deliberate promotion of rumor and innuendo to link Crisis Actors to the Sandy Hook shootings. We do not engage our actors in any real world crisis events, and none of our performances may be presented at any time as a real world event."

Cooper aired a brief interview with Tracy by a reporter from Florida TV station WPEC, who found Tracy on the university campus.

"You had twenty families that were mourning, that buried children. Are you concerned about that at all?" the reporter asked.

Tracy, in jeans and a white button-down, a battered cloth messenger bag slung over his shoulder, struck the classic conspiracist's pose: arch, secure in his superior knowledge. "Once again, the investigation that journalistic institutions should have actually carried out never took place, as far as I'm concerned."

He scratched a graying temple. "We need to, as a society, look at things more carefully. Perhaps we as a society have been conditioned to be duped."

Cooper aired an outraged statement from Pat Llodra, Newtown's first selectman: "Shame on you, too, FAU, to even have someone like this on your payroll. I can assure you, sadly, that the events here in Newtown unfolded exactly as they're being reported."

CNN sent Florida-based reporter John Zarrella—its space reporter, appropriately enough—to Tracy's house. The professor refused to be interviewed and submitted a statement.

"I apologize for any additional anguish and grief my remarks—and how they have been taken out of context and misrepresented—may have caused the families who've lost loved ones on December 14. At the same time I believe the most profound memorial we can give the children and educators who lost their lives on that day is to identify and interrogate the specific causes of their tragic and untimely demise."

That sounded more reasonable than what Tracy posted on his blog: "Does Anderson Cooper want James Tracy and/or His Family Members Harmed?"

Tracy told me he said no to CNN due to work and family pressures. But on the same day he refused CNN's request, he was arranging to appear on Infowars.

TRACY WAS INTERVIEWED BY Paul Joseph Watson, who was sitting in for Jones. The professor's willingness to entertain any and all batshit claims about the shooting proved too much even for Infowars.

Watson was thirty at the time, and an up-and-comer in conspiracy circles. A native of Sheffield, England, he learned about Jones when he was eighteen, watching Jon Ronson's miniseries *The Secret Rulers of the World.*

Ronson called Jones out as a huckster in the series. But he inspired Watson to start his own conspiracy website from a computer in his childhood bedroom in Sheffield, where his anti-immigrant, Islamophobic, and misogynist content racked up millions of views.

Watson joined Infowars as a contributor two years later, in 2002. A fast writer and more tech savvy than Jones, Watson created the PrisonPlanet website and was soon earning a hefty six-figure salary from Infowars, internal documents suggest.

Watson disagreed with Jones on Sandy Hook. An email from him to his Infowars colleagues unearthed during court proceedings suggests this wasn't on moral grounds. Watson thought questioning the murders of children would blow Infowars' chances of getting its content posted on the wildly popular Drudge Report website, the holy grail for click-thirsty conservatives.

When Watson interviewed Tracy on January 18, 2013, he told the professor, "I think it was a real tragedy. It really happened."

He prodded Tracy to focus on the gun debate and the media's role in demonizing assault-style weapons. He wanted Tracy to debunk some of the weirder Sandy Hook theories, like one that speculated Robbie Parker was actually skateboarder Tony Hawk, or "the really bizarre reports of people dressed as nuns fleeing the scene in a purple van."

Tracy didn't see that theory as weird. He bought it. "I believe there was a photograph taken of these individuals and they're wearing footwear similar to what police officers would wear," he said. Watson, who spews facile nonsense for a living, went temporarily mute.

By the end of the interview, Watson was debunking the professor.

"There's a plethora of claims that are obviously either completely baseless or are woefully reductionist," he said. "I think the whole thing about Emilie Parker supposedly not being killed and then being included in that photo with Obama is clearly one of them. Because if you look at the high-res images, you can quite clearly prove that it is in fact her sister, who obviously wasn't one of the victims.

"What do you say to the idea that some people are really making huge leaps of logic and in doing so, detracting from any real understanding of what happened?"

The unflappable Tracy, speaking as if to one of his students, saw a conspiracy behind that too.

"There is definitely a program to sow misinformation in the stream of information in order to muddy the waters and in the process, discredit the research."

THE CONSPIRACY WORLD HAS always been well populated with academics and self-styled free thinkers. In the past, they often remained in low orbit, a self-affirming and obscure collective murmuring among themselves. But the internet provided deluded scholars like Fetzer, Chang, and Tracy with tremendous reach. Their suggestion of secret insights, deep knowledge, and an amorphous enemy just outside the gates attracted a global crowd of disciples, all sidestepping the crucible of open, fact-based debate. This mind-meld of profs and proles is emblematic of the hybrid online linkages that hasten the spread of misinformation. Similar happened in 2020, when liberal yoga teachers and anti-globalist doomsday preppers unacquainted in the real world joined forces online to boost anti-vaxx nonsense during a pandemic. Influential online sites like Infowars amplified their messages to an angry, gullible audience of millions.

Secretive government plans, which by design or by default harm Americans and others, punctuate our history and certainly warrant exposure and serious study. But some academics' increasing willingness to lend their credentials and their institutions' reputations to promoting obvious harmful falsehoods around mass shootings, vaccines, voter fraud, and other issues presents a new set of challenges for academia and the rest of us. It's the role of a professor to teach critical thinking, but there's a consequential difference between questioning aspects of known historical events in a classroom and flat-out denying the reality of the mass murder of children at a school.

Tracy's views on Sandy Hook would cost him his job, but that did not stop his theorizing. In my interviews with him he doubted the official account of every major event we discussed. As we ended a conversation in early 2021, he stopped me from hanging up because he had forgotten to air his suspicions about COVID-19. To this day he believes his firing was unconstitutional. He was the canary in the coal mine, he told me, one of the first free thinkers gagged by woke academia.

Three of Tracy's former colleagues denounced this thinking in a letter to the editor.[7]

"He and his supporters quickly reference his First Amendment right to express his ideas," Jeffrey S. Morton, Patricia Kollander, and Thomas Wilson wrote in the *Palm Beach Post*. "What James Tracy does not understand is that ideas represent the end product of the intellectual process. Before they can be publicly espoused, ideas must be subjected to rigorous and intensive examination.

"Academics expose their theories to other academics; conspiracy theorists blog them to each other. Academics build on a rich intellectual tradition; people like James Tracy spin tall tales out of nothing."

7

Of all the Sandy Hook relatives, Lenny Pozner best understood the origins of the false narrative growing online, and the dangers of its spread.

Lenny's work made him well familiar with the mechanics of social media and its downsides; some of his clients had been victimized online. He may have been the only family member who was a regular Infowars listener. Like most Americans then, many of the other victims' relatives had never heard of Alex Jones. But in the days after Noah's death, Lenny, in deep grief, was only vaguely aware that conspiracy theorists were gathering online and in Newtown, and had little if any idea of what Jones was up to.

That all changed on January 27, 2013, when Jones turned his attention to Veronique. Lenny would appeal to Jones directly for the first time, asking him to leave grieving families alone. Jones refused, a decision that in time would prove costly.

For several weeks Jones had been hearing from Infowars followers hungry for more Sandy Hook hoax content. He'd been promising on his show to staff up, to do more, to send people to Newtown. Meanwhile random amateurs were racking up millions of views with videos like "The Sandy Hook Shooting—Fully Exposed."

Jones's video "Why People Think Sandy Hook Is a Hoax" was his bid for a bigger piece of the online action.

He began his bogus exposé with Robbie. "One of the big issues out there that has people asking questions is Robbie Parker, who reportedly lost one of his daughters," Jones said. "I know grieving parents do strange things, but it looks like he's saying, 'Okay, do I read off the card?' He's laughing, and then he goes over and starts basically breaking down and crying."

He played footage of Robbie's press conference yet again, this time with the chyron "Odd Parent Reaction from Sandy Hook."

"Now, ladies and gentlemen," Jones said, milking the drama, "the finale."

Jones played a copy of Veronique's CNN interview with Anderson Cooper. Jones stopped the video after Veronique said, "If I asked everybody in this world who has ever loved someone, who has ever had a human being in their life who was essential to their well-being, to raise their hand, I don't think there would be many hands down in this world. And every one of those hands is a reason why those weapons should not be out in the general public."

Jones wasn't interested in Veronique's words. He zoomed in on a split-second glitch in a copy of the CNN broadcast in which Cooper's nose blurs, then seems to disappear.

"He's supposedly there at Sandy Hook in front of the memorial, and his whole forehead and nose blurs out," Jones intoned. "I've been working with blue screen, again, for seventeen years. I know what it looks like.

"Anderson Cooper has got some explaining to do, because I know blue screen when I see it." A classic Jones insinuation: carefully hedged calumny plus braggadocio. Blue screen, like green screen, is a plain background that allows a TV production team to drop in whatever background images they choose behind a person speaking. Jones was implying that the white-columned Edmond Town Hall, with its heaps of flowers behind Veronique as she spoke with Cooper, were projected onto a plain blue backdrop, like the map behind a TV meteorologist.

Jones was relying on carnival showmanship to sell the verifiably phony claim that Veronique Pozner was performing from a CNN studio.

The glitch, Jones's "proof," was created by Infowars, Grant Fredericks, a forensic video analyst, told me.

Fredericks's clients include the FBI and police departments across the country. He and his team work in a secure facility in Spokane, Washington, using sophisticated image-processing tools to interrogate visual data for the purpose of revealing hidden details in video evidence.

But Fredericks didn't need any of that to debunk Jones's garbage claim. It took him less than five minutes to find a clean copy of Veronique's interview on YouTube, without the glitch.

In a sworn affidavit years later, Fredericks would testify to what Jones and his team did. In our conversation, he walked me through it.

When Jones's staff copied Veronique's interview from the web, they had to change its original format to the one used on Infowars' platform.

"Transcoding—changing it from one format to another—is very common, and pretty much always results in loss of detail and changing of the visual information," he said.

"Given what their accusations were, the person who did the work either ought to have known or did know," Fredericks told me.

"Jones *created* the anomaly that he complained about," he said.

"He created that issue and then reported it . . . And he can't defend it."

THE SANDY HOOK CONSPIRACISTS did not question Jones's claims because they reinforced their views. "Why People Think Sandy Hook Is a Hoax" grew into one of the more popular offerings on Infowars' YouTube channel.

Immediately Lenny began noticing comments attacking Veronique on their family's social media pages. He watched the Infowars video. He read further online and saw how the hoax had metastasized.

A poster called MoboEarth wrote in a lengthy comment:

> This interview was the point where I KNEW that Sandy hook was a Hoax. When I saw Veronique Pozners [*sic*] shawl and perfectly matched earrings . . . Even her shade of lipstick is the exact red shade for the color of her olive green outfit.
>
> No mother who has just had her baby shot through with holes would take such care to match an outfit . . . A Mother (and Father) would be in agony, imagining the terrible fear that their baby must have felt..I am a Mum of five children and KNOW that taking such thought into an outfit, JUST after having your child MURDERED . . . would NOT be possible.

Noah's death and their need to care for their girls had drawn Lenny and Veronique back together. Veronique wasn't online in the weeks after Noah's funeral, thank God, but most days she had trouble getting out of bed. Lenny needed to protect her.

TWO DAYS AFTER JONES aired the video, Lenny emailed him. It would be Lenny's first contact with Jones, sent nearly six years before the damage caused by Jones's actions dawned on the wider world. Lenny made a plaintive request asking that Jones stop.

> Alex,
>
> I am very disappointed to see how many people are directing more anger at families that lost their children in Newtown. Accusing us of being actors.......
>
> Haven't we had our share of pain and suffering? All these accusations of government involvement, false flag terror, new world order etc.
>
> I used to enjoy listening to your shows prior to 12-14-12.
>
> Now I feel that your type of show created these hateful people and they need to be reeled in!
>
> Lenny Pozner

Three hours later, Aaron Dykes, an Infowars producer, responded:

> Mr. Pozner,
>
> Thank you for contacting us to share your point of view. Alex has no doubt that this was a real tragedy, and sympathizes with the victims and their affected loved ones.
>
> Alex would very much like to speak with you. How can we confirm that you are the real Lenny Pozner? Please let us know how we can best be in contact with you, and we'd love to discuss your concerns and hear your side of the story.
>
> Sincerely, Aaron Dykes/Alex Jones

Separately, Lenny received an email from Watson:

Sir,

We have not promoted the "actors" thing. In fact, we have actively distanced ourselves from it.

We have said on numerous occasions that it was a very real tragedy with very real victims.

I hope we can continue to count you as a listener, and I am deeply sorry for your loss.

Best regards, Paul Joseph Watson

Infowars invited Lenny onto the show. Lenny told them he would speak with Jones privately. He was not interested in being an on-air attraction, further exposing his family to hate. He and Jones never connected.

8

BY 2012, AMERICANS HAD WEATHERED scores of mass shootings, including at schools. But Sandy Hook tore at people's basic sense of order and safety. Such wanton, sweeping violence visited on children barely older than toddlers struck most Americans as a new level of senseless savagery. One almost impossible to believe.

Parents across the country pulled their kids from school that December 14. For days afterward they kept them home or they camped in their schools' parking lots, afraid to abandon them to the fate met by the children in Newtown. Every anniversary since then, Americans have gathered on social media, sharing their recollections from that morning.

"I was teaching. Rumors started flowing like wildfire. The truth turned out to be even worse than the rumors. My students, teenagers, were shocked. That day we took the first steps down the road to lockdowns, active shooter drills, and ever more horrors."[1]

"At work, realizing it was almost Christmas, and crying with my male colleague who had just become a dad."[2]

Many recall not believing the news.

"I was eating lunch at the MFA in Boston, looking at my phone, when I saw the news. I remember thinking, this just can't be true."[3]

"Sat and cried. Couldn't believe the horror."[4]

"in office. had to leave, sat in car and wept."[5]

"I picked up my daughter from preschool that afternoon and burst into tears in the hallway as soon as I saw her teachers. We couldn't even get words out."[6]

People often greet random trauma with disbelief or magical thinking, says Joanne Miller, a professor at the University of Delaware whose research focuses on the causes and dangers of conspiracy theories, misinformation, and political propaganda.

"Seeking out explanations for negative, scary events that make us feel powerless is kind of a natural thing that we all do," Miller said. "If we can understand why and how the event occurred, we can maybe mitigate its negative effects and try to prevent something similar from happening in the future."

Miller categorizes conspiracy theorists into three groups. First are profiteers like Jones. Second are people who grasp at conspiracy theories to meet a need, either psychological or political. The third group encompasses people whose conspiracy beliefs spur them to action, whether it's setting fire to 5G towers, protesting in front of vaccine clinics, or harassing the families of shooting victims.

"I don't think we have a good sense of what takes someone from just passively believing to actively acting on some of these beliefs," Miller said. "The problem is, there's such a small number of people who actually act on these beliefs that it's hard to study them."

Soon there would be more of them.

In the desolate months after Noah's death, Lenny created a Google+ page as a tribute to his son. He posted a video of Noah, doe-eyed, saying "Hi" in his toddler's voice, and one of him in his pajamas, gliding backward on his belly down carpeted steps in the house on Kale Davis Road. Noah with his twin, Arielle, celebrating what would be his last birthday, less than a month before his murder.

The page drew expressions of grief and care from family, friends, and total strangers.

"Wishing you peace. I will never forget . . ."

"Noah is watching over all of you and his sisters."

"I'm happy to know that in his short time on earth that he was so loved."

"Reading Noah's story has inspired me to be kind to others."

Lenny found the messages a priceless source of support. "I get my strength from all of you," he wrote back.

Then he watched in numb disbelief as the page grew into a magnet for sentiments of an entirely different sort.

"There was no Sandy Hook shooting."

"Let's just dig up all the little bastards."

ONE OF THE MOST shocking messages to Lenny arrived from a woman with the online handle "gr8mom":

> I want to hear the "slaughter" and I won't be satisfied until the caskets are opened (hopefully by Geraldo).

"Gr8mom" is Kelley Watt, who by mid-2014 had become a daily visitor to Lenny's page. A divorced mother of two children in their thirties and a grandmother of two, Watt is the proprietor and sole employee of Maid in the USA, a housecleaning business in Tulsa, Oklahoma.

Women, particularly mothers, made up a large proportion of Americans airing questions about Sandy Hook during the first year after the shooting. Most moved on when confronted by the facts, embarrassed to have entertained such notions even briefly. Not Watt. Suspicious by nature and distrustful of government, she believed "proving" that anti-gun politicians faked the shooting was an end that justified all means. "Sandy Hook is my baby," she told me.

Some who pushed back on her in those years dismissed Watt as a vicious troll. But Lenny made a distinction between commercial conspiracists like Alex Jones and relative unknowns like Watt. "Jones was not interested in getting to any sort of destination or truth," he said. "He's about doing his monologues and keeping his listener entertained. His show needs to be filled with his gravelly shit." But some of the others struggled "to carry the pain of women and children being executed," Lenny said. Maybe their questions sprang from a genuine desire to understand how this could have happened. Lenny hoped that by walking these people through the reality of Noah's life and the hell of his death, he could make them believe. Or at least make them stop.

SHORT AND SQUAT, with chestnut hair, Watt is in her early sixties but retains the energetic air of the gymnast she was in her youth. She leads a quiet life in Tulsa with Duke, her boyfriend of nearly two decades, a retired sales manager. When I told her I'd visited Oklahoma City a couple of times, she said she'd never seen the capital building: "Too busy researching Sandy Hook."

Kelley grew up as the daughter of an Oklahoma oil company engineer and a homemaker. When she was in middle school, her father was transferred to Tripoli, but the family had barely settled in when the 1969 coup brought Mu'ammar al-Gaddhafi to power. Her mother decamped with Kelley, her brother, and her sister to Bartlesville, Oklahoma, her mother's hometown. Together they lived in an apartment over the garage of Kelley's maternal grandparents' home. "I absolutely loved Bartlesville,"

Watt said, so much so that when her father returned from Libya and the family resettled in Tulsa, she remained behind to finish high school, then went on to Oklahoma State University.

"I was mainly going for the social life and partying," she said. She left OSU with an associate's degree and soon after married Jim Watt, whom she'd met on a blind date in high school. Popular and adventurous, Jim interrupted college to compete as a motocross racer, then earned electrical engineering and law degrees. He came from a "really good family," Kelley said, owners of a prominent electrical contractor in Tulsa, which Jim took over. The Watts bought a "cute little house" in Tulsa and a summer place. Kelley threw herself into volunteer work, serving as a "play lady" in a local hospital and an aide in a Christian school for special needs children. She ran a scout group even before her son, Jordan, and daughter, Madison, were born.

"My whole life has been about kids," she said. "Aw, I wish I was a teacher. I would have had a cute little school classroom. I would have decorated it so cute.

"That's my biggest regret in life. I should have been a teacher. I would have been a really good first-grade teacher."

Watt has a Pinterest board called "Beautiful children." She had posted more than one hundred photos there of babies, toddlers, and prepubescent girls, many of them twins. They wear fur-trimmed hoods, chic berets, oversized bows, earrings. Their hair is often flowing, framing enormous eyes with irises in unusual colors. They smile and hug, peek through doorways—a fantastical, eerie ideal for how children should look and live.

I wondered whether she doubted Sandy Hook because first-grade children being murdered in their classrooms was too hard for her to face.

"No. I just had a strong sense that this didn't happen," she said. "Too many of those parents just rub me the wrong way. The things they said: 'We're moving forward' and the laughing, the smiling."

She judged the parents as "too old to have kids that age." She found their clothes dowdy, their hairstyles dated. Where were their "messy buns," "cute torn jeans," their "Tory Burch jewelry"? She mocked their broken stoicism. Their lives had fallen to pieces, but in Watt's mind they seemed "too perfect," and also not perfect enough. Watt had read widely about the shooting and the families, choosing from each account only the facts that suited her false narrative.

She brought up Chris and Lynn McDonnell, parents of seven-year-old Grace, a child with striking pale blue eyes who liked to paint, Noah's classmate. Lynn McDonnell told CNN's Cooper that Grace had drawn a peace sign and the message "Grace loves Mommy" in the fogged bathroom mirror after her shower, whose traces her mother found after her death. She described the abyss she felt on seeing her daughter's white casket, and recalled how she, Chris, and Grace's brother, Jack, used markers to fill its stark emptiness with colorful drawings of things Grace loved.

Watt mocked her reminiscence in a singsong tone. "'Ohhhhh, Grace. She loved loved loved loved loved Sandy Hook, and we're glad she's in heaven with her teacher, and she's with her classmates, and we feel good about that,'" she said. "'She had a white coffin, and we busted out the Sharpies and drew a skillet and a sailboat.' NOBODY CRIED," she barked.

Her feral lack of empathy astonished me. I reminded her that a photograph of the McDonnells appeared on newspaper front pages around the world, openly weeping in the firehouse parking lot after learning Grace was dead.

"Not one person had a Kleenex," she retorted. "That's a dead giveaway. They didn't think about *that* prop. You've got hundreds of people and not one Kleenex? Impossible."

I thought about the tissues I had seen in parents' hands in the photos. The literal crates of tissues Walmart had donated to Newtown.

She continued, "And the way they made those kids bigger than life.

I'm sorry, Emilie Parker did not know Portuguese. And the Kowalski boy: 'Oh, we've lost a future Olympian,' with pictures of him swimming in a kiddie pool and riding a bike with training wheels."

Watt a few minutes earlier had boasted about her son Jordan's voracious reading habits and how well her daughter, Madison, played the piano. If Watt's children died, wouldn't she also speak highly of them and their gifts?

"No. This was to build up the sympathy factor. I do not own a gun, never shot a gun—I'm scared to death of them. But I think they're people with a gun control agenda."

Watt had been speaking to me from a hospital, where her son had been admitted with symptoms resembling appendicitis. She had been speaking to me for hours, from the hallway outside his room.

"If Jordan died, I wouldn't be in Washington lobbying," she added.

It was a strange thing to say. When Watt's children were young, she ignored them, obsessed with saving families from imagined government plots while her own family unraveled around her.

WATT'S CHILDREN WERE BARELY in grade school when a neighbor urged her to join a battle against the passage of Oklahoma House Bill 1017, a.k.a. the Education Reform Act of 1990. The proposed overhaul, including new curricula and testing standards, would cost more than $500 million over five years, funded through a tax increase.

"I didn't even know property taxes funded the schools," Watt said, but the cost wasn't the problem. She believed the reforms masked government's true intent: "dumbing down the population," asserting control. She threw herself into the campaign, speaking at meetings, picketing, making phone calls late into the night. She lost; the bill passed. But the campaign "changed my life," Watt said. "I kept going and going."

Watt and her allies formed Families Restoring Excellence in Education (FREE), a vanguard against plans by liberals in the Department of

Education to mold Oklahoma schoolchildren into pliant underachievers. Inspired by books like *Educating for the New World Order* and *Government Nannies*, Watt exposed what she claimed were examples of social engineering in reading texts, math problems, even the free-lunch program. Her group campaigned to ban a book titled *Earth Child* and its corresponding science curriculum, saying it taught children to worship "earth above humans." She compiled stacks of "research," pressing it on PTA parents and local politicians, hand-delivering it to the *Tulsa World* newspaper and the city's three network affiliates. She grew enraged when ignored, ringing people in the middle of the night, turning up at their offices and homes.

"Every education reporter during that era remembers her," Ginnie Graham, a *Tulsa World* writer who covered education at the time, told me. Well-spoken and fashionably dressed, Watt came off at first "like any active PTA mom," Graham said. "But it didn't take long to uncover more conspiratorial thoughts." Dodging Kelley Watt grew into a rite of passage for the *World*'s rookie reporters, Graham said. "Once a reporter gave her some time, she would then be a persistent caller to that reporter. Kelley would give up on the older reporters and try breaking through to the new ones."

This was before the decline of local media and the rise of Facebook's news feed, when trusted community outlets had the power to shut down verifiably cuckoo claims.

Watt viewed the rejection as evidence of the media's capture by government. Her 1993 bid for a school board seat failed after "they" labeled her an extremist, she said.

"The *Tulsa World* only wanted to write articles about fundraising events like Mommies and Muffins, Daddy's [*sic*] and Donuts, Beach Blanket Bingo, to distract parents from what was really taking place," Watt wrote in one of fifty text messages she sent me over the course of an hour one evening.

Parents in her group drifted away. "They were playing tennis and working and distracted and did not wanna know what was going on in their kids' school," she said.

Watt's battle for the minds of local schoolchildren blinded her to ominous developments at home. The family electrical business was racking up losses on Jim's watch, and consumed by guilt, he fell into a depression. "I had seven-hundred-dollar- and eight-hundred-dollar-a-month phone bills, and we didn't have any money," Watt recalled.

"I'd be up all night researching, and Jim used to go to this little bar up the street because he hated my research. So did I drive him to drink? I guess I did."

The business went bust. The Watts eventually lost their houses and cars and had to withdraw their children from private school. In the early 2000s, Kelley Watt left her husband and moved into a one-bedroom apartment, riding a bicycle to work or driving a "bomb" of a car she borrowed from neighbors. The couple divorced in 2004. Jordan lived for a time with his father, who continued to spiral, at one point cleaning out his son's college-money bank account to pay a DUI fine, Watt said. Jordan got into trouble with the law too, and served a six-month jail term, she said.

Watt did secretarial work, then switched to cleaning houses because she couldn't afford a full-time babysitter for Madison, then in second grade.

"You know what? The bad times teach you a lot—to persevere, be determined, and stick to your guns, seeking truth, seeking justice," she told me. "When you're poor, you have more time to do stuff. You clean two or three houses a day and you have a lot of time left over." She devoted most of that time to more research.

Madison's public school denied Watt a slot on a task force to review proposed new textbooks, so when picking up her daughter one afternoon Watt grabbed one from atop a stack. She wound up in a standoff with the principal, who called the police. Watt withdrew Madison from the school, and homeschooled her for a time. Her business slowly grew. In a bar one night she met Duke, striking up a conversation about the poor quality of the public education his children were receiving. He hired her to clean his house; they dated and eventually moved in together. Watt gave up her public schools campaign.

"Kelley's efforts never reached any mainstream acceptance or status," Graham, the Tulsa reporter, said. "These were days before the internet, so her reach was limited." Indeed, without the web to enshrine her research forever, Watt's crates of files moldered in her attic.

WATT'S DAUGHTER LEFT OKLAHOMA. She attended the Cooper Union in New York, sleeping in the butler's pantry of an apartment to save money. She works in Poland, teaching English to business executives. She is a smart and accomplished young woman who has challenged her mother's views, both in person and in comments on her mother's Facebook page. "She doesn't question things the way her mother does," her mother said.

"Madison was a cute little girl," Watt told me. The quiet-natured Madison loved to paint and yearned for piano lessons. Her mother envisioned a different ideal for her, entering her in kiddie beauty pageants and ice-skating competitions. "I have a picture of her in diapers with ice skates on. I got her this two-hundred-dollar rabbit fur coat, got her little ice skates on. She was like this power skater. But then came the divorce, and I couldn't afford it, so that was all gone."

Jordan "struggles," Watt said. "He's very, very smart, extremely smart with a very high IQ, but he doesn't work. He's got every excuse in the book as to why he doesn't work."

Jordan recently told her, "'Mom, you ignored us so bad,'" Watt said. "He'd come in when I was on the phone and say, 'Mommy?' And I'd wave my finger and tell him to go play in the backyard."

Her voice broke into a sob. "Oh my God, yes. I have so much guilt."

Watt's ex-husband, Jim, continued to struggle with alcoholism and homelessness. In summer 2020 he was living a half hour away in Claremore, Oklahoma, behind the Claremore Motor Inn, "in a storage unit where they used to keep the lawn mower," Watt told me.

"He gets a free donut and coffee for breakfast, and he's happy as a lark," she said. "He tells me all the time, 'I owe you the biggest apology for everything.' Because I was right."

Watt was ebullient. President Trump had just made a speech at the National Archives, condemning "decades of left-wing indoctrination in our schools."

"Our youth will be taught to love America," he said on September 17, 2020. He pledged to end what he said was liberal educators' portrayal of America as "a wicked and racist nation," and create a new panel, the 1776 Commission, charged with "restoring patriotic education in our schools."

"I'm gonna go listen to what Trump has to say about them teaching critical race theory," Watt said. "I'm so glad I have all that stuff in my attic. I might need that!"

Two months later, Jim's body was found in his storage-shed home. He had died of natural causes, likely related to his alcoholism. His children took up a GoFundMe collection to cremate him. The state would have paid for it but would have kept and disposed of his ashes. I contributed $35.

"Our secret," Watt said when she heard about it. No, I told her, repelled at the idea. My donation was to her children. I felt terrible for them.

SANDY HOOK HAD GIVEN Watt a new cause; and social media, a global audience.

Three weeks after the massacre, Watt worked a late-night office cleaning job and headed to her and Duke's cabin in Grand Lake, Oklahoma. It was too cold to go fishing, so she flipped open her laptop. On CNN, she watched Anderson Cooper "railing about James Tracy," she said. "He was saying that this nutty professor from [Florida] Atlantic University who has a blog site called *Memory Hole* doesn't think it happened.

"Being a curious person, I immediately clicked off of Anderson Cooper and logged on to *Memory Hole*."

There she read Tracy's post about the December 15, 2012, press conference with H. Wayne Carver, the Connecticut chief medical examiner. An imposing man, his jaw and bald pate fringed in white, Carver had worked on some of Connecticut's most grisly crimes over his thirty-year career. In the late 1980s, he helped convict Richard Crafts, who murdered his wife, Helle, stuffed her body inside a freezer chest in their Newtown home, then ran it through a wood chipper to conceal the crime. Investigators found a fingertip and bone fragments, bearing unusual marks. Carver fed a frozen pig carcass through the same equipment, replicating the marks. Crafts's was one of Connecticut's rare murder convictions won despite the absence of a body.

Carver's offbeat earthiness—he often joked that his name suited his profession—served him well with juries. But that day, punchy with exhaustion, fielding reporters' repetitive, sometimes cringeworthy questions, his manner unwittingly fueled the suspicions of people like Watt. "I've been at this for a third of a century," he told a reporter who asked whether this case was "over the top." "My sensibilities may not be the average man," he said, with an awkward half laugh. "But this probably is the worst I have seen, or uh, the worst, uh, that I know of any of my colleagues having seen." He offered his gratitude and concern for them, adding, "I hope they and I hope the people of Newtown don't have it crash on their head later."

Watt watched a YouTube clip of the news conference three times. "I counted one hundred and fifty 'uhs' in twenty minutes," verbal tics she took as "a sign that somebody doesn't know what they're talking about," she said. "I'm watching this in total disbelief, that a coroner when all these children had died was acting like this," she said. "This professor was saying this didn't happen, and something just clicked.

"That's when I called to ask them who got the contract to clean up the blood."

Watt made hundreds of phone calls to the Newtown Town Clerk's office and city hall, the Connecticut State Police, state environment protection officials, and Carver's office, demanding the name of the company that cleaned the crime scene. When they told her off or slammed down the phone, she named them as participants in the cover-up.

Had she ever cleaned a crime scene?

"Oh no, no no. I'm just a janitor," she said. "It's a real specialized field and they test you psychologically, even. Somebody like me wouldn't be allowed in there—I looked it up online."

Watt believes Noah never existed. She insists Noah is actually Veronique Pozner's son from a previous marriage, born nearly a decade before Noah. Holding out her iPad like an evangelist wields a Bible, she toggles between Noah's photographs and photos of the young man that she lifted from his Facebook account. Compare, she urges, "the angel bow in the lip, the roundness of the face, the color of the eyes, the bushy eyebrows" of the dead child with those of his living older brother. "Pictures do speak," she told me, her voice low and brimming with import. Watt watches true crime shows on TV, and when she talks about Sandy Hook, she adopts their narrators' sonorous tone.

Watt would chat online with "mostly women, because women are more into social media and Facebook, and they'd say, 'I heard that was a hoax, and then they'd ask me, 'Why would the government do this? What was the purpose behind it?' And I'd say, 'They were trying to ban the AR-15 and of course that was the weapon of choice by this kid with Asperger syndrome,' and point out all the anomalies."

Her preoccupation migrated into the real world.

"We'd be filling up our boat at a boat dock, and I would have crowds captivated. People would be like, 'Oh my God.'

"I told my clients, people in elevators, waitresses. Duke would say, 'Come on, we've gotta go. Don't bring up Sandy Hook—we're not gonna be here that long.'

"But I felt like this was a global story that needed to be told."

In December 2013, Connecticut released the first recordings of 911 calls from inside the school. Todd Schnitt, an AM radio host whose syndicated program, *The Schnitt Show*, aired on WLAD in Danbury, played excerpts from the calls, inviting people from Newtown to weigh in.

"Kelley in Tulsa" called in on line three. She urged him to watch a "documentary that's gone viral" called "*Unraveling Sandy Hook in Two, Three, Four, and Five Dimensions.*"

Schnitt snorted. "Whoa, you're calling my show, and you're gonna sit here with a straight face and say the Sandy Hook tragedy—that twenty kids were not shot and killed? What the hell—what are you talking about?"

"Absolutely," she said. Watt told him about her calls to Connecticut, where "there was no blood to clean up, all right?"

"Oh my God," Schnitt erupted. "You know what scares the crap out of me? People like you, psychos and kooks who believe this conspiracy crap."

"It's kooks like you that are spreading this," she replied. "It did not happen. It was a drill and you know it and everybody else. The people inside Connecticut have had a major brainwashing job. Nobody outside of the state of Connecticut believes this crap."

They talked over each other for more than a minute, Watt repeating the name of the documentary and a list of reasons why the shooting was fake. "You're just splattering talking points from some kook-job video," Schnitt told her.

"Quit lying to people," she said. "It was a drill. It was a drill and nothing more."

"Kelley, you sound like a robotic nutjob . . . What do you do for a living? I'm curious . . . How old are you?"

"That has nothing to do with it," she told him. "It was a drill." Schnitt hung up on her, furious.

Watt was elated.

"I knew because of my education research," she said, that Schnitt "was using Delphi, which is a technique where you take a diverse group of people and dialogue them to consensus, using laughter and all these techniques to try and distract.

"I just kept saying over and over and over, '*Sandy Hook in Two, Three, Four, and Five Dimensions.*' All I cared about was people listening to that and googling it."

James Fetzer heard her. The retired professor contacted Watt online and interviewed her for his website. Watt was thrilled by the attention from a "researcher" with a doctorate and a two-decade history of bucking government narratives. She fed her biohazard theory to Fetzer and his online pal Wolfgang Halbig. Halbig added Watt's query to his blizzard of public records requests. He wanted receipts for the cleanup of "bodily fluids, brain matter, skull fragments and around 45 to 60 gallons of blood."

THE CONSPIRACY FRENZY surrounding Sandy Hook drew the notice of Charles Frye, a court clerk in Northern California who has since become a lawyer. In those days Frye used the online name "CW Wade." He originally took a professional interest in the shooting.

"I had some questions when it happened," he told me. "Like everyone else, I was completely shocked at the time. I did surface level watching of it, but I didn't get involved too much until the reports came out"—the November 25, 2013, release of the final report on the investigation into the Sandy Hook shooting by Stephen J. Sedensky III, state's attorney for the judicial district of Danbury, and the online release of the Connecticut State Police's investigation a month later.

"At that time YouTube was literally littered with hoax videos," Wade recalled. "I was like, 'I gotta know what happened in Sandy Hook. How the heck did that guy walk into a school building and kill so many children in so few minutes?'

"At first I gave the hoaxers the benefit of the doubt, thinking they're just asking questions" like he was, he said. "And I thought early on that if you were honest and you engaged them in an honest debate and presented evidence, that would persuade them," he told me.

"I couldn't debunk all the little guys, the low-hanging fruit, so I would go for the main guys," like Tracy, Halbig, and Jim Fetzer, whose theories gained traction on Infowars in those years.

Wade studied thousands of pages of records and reports, photographs and heavily redacted crime scene videotape from Sandy Hook and the gunman's house, which authorities released with images of the victims redacted. Wade re-created the shooting, the weapons used, and the gunman's movements that day.

Early on Wade disproved the Sandy Hook deniers' theory that it was physically impossible for Adam Lanza to shoot 154 rounds from his Bushmaster XM15-E2S in five to eleven minutes, killing twenty-six people before turning a handgun on himself. Wade presented his evidence as the hoaxers did, in videos and a blog he started in early 2014 called *Sandy Hook Facts*.

Wade posted documents from the state's attorney and Connecticut police reports, indexing them for others and using them to dispel falsehoods.

In 2014, Wade surfaced a Newtown Police Department report that answered Watt's question about cleaning the school. He posted the document on the *Sandy Hook Facts* blog with a scorching rebuke.

"Hoaxers are donating Halbig thousands of dollars so that he can ask this question for them!" Wade wrote. "Is this question a sick, disgusting, mean spirited, absurd, and irrelevant question designed to hurt [and] re-victimize families? . . . Yes, of course it is."

The report filed by Newtown Police Officer Jeff Silver attests to Newtown's determination to shield the grim operation from prying eyes. "On 12/31/2012 I met Wilson of Clean Harbors"—a national industrial cleaning company—"at the school. All members of his team were instructed not to bring phones, camera, or other electronic devices into the school. They all stated that they understood and complied."

The cleaning took four days. Police officers escorted a dumpster of "combustible materials" from the school to Wheelabrator Bridgeport, where "they witnessed the destruction of the material at 1305-1315 hours on 1/3/2013." The next day, officers accompanied and witnessed the destruction of a load of metal desks, chairs, and other metal objects at Sims Metal Management in North Haven, Connecticut.

Finally, "on 1/4/2013 one box truck was loaded with 63 boxes of biological waste removed from the school, computer equipment and the remaining items removed from the school. Clean Harbors indicated that these items would be secured at their facility and then shipped for disposal in Arizona and Oklahoma."

Watt, Fetzer, and the others ignored the evidence.

Watt told me she had visited Clean Harbors' website. "Now, it might have changed, but it said, 'We clean up harbors and waterways from oil spills.' Which told me they don't clean up urine and blood and guts from elementary schools."

That's wrong, I told her. It had taken me one phone call to confirm that Clean Harbors had cleaned Sandy Hook. The company had performed biohazard cleanup for decades, including of Ground Zero after the 9/11 attacks.

"You told me you made hundreds of phone calls looking for the cleanup information," I said, working to keep my voice neutral. "Did you ever read the one-page Newtown police report that spells out what you spent years trying to get?"

Watt was momentarily silent. "I haven't seen that document," she said blithely. "But where are the receipts?"

WADE WANTED TO DEPLOY his skills to help the families. Gingerly, he reached out to a few, offering to debunk false claims piling up on their social media accounts and charitable websites.

Some told him they refused to publicly acknowledge the conspiracists or didn't have the energy to engage them.

Wade concluded that "no one in Newtown was going to give these people any attention whatsoever. Early on they thought that even talking to them would lend them legitimacy."

In Lenny Pozner, however, Wade found a parent with a sophisticated understanding of the internet determined to counter falsehoods, whether spread by mainstream media, Infowars, or random people on Facebook.

"To this day there are very few parents who would associate with the dirty little pool of debunking and hoaxing. He was pretty much the only one willing to be there," Wade said.

The two grew into online friends.

"CW had this concept that the hoaxers wanted evidence, so if we can just show that Noah was real, showing proof," they would be satisfied, Lenny told me.

Lenny could not yet bear to read the official reports. "I didn't have any evidence of what the state police did, or any of that," he said.

"All I had was my experience of that day, and what happened to me."

He resolved to embark on a surreal endeavor. He would use his authority as a father and the records of Noah's life to prove his son had been a living little boy, murdered in his first-grade class.

9

IN EARLY SPRING OF 2014 Lenny Pozner began to assemble the records of Noah's life and death. Compiling data came naturally to him. He had already collected the most important material the previous year, and he knew where to find the rest. The emotional aspect of the task would prove more difficult. Asserting the truth of Noah's life would force Lenny into a public role he had never sought for himself, certainly not during the worst period of his life.

Lenny was in his mid-forties then, with short dark hair and eyes like Noah's. He speaks deliberately, weighing his words, in a timbre so low a friend nicknamed him Eeyore. It takes a long time for Lenny to reveal himself. In conversation he is freely argumentative, with a dry, ironic wit. He has a teenager's love of internet memes and cartoons.

When saddened, like on the shooting anniversary, he is more likely to send a broken-heart emoji than any words. Lenny's first email to me read simply: "It was nice talking with you too. I will create a google doc."

The Google doc contained what Lenny did not express aloud. A photograph of Veronique in the hospital, nuzzling newborn twins. The babies' footprints. Lenny holding Noah tight against his chest after a fall, tears clinging to the toddler's lashes. A link to a private videotape of Noah's funeral service, his father sitting stolidly in the front row.

Lenny did not speak or emote in public after Noah's death. He told me strangers are not entitled to that part of him. This reserve can prompt people to misapprehend him, or assume the face he chooses to show them is all there is. Of many wrong assumptions by the conspiracy theorists, the most inaccurate was that Lenny lacked intensity, that he wasn't angry enough to be a grieving father. He would soon demonstrate how foolish that was.

Delivering Noah's records to the hoaxers would plunge Lenny into a subterranean world that repelled the other victims' families. A year earlier, some thirty Sandy Hook relatives had successfully lobbied Connecticut lawmakers for legislation limiting public access to any record depicting the physical condition of a homicide victim without surviving family members' permission.

The other parents were the only people in the world who knew how it felt to lose Noah this way. The thought of alienating them bothered Lenny more than he let on.

But as he contemplated what he was about to do, he thought about the dedication in *Reclaiming History*, Vincent Bugliosi's tome throttling JFK assassination conspiracists: "To the historical record, knowing that nothing in the present can exist without the paternity of history, and hence, the latter is sacred, and should never be tampered with or defiled by untruths."

"We can't protect them. They've already been killed," Lenny told me. "This is Noah's story. All you can do to honor him is reveal more about his life and his death, to protect his story from being diluted or hijacked for someone else's cause."

Lenny made sure Veronique was on board. She understood Lenny's desire to make these people see the light, but did not share it.

"If you think about it, you're grateful that you don't understand them," she told me about the conspiracists. "You don't want to relate to that. It's abhorrent. And twisted."

Still, at the time, she had hoped "reason would prevail, and they would say, 'Oh, okay. Yeah, you've made your point. I'm going to come around.'"

CW Wade worried about the toll this endeavor could take on his friend. Worse, he suspected that it wouldn't change any minds.

"I said, 'Lenny don't do that—they're just going to tear it up,'" he told me. "But he was trying to debunk them with evidence."

Lenny started slow, posting Noah's kindergarten progress report on the Google+ page, with his teacher's praise for his reading skills and assessment of him as a "sweet, inquisitive boy."

He posted the copy of Noah's death certificate that he had gotten from the Newtown records clerk in early 2013. He used a marker to blot out Noah's social security number and the location of his grave in Monroe, because Halbig and others had threatened to dig him up to prove that it was empty.

Lenny had gotten a copy of Noah's postmortem report in late 2013, for a reason that still agitates him. An article published in the Jewish *Forward* after Noah's funeral said he was shot eleven times, a greatly inflated number. The source of the error remains unclear. But Sandy Hook deniers seized on it. "Tell me how you had an open casket funeral when Noah was shot 11 times?" one person wrote on Noah's Google+ page. "There'd be NOTHING LEFT OF HIM! You'd be picking up body parts and tossing them in a casket! You people are SO DUMB!"

Several months after the story ran, Lenny asked for a correction. Before acceding, the paper asked him for official "documentation" of the number of times Noah was shot. That angered him. How could they have trusted an anonymous, erroneous source over him? It was a question he would ask himself about the Sandy Hook deniers too.

EMERGING LEAVES HAD CAST a green veil on the trees along I-84 when in the spring of 2013, Lenny drove to Farmington, to the brutalist concrete building housing the office of the chief medical examiner. He met H. Wayne Carver there, and they sat down in a nondescript conference room to discuss Case No. 12-17604, the postmortem examination of his son.

Harold Wayne Carver II was born in St. Louis and grew up as a gregarious, self-described "geek" gifted in science and music. While pursuing his undergraduate and medical degrees at Brown, he played in three orchestras and marched as a drum major in the university band. He met his wife, Deborah DeHertogh, also a physician, in anatomy class.[1]

Carver returned to his native Midwest to train in the medical examiner's office in Cook County, Illinois, encompassing Chicago, whose crime provided plenty of training. On weekends Carver loved to cook, a talent betrayed by an ample frame that he draped in loud Hawaiian shirts on his off-hours. Carver thought deeply about the suffering families whose loved ones' bodies arrived at his office. He read on his own about their religious beliefs and cultures, training himself and his team to respect them. Carver's government superiors appreciated that. Carver had ascended to chief medical examiner in 1986 after his predecessor was fired, in part for letting her pet Doberman pinschers roam the office during autopsies. Decades of daily encounters with death had armored the pathologist with a dark sense of humor that some people found off-putting. Lenny was unperturbed. He wanted what all the families sought, some right away, some not for years: the details.

In his flat accent Carver chatted about unrelated things for a while. He told Lenny about one of his two sons, who was tall like him and played college basketball. Lenny had trouble focusing on the words but recognized their warmth. Then Carver asked what questions Lenny had about the five-page report.

Their teachers were already dead when the gunman found Noah, Joey, Charlotte, Emilie, Grace, James, Sammy, Ana, Caroline, Chase, Ben, Maddie, Jack, Catherine, and Jessica hiding in the bathroom. He shouted and laughed as he fired more than eighty rounds, shattering drywall, wood, porcelain, flesh, bone, their lives, Lenny's, Veronique's, all the families'.

They found Noah, sixty-one pounds and an inch shy of four feet tall, lying faceup on the floor. He wore black, red, and gray superhero-themed sneakers, black pants, and the red Batman hoodie. From its folds Carver retrieved a small-caliber bullet, deformed from impact. He inscribed it with the number 852—the 852nd bullet retrieved from a body by the Connecticut medical examiner's office.

Lenny wondered whether it was Noah whom Sergeant Bill Cario described when he wrote in his report: "As I stared in disbelief, I recognized the face of a little boy . . . I then began to realize that there were other children around the little boy, and that this was actually a pile of dead children."

Lenny and Veronique had asked Carver not to perform an internal examination. So on the third page of the report the coroner described Noah's injuries. Lenny agonized about letting people see this page.

The bullet that ended Noah's life entered his right shoulder. X-rays revealed minute bits of metal and bone along its path through his chest and left arm. A second shot took most of Noah's lower lip and jaw. A third struck his left hand, destroying his thumb.

> Needle aspiration demonstrates hemothorax in both chest cavities, as does x-rays.

"His lungs, his entire torso filling up with blood. I thought about that and thought about what that meant" for a long time, Lenny said.

> There is a gunshot wound to the extensor aspect of the left thumb . . . It passes through the thenar eminence for a distance of 3/4" and leaves the thumb through a 1/2" irregular stellate laceration.

There were soot deposits around the wound to Noah's hand. Lenny asked Carver, "Was that a defensive wound?" Had Noah, at close range, tried to shield himself?

Carver told him no. "I thought maybe he was trying to protect me," Lenny said. "Because no parent wants to think of their child in that kind of terror at the end."

Carver reassured him that it had all happened chaotically and swiftly. An isolated confrontation like that wasn't likely in "a dogpile of children."

Lenny wondered numbly at Carver's blunt metaphor. A dogpile of children. "I didn't understand the visual. I guess puppies pile on top of each other.

"Maybe that was toward the end, kids still wiggling on the bottom of the dogpile and to get at them he would have had to shoot through Noah's thumb."

Lenny stopped speaking for several seconds.

"I don't even know why it was important," he said.

"The gunman was blocking the door. It wasn't like anyone could have gotten out."

By mid-May in 2014, Lenny had posted all the documents to Noah's Google+ page. He couldn't tell whether they had satisfied anyone, because his antagonists tended to dominate the comments.

Wolfgang Halbig, for one, had continued trumpeting his doubts on Infowars.

That same month, Lenny scheduled a business meeting not far from the gated golf course community in Sorrento, Florida, where Halbig lived with his wife, an educator, and their dog, a moplike Havanese named Coco. Lenny hoped to visit Halbig at home to show him Noah's records.

Halbig wasn't hard to find. He plastered his home address on his website so fans could send him money, and includes his phone number on every email he sends. Halbig's voicemail was full, so Lenny emailed him.

"I would like to speak to you on the phone. Len Pozner (father of Noah Pozner)."

No answer.

Halbig had just returned from Newtown. He'd spent a good chunk of the month there, demanding the very proof Lenny offered to show him. On May 6, Halbig, Jim Fetzer, an Infowars camera crew, and about a dozen other hoaxers visited Newtown for a Board of Education meeting. "Board members, these are your children," Halbig pleaded, playing to the cameras. "We want answers. We want truth." An array of Sandy Hook deniers sat around him, including Fetzer, in his usual khakis and orthopedic hiking shoes, and a man dressed as a Revolutionary War patriot in tricorn hat and lace cravat, who earlier had hung an effigy of Governor Malloy from a tree branch with a sign reading, EXPOSE ME W/ SANDY HOOK . . . I'LL GIVE YOU BACK YOUR 2ND AMENDMENT RIGHTS! The group strode into the offices of the Danbury *News-Times*, seeking publicity. "I go for the throat," Halbig boasted to their video cameraman as they entered, then turned tail after an elderly receptionist told them to shut off their cameras and get out.

A week later, on May 13, Halbig called in to Infowars to tell Jones about his trip. On its webcasts, Infowars superimposed the link to Halbig's Sandy Hook Justice website, to help him raise money. "What's your bottom line? What do you really think happened at Sandy Hook?" Jones urged. "I can tell you, children did not die, teachers did not die on December 14, 2012," Halbig replied. "It just could not have happened." Jones nodded sagely. Halbig followed up with another trip to the Newtown police station in late May.

But when Lenny, an actual Sandy Hook parent, offered to hand-deliver the truth Halbig sought, he cowered.

On May 30, 2014, Kelley Watt emailed Lenny:

> Wolfgang does not wish to speak to you unless you exhume Noah's body and prove to the world you lost your son.

Hiding in his home, Halbig called the Lake County Sheriff's Office repeatedly over three days, claiming that Lenny was outside the gate of his community and/or stalking him by phone. "I was never there," Lenny said.

Lenny had already endured hours on the phone with Watt, calmly answering questions about Noah, and her suspicions that a "real" father would be more distraught. Giving up on Halbig, he asked Watt to relay a message to Tony Mead, who ran the Sandy Hook Hoax Facebook group, a rookery for deniers who gathered nightly, chattering into the wee hours. Lenny wanted to join them online and answer their questions.

Lenny thought that as someone who had entertained conspiracy theories himself in the past, he might be able to talk these people off the ledge.

"If I was on the opposite side of that perspective, and someone who was part of the tragedy made themselves accessible, I probably would have wanted to get some information from them," he said later, recalling his thinking. "I thought that I'd kind of bridge that gap and cross over and see what I would find."

Mead brought Lenny's proposal to the group.

"Lenny Pozner has asked to join the group!! All in favor say 'Aye' !!" Mead crowed in a post summoning the page's several hundred regular visitors.

Sandy Hook Hoax was a closed Facebook group; you needed an invitation to join. Its page was topped with the image of a child ghoul, her eyes ringed in black, a mud-encrusted finger pressed to her lips.

Its nightly discussions drew from a dozen to a couple hundred people. Some still struggled with the enormity of the crime. Other, darker types obsessively posted photographs of the young victims, comparing them with living children in an ostensible effort to "prove" they were still alive, that they had attended their own funerals. Many Sandy Hook hoaxers didn't tell family members about their membership in the group. Others admitted, with an "ugh" emoji, that their families questioned their sanity.

Tony Mead experienced no such qualms. He was in his mid-fifties, with a florid complexion and a broad, thin-lipped mouth, the pouchy slope of his neck from chin to collarbone giving him a vaguely reptilian profile. Mead runs Absolute Best Moving in Tamarac, Florida, schlepping home and office furniture. "All my business is referrals and repeats," he told me. "I don't have to advertise, and that's good, because I've gotten so much bullshit," he said, meaning people who connected his business with his online pursuit and dinged him on Yelp or on RipOffReport.com.

"People know me to be fair and honest and just," Mead told me, which sounded like protesting too much. "I have integrity. People trust me with their belongings; they trust me with their families."

The Sandy Hook Hoax group debated for more than an hour about whether to admit Lenny. Members seemed worried, even threatened, by his overture.

(I have used first names only to protect unrelated people with the same names or pseudonyms as the participants quoted, as well as those who have since repudiated the group and its leadership.)

"I say nay, because I don't see what will come out of it," wrote Thomas. "It is just us voluntarily giving them more information on what we are doing."

Pamela wrote, "Does he want to trash us or is he on our side?"

Erin: "How do you even know it's him?"

"Of course you should let him join. Why be afraid?" posted Steve.

"It's not really fear, Steve," replied Kathryn. "He will have info available to him so they can spin the story to fit their agenda. Things they don't know we have picked up on."

"Lenny Pozner is a professional shill," said Tony Mead, who during the discussion had posted Lenny's home address, LinkedIn page, and photos of his surviving children. "He gets paid to troll the internet."

Reading the exchange, I was struck by the group's determined defensiveness. They were a ragtag army of errant thinkers holed up in a Facebook fortress, fending off intrusions of truth.

Political scientists began studying the spread of conspiracy theories about thirty years ago. Psychologists joined them less than a decade ago, partly in response to a surge in outlandish claims and mass delusions that spread on social media during the Obama and Trump eras. While the bulk of studies focus on understanding why *individuals* believe conspiracy theories, researchers like Nicholas DiFonzo delve into the role of conspiracy theories within groups, the prime vectors for their spread.[2] DiFonzo, associate professor of psychology at Roberts Wesleyan College with a psychoanalyst's empathic style, tapped eight decades of research into the spread of rumors to understand what motivates and sustains groups like Sandy Hook Hoax.

Assessing and vetting rumors builds trust among group members who may be genuinely trying to learn the truth, as well as those who tend to distrust traditional sources of information. "Conspiracy theories are actually a subset of rumors," DiFonzo told me. As distrust in American institutions like the media, government, and religion has risen, so too has the spread of conspiracy theories.

Conspiracy rumors can be a collective response to psychological threats, which DiFonzo defines as attacks by outsiders on a group's identity, values, faith, or politics.

Conspiracy theorizing is an enduring feature of American life. Since the country's founding, Americans have proved willing to believe that an ever-shifting panoply of internal villains, from Catholics to financiers, threaten the American project. Kathryn S. Olmsted, professor of history at the University of California, Davis, writes that scholars generally agree that "American conspiracism stems from the difficulty of defining a national identity. In a land of immigrants, some Americans have resorted to demonizing outsiders as a way of bolstering their own sense of self." What historian Richard Hofstadter called the "paranoid style in American politics" in his seminal 1964 essay[3] springs partly from "the fluidity and diversity of American life, which led Americans to have 'status anxieties' that they assuaged by striking out against hidden enemies," Olmsted writes.[4]

"Americans define themselves by defining the other," Olmsted told me.

Although American conspiracism has always been with us, Olmsted and others argue that the current wave of political conspiracism and government distrust has its roots in the Cold War, when our intelligence and security agencies spied on a range of groups perceived to be national security threats, from the American Communist Party, to the Students for a Democratic Society, to Martin Luther King Jr.'s Southern Christian Leadership Conference. Alex Jones's conspiracies about poisoned drinking water and CIA-controlled mainstream media are a plausible if often incoherent and self-serving reaction to historical events. The sweeping COINTELPRO domestic surveillance program was launched by J. Edgar Hoover's FBI. In MK-ULTRA, CIA scientists carried out LSD and other mind-control experiments, often on unwitting human subjects, to create a defense against similar actions by foreign adversaries. In Project Mockingbird, the CIA targeted two American journalists using warrantless surveillance in an effort to track down the theft of classified national security documents. Jones rightly decries those illegal programs while at the same time adopting the very same nativist paranoia that fueled some of the government's extremist reactions back then.

American anti-intellectualism provides a rich cultural agar for growing these theories. Americans distrustful of institutional expertise have increasingly been swapping facts for gut feelings, a trend accelerated by social media algorithms that feed us "information" tailored to our prejudices, and by President Trump, who blithely denigrated facts and science. This collective pivot away from established knowledge undermines efforts to understand and battle global challenges like climate change, pandemics, and societal violence.

Conspiracy theories gallop across the political spectrum. The University of Miami's Joe Uscinski said insecurity among members of the political party out of power drives conspiracies centered on the party in charge. Still, it's quite a leap for some pro-gun conservatives to believe that

Obama pulled off a seamless, staged mass shooting in Newtown a month after a majority of its residents voted for Mitt Romney. Obama left office in 2017 with Congress having failed to pass any substantive gun safety legislation. Yet not only do false flag claims around Sandy Hook persist; they have metastasized to virtually every mass shooting since.

Research suggests three main psychological motives behind the spread of conspiracy rumors. *Fact-finding* drives collaborative efforts to reach the truth within groups that, sometimes with good reason, distrust official sources. *Self-enhancement* involves spreading rumors to boost group members' self-esteem, as possessors of superior knowledge. In *relationship enhancement*, group members share rumors to deepen and cement the bond among them.

DiFonzo believes relationship enhancement to be the most powerful of the three. Just as membership in a group enhances an individual's self-esteem, he said, "sharing conspiracy rumors and stories tends to solidify the identity of the group."

But what explains the dehumanizing cruelty these groups exhibit toward people like Lenny?

DiFonzo sighed. "It's hard for us outsiders to wrap our heads around it, but for this particular group that disbelieves Sandy Hook happened, that poor parent was threatening a very central cherished belief."

The Sandy Hook Hoaxers were bound by the powerful, false belief that their freedoms were under siege. Group survival demanded they nurture that falsehood, and protect it. To them, the people who died in the gunfire, their families and all who tried to help, weren't victims, grieving survivors, or heroes.

They were threats.

As THE DEBATE WITHIN the Sandy Hook Hoax group continued, members' suspicions gave way to a cautious curiosity.

Evee wrote, "I say you make it clear to him that if you allow him to be a member he must agree to use the information solely for the benefit of all the other members. I think he should even sign a non-disclosure waiver.

"I change my answer to aye, only for one reason," wrote Jason. Lenny's sharing of Noah's records "either shows he is trying, or he is breaking."

Stephen wrote, "I think he's the loose cannon of the bunch and he'll be their downfall. Let him talk."

"Like the ole saying," wrote Celena, "'keep your friends close and your enemies closer.'"

LENNY JOINED THE GROUP on the evening of June 28.

Unbeknownst to the moderators, Wade was there too, screenshotting the conversation as it rolled downhill. This is a sampling of the hours-long online conversation, drawn from that record. The typos and abbreviations are unedited.

"Lenny Pozner has been accepted into the group. I look forward to hearing what he has to say," typed Erik, Tony's fellow administrator from Yakima, Washington.

"What verification we have that it is hime?" James, a.k.a. "Zomblie," asked Pearson.

Lenny, anticipating this, had sent a photo of his driver's license to Watt.

"I have no goal in mind," Lenny typed. "I respect your right to question public events. I am not an authority on the event. What I can share is my experience before and after the event."

The group at first seemed open to that, even welcoming. Several offered Lenny their condolences.

"Why are you here and how can we help you cope," Kathryn wrote.

"I would like to see less abusive videos on youtube," Lenny responded. "They are irresponsible and downright mean."

Members objected to Lenny's ask that they stop sharing hoax videos. Videos were their version of DiFonzo's rumors—sharing them was a bonding exercise. "by getting videos removed, we might as well stop trying to find the truth . . ." wrote Melissa.

Mike wrote: "Why would you watch them? I wouldn't. I would want to tell the world that they are full of shit, and it did happen and show that it did happen to prove to these people that it wasn't a Hoax. Lenny please HELP US UNDERSTAND."

"I cannot be the convincer for people who have a conspiracy spin on everything. I want to keep my Sons memory pure," Lenny replied. "I cannot help anyone understand until they are ready to carry the pain of women and children being executed."

The group asked Lenny whether Noah had been taken to the hospital (no), whether he'd seen him after his death (yes), and whether he keeps a "Kosher home" (he ignored that one). They posted news photos from the scene outside the firehouse that they had pulled from the internet and asked Lenny to identify people in them (he could not).

James: "Do you accept the idea that there was one shooter when witness reports contradict it"

Lenny: "I believe there was one shooter. Witnesses in shock get confused."

Melissa: "what do you think about teachers saying there were 2 men running past the outside of the gym with guns shooting . . ."

Lenny: "I think the teachers were scared, and maybe saw police moving about with weapons."

Melissa: "they reported this BEFORE cops arrived."

Lenny: "The questions you ask are in the report. I am not part of the investigation and dont know the answers."

Kathryn: "Could you please stop pointing to reports and talk to us as a father, as human, so we can defend your son as well.

"tell us how he felt toward his first dental appointement, tell us what he liked

"what was his favorite ice cream

"tell us what made him happy . . . tell us what made him sad, we need to feel him to understand and make this go away"

It grew clear that Kathryn wasn't trying to gather details in order to "defend" Noah. Like many of the members, she doubted Noah's existence and was trying to trip Lenny up.

Kathryn's line of questioning was too tame for a member named AJ, who jumped in to reassert the beliefs that bound them.

"Its just lie after lie. Proven actors, photoshopped pictures, green screens, all the obvious stuff. If this actually happened, why go through all that work? . . . Its not ONE little piece of misinformation that the media did a bad job of reporting and saying, oh, that's a logical error. Errors happen. THIS is HUNDREDS of issues that simply don't add up. Except in some bank accounts of course."

A member named Jeff chimed in. "when did you set up a donation page for your 'loss'? and . . . how much have you collected?"

The tone darkened further when Lenny couldn't recall the first name of Noah's first-grade teacher. Amanda D'Amato was absent on the day of the shooting. Her substitute, Lauren Rousseau, was killed.

Melissa: "how don't you know your own kids teacher? You would know if your kid got shot through . . . You would know it from all the talk and Im sure the teacher would have offered condolences to you."

Tony Mead, who'd been lurking in the background, sensed a vulnerability. "You don't know the name of the teacher that had your son in her custody on the final day of his life? Seriously?"

Mead was heavily invested in discrediting Lenny. He got an enormous ego boost from his leadership of the hoax group. Its members praised his findings and his writing, and laughed at his jokes. He spent up to six hours a day on it. No longer just a house mover, he had begun calling himself an investigative journalist.

AJ brought up the Infowars video of Veronique. "Where was Veroniques interview done with Anderson Cooper? I mean, since it was a green screen and all. Was it in a studio?"

Lenny: "I beleive the interview was done in Newtown, outside. I don't think V travelled outside of Newtown"

AJ: "I believe that's GARBAGE!"

"I got that AJ," Lenny typed back.

AJ: "You would have been better off just saying, I don't know. Man, Im gonna have to coach you up if you wanna go on tv and make money Lenny."

Another member named Mike chimed in. "They don't care Mr. Pozner. They are completely delusional, and have completely bought into this nonsense because they can't rectify the fact that every day they put THEIR kids on the school bus and that this could happen to them. This they cannot handle so they deny. It's pathetic to look at from anyone possessing a rational mind. I apologize for the classless segment of humanity above. They will never believe. You are the proof they claim doesn't exist, and yet they still deny."

"There is a reason I am here," Lenny wrote. "I used to argue with people about 9/11 being an inside job. I listened to Alex Jones podcasts in my car, i entertained that we didn't go to the moon. My world changed . . . Try it on for 10 minutes . . . just 10 minutes. Beleive it and see what it feels like and then tell me to fuk off."

For hours Lenny answered questions that seemed designed to catch him in a lie, while Mead and his fellow trolls drove off anyone—and there were several—who gave him the benefit of the doubt. Lenny began blocking people who abused him, and those he blocked boasted that he'd done so.

They asked if Veronique represented Switzerland, lobbying in Washington for gun control. ("I believe that is one of those youtube lies.") Was it true the school hadn't been used in years? Did Noah's sister Sophia really see a dead body on the floor? Was Lenny "data mining" them? Outing them to the government?

"I beleive its easier for people to blame a group for this rather than face reality of this sucky unfair world," Lenny told them. "Thats my theory."

"Are you serious?" replied a member named Jen. "We know there is CRAZY in this world. BATSHIT CRAZY. We get that. A lot of us have seen it first hand. We are perfectly capable of accepting that a psychopath exists that could do this. It is FAR scarier to think that your government is involved! I WISH I believed this was the doing of "Adam"! That would be SOOOOO much easier to accept!"

"if you say so," Lenny replied.

His calm rebuttals riled them.

"I've never in my life witnessed a parent not lose their minds at the death of a child, and in Newtown, EVERYONE was composed," Jen wrote.

"everyone?" Lenny replied. "i shed tears a few hours ago . . . its not a reality tv show"

Jen retorted: "It's not reality? So it's staged?"

They demanded Lenny post a photo of himself. He refused.

"Lenny do you have any kind of proof or evidence that this whole shooting is legit. Any at all would help everyone out," Nick wrote.

"Nick, I buried my son, thats my story"

"Let's Bump this dude, we have no proof that he is real," said Mike, who three hours earlier had told Lenny, "I'm sorry for your loss."

Lenny had had enough too. "g'nite," he wrote.

Jen responded: "If you're the real deal, thank you for letting us abuse you. If you aren't, then fuck you very much."

Mead kicked Lenny out of the group. The hoaxers praised one another, reveling in their takedown.

"LOL Jen! Way to cap off the night," wrote one. "Take care all . . . Man, we could have some party if we were all together."

Wade logged off, sickened. "After releasing that autopsy report and everything, he put himself out there to try and answer their questions, and all he got was vitriol and abuse. He was willing to talk to people when no one else was. But they didn't hear anything he had to say."

LENNY KEEPS NOAH'S BATMAN pajamas in his top dresser drawer. When he presses them to his face, they remind him of how, late at night when Noah was alive, he "would pass by his room and sneak in just to take a breath of his scent," Lenny told me.

"I can smell it even now," he said, the sweet essence of salt and minerals that collects in sleeping babies' hair.

After Noah died, "I went through so many boxes of his clothes looking for anything that hadn't been washed, that had something, anything Noah on them," he said. "But nothing."

After his experience with those who doubted Noah's story, watching their questions distort into accusations, their suspicion into hate, Lenny understood that to them, he and Noah were villains. Bound by a shared delusion, they reserved any sense of respect or humanity for those in their group.

Lenny called his hope that the facts would prevail "naïveté." But his disappointment ran deeper. He believed in the power of the web to expand human knowledge and experience. The internet had allowed him to virtually sit with the group as witness and participant and to share what he knew. That they were not curious or interested signaled to him a profound shift.

"We thought the internet would give us this accelerated society of science and information, and really, we've gone back to flat earth," Lenny told me. "Online hoaxers are not the conspiracy theorists of forty years ago. All they see is this stupid, inconsistent reporting that doesn't mean anything."

Powerless to counter the political subtext of Noah's murder, to push back against the cacophonous attention it drew, Lenny felt like a spectator to his own loss. Now he feared similar forces threatened Noah's memory. It horrified him to see the image of his son, smiling in his bomber jacket, passed around by an online mob attacking Noah as a fake, a body double, a boy who never lived.

Lenny realized that "if this hoax continues to spread, in thirty years it will be past the point of no return." He resolved to fight. He had to.

People were a big part of that—regardless of how he was received, he would keep talking, insisting on the truth of what happened.

But machines were another part of the story entirely. Inside them, across the internet, relentless algorithms would push those human lies to the top. In the eyes of the internet, all-powerful booster of outrage and denial, Noah could fade, like his scent, to nothing.

10

Everywhere in Newtown, Veronique saw reminders of Noah's death. The route she took to the school; the strip mall where she parked, then ran to the firehouse; the school itself, fenced, locked, and guarded. The house where Noah had spent his life, sliding down the carpeted stairs on his belly and tearing through the kitchen, was tainted too. That same winter, its pipes froze and broke, the damage rendering it unlivable. Veronique abandoned it.

It had taken Veronique about two months to learn that Jones had aired his botched copy of her CNN appearance in front of the Edmond Town Hall and accused her of participating in a staged studio interview. She noticed people holding up their cell phones toward her on her errands around town. Were they watching her? If so, were they merely disaster tourists, or stalkers? She logged on to Facebook to read notes of solace, and also found messages calling her a fraud, attacking her and mocking Noah.

"It's like you've entered the ninth circle of hell," she told me. "Never even in your wildest, most fear-fueled fantasies would you have guessed you'd find yourself having to fight not only through your grief, which you know is at times paralyzing, but to even prove that your son existed?

"I think that's the dark underbelly of technology. It really can be used to highlight the worst demons of human nature.

"When anybody's behind a machine, whether it's a gun, a computer, or a car, there's a dehumanization of the other," she said. "It's easier to commit an act of violence. And it's easier to desecrate my son's death and my grief as a mother."

In June of 2013, the Pozners escaped the miasma that hung over Newtown. Veronique and Lenny had decided to eventually divorce and co-parent the girls. They moved together to Boca Raton, Florida, establishing separate households about fifteen minutes apart.

By mid-2014, about the same time Lenny had gone onto Facebook to talk with the Sandy Hook Hoax group, he had moved into a rose-colored-stucco apartment building in Boca's Mizner Park area, near the beach. The building was near a big outdoor amphitheater, and his apartment had a sweeping balcony. "It was loud as hell," Lenny told me, and he liked it that way. He pulled his favorite seat, a hulking Sharper Image massage chair, next to the open balcony doors. The music practically shook the walls at night, cover bands booming the Eagles or the Grateful Dead. Anthems from his early youth, pulling him back to a more peaceful time.

Newly settled, Lenny began to plan his counteroffensive against the conspiracy theorists. He was still angry about how he had been abused by them. "I put my best foot forward and said, 'Here I am. What do you want to know?'" he told me. "I treated them like my peers, like one of my cyber communities. But they would not recognize my pain."

Stretched out in his recliner, his face ghostly in the light of his laptop, he "was ready to address what the hoaxers were doing," Lenny told me. Hoaxers: Lenny repurposed the word after his evening with the

Sandy Hook Hoax group, as a satisfying bastardization of their preferred moniker, "truthers."

Many Sandy Hook family members had channeled the contributions they received into foundations, public service campaigns, acts of charity, and community places of sanctuary dedicated to their loved ones. Fighting online abuse ignited Lenny. Even for someone in his profession, he was fast and sharp on a computer. Few understood as he did in 2014 how online misinformation grew and replicated, spreading like a noxious vine, germinating on one platform, throwing out runners that sprouted on others. At that time, not many people cared.

Lenny tracked Noah through cyberspace. In those days, any basic Google search conjured him up, smiling in his bomber jacket at school, his image used by the hoaxers in a grainy video, or pasted into a Facebook post labeling him and the twenty-five others "so-called victims."

Lenny was shocked at the major social media platforms' wholesale failure to protect vulnerable people. While Facebook and its wunderkind creator, Mark Zuckerberg, preached the virtues of "community" and "connectedness," Lenny's daughters, not yet ten years old, checked the locks on their bedroom windows at night, their parents targeted by an unseen community of millions. Was that what Zuckerberg meant by "bringing the world closer together"?

Lenny looked up the big platforms' community standards. Under its "image privacy rights" standard, Facebook said it "may" remove "a reported photo or video of people in which the person depicted in the image is: a minor under thirteen years old, and the content was reported by the minor or a parent or legal guardian."

But appealing to Facebook to remove Sandy Hook hoax posts or images of Noah was a maddeningly quixotic endeavor. Lenny struggled to find even a number to call, or an email address that didn't kick back an unmonitored autoresponse.

"You're powerless," he told me. Trying to reach Facebook in those early years reminded him of a scene in the 1960 film adaptation of H. G. Wells's classic *Time Machine*, in which Rod Taylor as Wells lands his craft in an exotic future, only to be confronted by the impervious metal doors of a stone fortress. "You're knocking to get in, and there's nobody to answer it," he said. "That's Facebook."

Facebook had allowed the Sandy Hook Hoax group and dozens of others to operate virtually uninterrupted for two years by the end of 2014. Hundreds of videos on YouTube, owned by Google, wallowed in the hoax, drawing thousands of people who chatted and made threats against the families in the comments. Jonathan Lee Riches, the prison litigant who broke parole by traveling to the Lanza home in Newtown, kept a pseudonymous YouTube channel where his video of that visit and others he made in Newtown remained. Halbig used PayPal to receive thousands of dollars from fellow believers to fund his incessant public records requests and travel. Alex Jones posted excerpts from his show on every platform, from Twitter to Pinterest, hawking Infowars merch while broadcasting false claims about the tragedy.

"Facebook and Twitter are monsters. Out-of-control beasts, run like mom-and-pop shops," Lenny said.

He recalled how for five years, beginning about the same time as the shooting, hundreds of military personnel in Myanmar used Facebook to incite genocidal attacks on the country's mostly Muslim Rohingya minority,[1] seeding troll accounts and fake celebrity fan pages with false stories of Muslim-led violence against Buddhists. Given the enormity of that failure, how could Lenny expect Facebook's Mark Zuckerberg to protect families of dead women and children from hoax-mongers who lurked near their homes and phoned them in the middle of the night?

Facebook is by far the largest of the "monsters." The year Noah died, the social network Zuckerberg founded in 2004 in his Harvard dorm room became the first platform to surpass one billion active users. By the end of 2021, that number was close to three billion. It enraged Lenny that Zuckerberg, Twitter's Jack Dorsey, and YouTube's Susan Wojcicki shaped what people read, bought, and got angry about. How their platforms fostered personal attacks by refusing to remove vicious content in the name of "free speech," then threw up their hands when what any half-savvy user could predict would happen, did.

"That kind of influence on human lives has to be government-regulated. It needs law enforcement and laws and human ethics that come from our society," not from the monsters' errant creators, Lenny said.

"It's like leaving a bunch of teenagers alone in a house," he told me. Without supervision, chaos.

KARA SWISHER, TECH JOURNALIST and force of nature, well understood the damage Lenny described. Swisher has been covering technology since 1994, when Mark Zuckerberg was ten years old. Some tech writers pull punches, worried about access or awed by the brainpower of Silicon Valley entrepreneurs. Swisher, a colleague at the *Times*, is unsparing in her criticism of Facebook and Zuckerberg, whom she calls its "nerd-god." She describes Facebook as a city run by people who collect plenty of rent but fail to provide their citizens with police, firefighters, road signs, sewage treatment, or trash pickup. She routinely calls out Zuckerberg's lack of accountability for the horrific abuses perpetrated on his creation.

"He's uneducated. I don't know how else to say it," she told me. "Everybody's job depends on him, and he has everyday decision-making power. It is a prescription for bad decisions."

Facebook can and has been used for good connections and causes. The Sandy Hook families received thousands of messages of support and pledges of help through the platform. Facebook groups have connected lost classmates to weekend hobbyists: one of the first Facebook groups was for knitters. But the system is also tailor-made for toxic groups—anti-vaxxers, white supremacists, conspiracy theorists—to find and recruit new members. Swisher says far-right demagogues and groups shunned by mainstream media found an early new home on Facebook. So did Alex Jones, a savvy user who embraced Facebook as an efficient channel for his online outrage.

Facebook's business model focuses on growth, to the exclusion of most everything else. Its algorithms are designed to keep users on the platform for as long as possible, while the company collects personal data it sells to advertisers. The algorithm operates with relentless, sometimes murderous neutrality, rewarding whatever horrible behavior and false or inflammatory content captures and retains users. "Enragement is engagement," Swisher told me. "You can't shut it down, and you can't fix it. It's like air. You can't stop it. Even if some of it's poisonous."

The other big platforms operate similarly. When confronted over their role in spreading hate or inciting violence, their creators typically cite free speech principles. Zuckerberg has for years protested that Facebook is not an "arbiter of truth," while it devolved into a spreader of hate.

"These people," Swisher said, meaning the Sandy Hook families, "are not going to win against the internet unless they shut it all down. And then what? Does that really solve the problem?

"The problem is humanity. It really is. But humanity with the internet is like—aw, Jesus."

AND YET, AND YET—there was one significant bright spot to Lenny's encounter with the Sandy Hook Hoax group. As the conversation slid downhill on that evening in June 2014, several group members, including the ones who defended him in the Facebook chat, started sending private messages to him, asking earnest questions.

Lenny detected something different among the private messages they sent him that night. "It's not that they didn't believe that it happened. It's that a lot of them *needed* to believe that this didn't happen," he said.[2]

After Tony Mead booted him from the Sandy Hook Hoax group, Lenny created a closed Facebook group he dubbed Conspiracy Theorists Anonymous. He invited some of the people who had been messaging him with sincere-sounding questions to join him there. The conversations continued for months, growing into the foundation for his hoaxer counteroffensive.

Jennifer Forsman was one of those who joined in the conversation. Bubbly and sociable, Forsman lived in Tacoma, Washington, with her four daughters, the eldest of whom was twelve, and the youngest who was Noah's age. Forsman was thirty-two at the time, and a shift manager at a Starbucks in Tacoma. Members of her sprawling family would drop in on her at work, chatting and laughing with her staff.

In her rare free time, Forsman moderated Sandy Hook Truth, a closed Facebook group of more than one thousand members who took a "true crime" approach to the shooting. Sandy Hook Truth was more sophisticated than Mead's group, but filled with deniers nonetheless.

Forsman joined the Sandy Hook Hoax group's grilling of Lenny but arrived late, about the same time one participant was writing, "Fuck you Lenny fuck off and fuck your fake family."

"Lenny, I'm sorry for your loss," Forsman wrote. "Thank you Jenn," Lenny replied.

After Lenny was ejected from the group, Forsman stuck around, gently cajoling the Sandy Hook Hoaxers. "So here's the thing. If u go look at Lenny Pozner profile (please don't think I'm a crazy stalker) his profile is legit," she wrote. "An open mind goes a long way."

"That's true Jennifer, but why did he bother coming on here? He didn't achieve anything," one of the hoaxers responded.

"Idk," Forsman wrote. "My guess is he was hoping we would realize he is a human being."

Then she joined Lenny in his Conspiracy Theorists Anonymous group and recruited other doubters who might be persuadable.

Early on, many of the doubters were young mothers who could not come to grips with the murder of so many children so young. A couple of them had lost children of their own; others simply feared they would.

Tiffany Moser was a young mom from California who not long before the shooting had accidentally struck and killed a child on a bicycle with her car. Her two sons were ages three and twelve in 2012. "I actually kept my kids home from school for a couple days" after Sandy Hook, she told me. In 2014, "I wanted to see how the families were doing and did a Google search for memorial or support pages. Emilie Parker's family in particular really resonated with me."

On Facebook she found the Sandy Hook Hoax group and asked to join.

"It's hard to understand evil, and it's hard to understand the fact that humans are capable of such evil. That really resonates with a lot of people. It resonated with me. The thought that those children suffered so much in the final moments of their lives is enough for me to break down," she said, pausing to settle herself.

Lenny saw disbelief among people like Tiffany as a form of post-traumatic stress. He let thirty or forty people at a time into his "CT Anon" group. Most arrived highly skeptical but open to hearing from him. No question was too weird or too painful. Lenny found that answering them helped him as well.

Over time, "most of them disappeared," he said, their questions satisfied. But some stayed on.

"Having conversations with Lenny helped me emotionally. I would never tell him that, of course, because what I wanted to do, what I was hoping, was that my conversations were helping him," Vanessa, an early member, told me.

Vanessa never doubted the shooting took place. Her true name, profession, and address are known to me. A successful entrepreneur based outside the United States, Vanessa, then forty-seven, had scarcely used social media before 2012.

Her youngest son was a year older than the children killed at Sandy Hook. Riveted to the television, she saw a montage of the victims' photographs. Noah Pozner looked exactly like her own boy, and she couldn't stop thinking about him.

"I almost felt guilty that somebody was going through this. My family was whole and my children were safe, and it just ate away at me," she told me. "I was emotionally just sunk, for a long time."

She became a visitor to Noah's Google+ page. Several other moms visited too, posting encouragement to Lenny, and they got to know one another. By mid-2013, many of the visitors to Noah's and other victims' memorial pages were deniers like Kelley Watt, "that indoctrinated moron," Vanessa said, whose comments seemed to Vanessa and the other moms like desecrating a grave site.

"We would chime in and we would fight with these hoaxers. Lenny didn't like that, because his Google+ was basically supposed to be about honoring Noah," she said. The Conspiracy Theorists Anonymous group was a better place for those discussions, at least for the people who weren't so far gone they could never be convinced.

THE NUMBER OF CONSPIRACISTS GREW, as did their brazenness.

In late 2014 someone named Andrew Vaessen posted threats against Lenny on YouTube: "Ha Lenny soon the world will all know what you did then [we're] going to come after you and kill you slowly."

Vaessen posted a link to Noah's Google+ page with the subliterate comment "Its his dad I tooled him I was going to fuck him up his a cunt."

In November 2014, Jonathan Reich, a twenty-four-year-old from Flushing, New York, with a history of stalking, was released into a pretrial supervised diversionary program for crimes in which mental issues may be involved.[3] Reich had called the office of the Connecticut medical examiner H. Wayne Carver again and again, demanding autopsy reports for the children. "Yeah, we're gonna believe he performed all these autopsies when Noah Pozner's mother said no autopsy was performed, saying that it was eleven shots at close range when no autopsy was performed? What kind of office do you run here?" Reich said in recordings he posted for several thousand YouTube fans.

Reich was a member of the Sandy Hook Hoax Facebook group. He was arrested and charged with harassment that year. Halbig leveraged Reich's arrest to solicit donations.

The same month, Lenny sent a plea to an agent he knew in the FBI's Fairfield, Connecticut, office, cc'ing Senator Chris Murphy, Newtown first selectman Pat Llodra, Connecticut governor Dan Malloy, and an aide to then vice president Joe Biden.

The subject line was "Online Threats made to me by unstable Conspiracy nuts."

> I have made several reports of threats against my life to your office yet I have no indication that your office has taken any action or interest in treating those reports seriously.
>
> I have no other direction to take other than to start to speak publicly on this very real lack of concern and interest in preventing something terrible from happening to any of the victim's families... I am praying that we are all safe after all the constant threats that continue daily.
>
> Len Pozner

The Connecticut FBI agent responded, telling Lenny to bring his concerns to local law enforcement or the local FBI office in Florida, since he lived there, and there was no indication that the hoaxers messaging him were in Connecticut. "I have explained this to you on multiple occasions," the agent added testily. Lenny was incensed; he'd been told the agent would help him.

He would eventually work with the FBI in Florida, but only after the threats against him escalated.

Lenny and Llodra tried to enlist help from authorities in Connecticut, and even appealed to its congressional delegation. While sympathetic, some of them thought confronting the "sick bastards," as Governor Malloy called them, would only encourage them, and detract from Newtown's efforts to recover.

Lenny felt the opposite.

"Grieving requires a calmness, and a silence. And all of this material was a distraction for me, and it was noise. And I needed to handle that noise, so that I could have the silence and calm that I needed," he said.[4]

The Conspiracy Theorists Anonymous group would become the core of Lenny's hoaxer pushback, a network of volunteers who ensured he didn't fight alone.

The HONR Network's website went live around Noah and Arielle's birthday in November 2014.

I asked Lenny how he had chosen its name. Before the twins were born, Lenny had been the chief technology officer for the Hotel on Rivington, a New York boutique hotel. Looking for a domain name, he snapped up HONR. The client ultimately chose another name, but Lenny kept HONR, figuring he would eventually find a use for it. In 2014, his life utterly changed, he did.

"It's not the best fit, but it sounded like honoring Noah," he told me.

The website explained HONR's mission: "We bring awareness to the cruelty and criminality of Hoaxer activity and, if necessary, criminally and civilly prosecute those who wittingly and publicly defame, harass, and emotionally abuse the victims of high profile tragedies and/or their family members. We intend to hold such abusers personally accountable for their actions, in whatever capacity the law allows."

On the home page, Lenny posted a plea. "Families in grief have the right to do so in peace and dignity," it said. "We need to band together as a community of caring individuals, and act to stop these abusers from inflicting any further harm to people who have committed no crime, and simply want to be left alone to grieve their loved ones and live out the rest of their lives in peace and anonymity."

Lenny made Jennifer Forsman the site's administrator. She left her Sandy Hook Truth group, crossing over from the dark side for good. Persuasive and unflappable, "she had the ear of a lot of people in that group, and as a result a lot of people's views were transformed," Lenny told me.

Five years later, Forsman became a victim of gun violence herself, shot to death in her home by her ex-husband.[5]

"She remained a volunteer, and an administrator until the very end," Lenny told me. "She was probably my first cyber friend." A conspiracy theorist who defended him before an online mob, Jennifer showed Lenny his coming battle was worth it.

11

LENNY'S NEW NETWORK OF ANTI-HOAXERS drew people with not much in common, except their determination to debunk the lies spreading unimpeded across the big platforms.

"Lenny had the motto, and I agree, that whomever we can get help from, we will take it, because the situation at that time was so, so bad," Vanessa, one of HONR's first volunteers, told me. "Just thousands upon thousands upon thousands of videos and people seeing those videos and spawning new videos of their own. It was like, 'Can this ever end?' It was intolerable."

More than a few of the early HONR volunteers had been conspiracy theorists themselves. Engaging with hoaxers required plumbing the caverns, and nobody knew that terrain like people who dwelled there.

Keith Johnson volunteered early.

Johnson, then in his early fifties, was a freelance writer with the *American Free Press*, based in a suburb of Washington, D.C. The publication was founded by Willis Carto, once an important backer of the late segregationist and Alabama governor George Wallace's 1968 presidential bid, later a neo-Nazi. Carto died in 2015, with his views experiencing a resurgence among the white supremacists supporting the presidential candidacy of Donald Trump.

Johnson didn't share Carto's views when he joined the *American Free Press* in 2010, but he had done jail time for car theft and other nonviolent offenses and needed a job. Johnson lives in the South now and holds a manufacturing job for a company I agreed not to name. He insisted I make the details of his "dark period" brutally clear. "I wrote articles that appeared on Infowars and PrisonPlanet, but I didn't get paid for them. The Southern Poverty Law Center calls *American Free Press* a hate group.[1] I got paid by them," he told me.

Johnson had known the conspiracy world since childhood. He grew up in the Los Angeles suburbs, son of a traveling salesman and a homemaker.

"My mother was extremely conspiracy-oriented because she was involved in the evangelical movement and they were all doomsdayers," he told me. "I would accompany my mother to church, and they'd be saying Jimmy Carter is the Antichrist."

As a child Johnson joined the Police Explorers, for kids interested in a law enforcement career. When he was eighteen, he became a police dispatcher, and a reserve police officer for a time. He got his private investigator's license and worked in the field in California for seventeen years. In his travels he listened to far-right talk radio, including Infowars. Substance abuse got him into trouble, and after serving his prison sentence he made his way east, to the *American Free Press*. There he tried, not always successfully, to sidestep the uglier topics.

"Everything pretty much changed for me over there when Sandy Hook happened," he told me. He had never seen a shooting draw so many bogus claims. "I tried to do some debunking of these theories. My stance, to try to maintain my audience, was 'this is just making us look bad.'"

Johnson reached out to Lenny, who to Johnson's shock agreed to an interview.

"That's the crowd I wanted to reach," Lenny told me. "Keith had been a very bad truther before that. But denying a real tragedy was where he drew the line. That's what a lot of hoaxers should have been doing, saying, 'This is giving a bad name to the truther world.'"

Lenny talked to Johnson for an article about his ongoing struggle to counter the falsehoods surrounding the tragedy. "Although these hoax theories primarily flourished on obscure websites and YouTube videos in the immediate aftermath of the tragedy, Lenny said they attracted a much larger audience after radio host Alex Jones virtually endorsed them in a video segment posted on his website," Johnson wrote.[2]

The article marked the beginning of the end of Johnson's *American Free Press* gig. By the end of 2014, he had become one of HONR's most enthusiastic warriors.

Johnson and CW Wade debunked scores of bogus hoaxer claims in YouTube videos, on Wade's *Sandy Hook Facts* blog, and on a website Johnson named the Newtown Post-Examiner.

Johnson took on James Fetzer, by then the leader of a motley conspiracy crew Fetzer called the "Sandy Hook Research Group."

Fetzer and Johnson had occupied intersecting circles in the conspiracy Venn diagram. In 2011, bored of his retirement in a bedroom community outside Madison, Wisconsin, Fetzer started freelancing as a "reporter" for *Veterans Today*. *VT* calls itself an "independent alternative journal for the clandestine services focused on U.S. foreign policy and military issues." In fact it's a conspiratorial, pro-Kremlin garbage fire that the Southern Poverty Law Center calls "an endless stream of Israel-bashing mixed in with some bona fide anti-Semitism."[3] None of that posed a problem for Fetzer, a Holocaust denier who used starkly antisemitic terms to falsely cast Lenny and Veronique as "monstrous liars" who raised millions by perpetrating a scam.

A few days after the massacre, Fetzer wrote that the shooting "appears to have been a psyop intended to strike fear in the hearts of Americans by the sheer brutality of the massacre, where the killing of children is a signature of terror ops conducted by agents of Israel."

The theory was so widely condemned and ridiculed that even Fetzer backed away from it. He pivoted to a new, equally absurd claim: no one died at Sandy Hook, because the shooting was a FEMA drill.

Sitting at the dining room table or in his spare bedroom office in his modest tract house in Oregon, Wisconsin, Fetzer plied his MacBook while his wife, Jan, who does not share his Sandy Hook delusions, watched MSNBC in the living room a few feet away. I visited there once, finding it the type of midwestern grandparents' house I remembered from my youth: colorful throw rugs, a granny-square crocheted afghan draping the couch back, a cartoon plaque in the bathroom depicting cats playing with a roll of toilet paper with the caption "Let's Potty!" The difference of course was that the grandpa in residence mocked the parents of murdered first graders. Corresponding obsessively with Halbig, Mead, Watt, Tracy, and others, Fetzer wrote more than thirty articles denying the shooting happened and collaborated on scores of others. Most of Fetzer's theories wound up on *Veterans Today.*

Fetzer broadcast an online radio show, *The Real Deal with James H. Fetzer*, imitating Alex Jones's format, live from his house. He built a library of more than two hundred episodes, which he posted to YouTube and his website. One of them was his initial interview with Kelley Watt.

Retirement had freed Fetzer, like Halbig, to devote his full energy to hateful conspiracy theories, cranking out falsehoods that Johnson and Wade raced to debunk.

In mid-2014, Lenny wrote a guest essay in the *Hartford Courant* alerting the public to the existence of "a small but obsessive faction of hoaxers" tormenting the Sandy Hook families.[4]

Relatively few Americans knew this was happening then. People who believed the truth of the shooting vastly outnumbered hard-core deniers. But the hoaxers derived outsized power from the unregulated megaphone of social media. There, baseless claims by anonymous people were shared and amplified by pro-gun local officials and activists with big followings, then amplified again by Alex Jones.

In December 2014, just before the second anniversary of the shooting, a "collaborative documentary" titled *We Need to Talk About Sandy Hook* appeared on YouTube. The film was a mash-up of false claims already debunked by HONR's volunteers and many others, produced by a collection of randoms who called themselves "Independent Media Solidarity." One of them was Tony Mead, house mover and self-styled journalist. Alex Jones invited a couple of them onto Infowars, and the film racked up some three million views.

Keith Johnson emailed Infowars: "I must warn you that you are travelling down a road that will ultimately lead to serious lawful consequences." He offered to appear on Infowars to debunk the hoaxers' claims.

Rob Dew, Jones's top lieutenant and a Sandy Hook hoaxer himself, agreed to put Johnson on.

Johnson's warning about legal consequences would eventually prove apt. But Infowars viewed his offer to appear as an opportunity to stage a spectacle timed to the second anniversary of the murders. Dew invited Wolfgang Halbig to debate Johnson.

Dew moderated the debate from Infowars' studio in Austin. Johnson and Halbig joined remotely.

"If you remember the last segment, we were talking with two documentary filmmakers for the documentary *We Need to Talk About Sandy Hook*," said Dew, who had traded his T-shirt and jeans for a brown suit and shiny gold paisley tie. "I'm now joined by two individuals who are both on the opposite ends of the Sandy Hook spectrum."

Johnson spoke first, his voice tinny over the telephone.

"Our government is involved in a lot of nefarious things," he said. "But Sandy Hook doesn't qualify as that. There's not one shred of evidence to suggest that there was a hoax. Quite the opposite," he said. "These were my fellow Americans who were slaughtered. And they need to be honored for the lives they lived, not cast in some delusional fantasy someone has."

Halbig spoke by phone while Infowars displayed a photo of him that was at least twenty years old. He got right to it, threatening to sue Johnson for calling him a fraud and a liar.

"Those researchers have dedicated their lives in trying to take a look at Sandy Hook for what it is," Halbig said, in his aggrieved, adenoidal whine. "So if we're going to talk tonight, let's only talk about the facts, only about the facts."

Dew, trying to keep the discussion on track, agreed. "I think the reason you have all the speculation, Keith, and I think you will agree, people do not trust the government.

"If you type in 'Sandy Hook conspiracy,' it's hundreds of millions of videos. Everybody and their mother out there is looking at this from a different angle. And they are—they're all bringing their collective expertise . . . That's the great thing about the internet."

Johnson focused on exposing Halbig.

"Mr. Halbig, you *are* a fraud . . . You don't have significant law enforcement experience. Even on your website, you admit that you have one year as a ticket-writing, Florida highway patrolman.

"I would like to see one person with some credible law enforcement experience come forward and say, 'Yeah, there's something wrong here' . . . You've collected what—thirty thousand, forty thousand dollars? And you can't even give us one expert to substantiate your claims? Ridiculous."

Dew led a discussion about the "anomalies" covered in the new documentary. He noted a scene with police "setting up a lunch table on top of the squad car, eating lunch during this scene. Do you think that's a normal response to this?" he asked Johnson.

Johnson dismissed that as silly. "If there's something out there that sounds compelling, we research it, my friend and I, CW Wade . . . A lot of these things are just ludicrous."

Dew asked Halbig for a response, but Halbig wasn't debating Johnson anymore. He was performing for the Infowars audience. In a folksy, Southern-accented ramble punctuated with plaintive exclamations of "Who does this?" and "Dear God," and he provided his rendition of the police officers' lunch caper.

"Twenty children shot three to eleven times? Teachers shot three to eleven times? Somebody goes shopping for bananas, sub sandwiches, Doritos, Gatorade, chips, and then bring it back to the crime scene?

"They actually took the food, if Keith will look, they took the food and had lunch inside the school! Now, that's a crime scene! Who in the hell is going to eat lunch with children's dead bodies, body fluids, blood spatter, brain tissue?"

Halbig, who had delivered this performance many times, talked over Johnson, who was fairly shouting that the entire vignette was "a lie."

"Dear God, Keith, I'm asking these questions using the Connecticut Freedom of Information Act. And for seventeen months they have refused to answer those questions."

The Johnson-Halbig debate dissolved into a long cross-talk argument. Finally, Dew broke it up.

"Well, guys, let me say this. It's been about forty-five minutes we've been going at it. And I'm still not convinced one way or the other about what has happened."

"There's not one shred of evidence," Johnson told Dew. "You cannot smear these people who just lost their children to a massacre two years ago and say that they're characters in some kind of conspiracy theory unless you have solid proof!"

Dew agreed—and gave Halbig the last word.

"I think it's a great debate," Halbig said, triumphant. "All I care about is the truth."

WOLFGANG HALBIG ILLUSTRATED the limits of debunking to Lenny and the HONR volunteers.

By late 2014 the people still questioning Sandy Hook were on "a one-way journey," Lenny told me. Lying about Sandy Hook had afforded them online recognition and identities that were much more interesting than their real lives.

Their viciousness when challenged convinced Lenny that "what drives most of these people is the demonic part of them. It's the evil in them. Goodness and ethics is not their true north. Causing pain is."

Like many of the HONR volunteers, Vanessa's anger found locus in Halbig. Infowars had helped make him a hoaxer hero, spurring him to new heights of abuse.

"Lenny would tell me, 'Hatred does not get things done,'" Vanessa said. "'When you're angry, that puts you all over the place and you can't be effective.'"

But Halbig's blend of small-time con man and sadistic troll got her every time.

"He's horrible. He sounds like this wise old man just doing things for the good of society, just trying to save what the U.S. stands for, you know?" she said. "But no. He was making money."

Listening to Halbig, you could almost buy his portrayal of himself as a homespun Columbo, a Mr. Smith goes to Newtown, his humble citizen's requests thwarted by a gray phalanx of bureaucrats.

Initially, Halbig pressed for responses to "sixteen questions" he had about the event. Some, like his question about why rescue helicopters weren't deployed, had easy answers. The hospital was less than a half hour away, less time by ambulance than it would have taken to summon a helicopter.

Newtown officials had given Halbig invoices and receipts, meeting minutes and police dash-cam video in response to his requests. They couldn't find other records he wanted, like receipts for porta-potties, whose appearance at the school that day Halbig took as a smoking gun, evidence of prior planning. Dissatisfied with what he was given, Halbig filed complaints with Connecticut's Freedom of Information Commission, securing public hearings, which he attended with an Infowars camera crew and an entourage of hoaxers.

Halbig didn't acknowledge most records he received, except to raise more questions about them, Tiffany Moser, the young mom who had converted from conspiracist to HONR volunteer, told me. Moser found records online disproving several of Halbig's claims, including that the school was defunct at the time of the shooting. In a text exchange and a phone call she recorded, Moser pushed him to publish the records he received and account for the money he'd collected. He affably shined her on—"I'm publishing a book, and you know what? That is the biggest surprise that you're gonna get!"—then angrily shut her down, calling her dumb and immature and part of the cover-up.

"These are society's discarded, failure-to-launch, underachieving grown adults who are socially awkward and isolated, who want to belong to something. They found each other, and this gave them a voice and made them feel they're part of something big," Moser told me.

"Debunking wound up being a good effort, because it proves them wrong. But it also gives them this delusional importance, like 'We must be onto something. We're getting closer to the truth and they want to silence us.'"

Halbig graduated from Abilene Christian University in Abilene, Texas, in 1973, after a stint in the U.S. Air Force. He spent a year as a Florida state trooper, from 1974 to 1975, highway patrol experience he has inflated to homicide investigator. In an appeal to the NAACP seeking justice for arrested conspiracist Jonathan Reich—seemingly cc'ed to every African American leader whose email Halbig could find, including Dean Baquet, executive editor of the *Times*—Halbig said he had driven Martin Luther King, Jr., in a protective motorcade. That duty, he wrote, was "the greatest honor this old man of 70 ever had." Indeed it would have been, had King not been assassinated in 1968, six years before Halbig joined the police.

Halbig bounced around the Florida public school system, working as a teacher, a coach, and an assistant principal. From 1995 to 1999, he was Seminole County Public Schools' security director; an *Orlando Sentinel* article quoted him after Columbine as saying a school shooting "can happen any time [*sic*], anywhere."[5] Halbig worked as Lake County Public Schools' risk management director from 2005 to 2009; when his position was eliminated, he said it was because he'd blown the whistle on mold issues.[6]

In retirement Halbig cast about for a second act. He spent years failing to parlay his school security experience into steady work as a consultant or an expert witness. That quest prompted him to email Newtown Public Schools after the shooting, and after his pitch fell through the cracks he decided the town had conspired to stage the crime.

Halbig's son Erik grew up admiring his father's public service. A man in his thirties in the mortgage business in South Florida, the younger Halbig speculated that maybe his father "started with a good angle on Sandy Hook and felt put in a defensive spot." Maybe, angered at being dismissed, his father perhaps started "lashing out with statements to catch people's attention," Erik Halbig told the *Orlando Sentinel*.[7] But he couldn't be sure. Halbig was too preoccupied with Sandy Hook to spend much time with Erik or the grandchildren Halbig claimed were inspiration for his quest, and his reason for wanting to crack the case and get back to retirement.

"If he's awake for 12 hours, he's working on Sandy Hook for 11 of them," Erik Halbig told the *Sentinel*. After Halbig's son warned him that his obsession with Sandy Hook was ruining his relationship with his family, Halbig persisted. His wife eventually left him.

Halbig's approach suggests issues deeper than his son surmised. Lenny's take: "He wants to be caught, to be stopped. But nobody ever does anything."

Halbig has called victims' families repeatedly on their cell phones. He wrote the mother of a disabled child who was killed, telling her the child "expected more from you" than to send her to a "filthy" school. "I have some questions which in a deposition from you I will learn the truth," he wrote to another child's mother. He repeatedly tried to contact J. T. Lewis, who was twelve when his half brother, Jesse, was killed.

Early on, Halbig assured Jones he was willing to give his life for the cause, and reported many of his activities to Infowars. In a 2019 deposition, Jones acknowledged that Halbig by that point had sent him about four thousand emails,[8] and that by around 2015, Halbig "seemed to get agitated." But Infowars still assigned a camera crew to travel to Newtown with him that year.

LIKE KELLEY WATT, who insists that Noah is actually his half brother, Michael, Halbig fixated on "proving" individual children killed at the school were still alive and in hiding. He dashed off hundreds of emails to a rotating list of victims' families, Newtown and Connecticut officials, and local and federal law enforcement, circulating photos of living people he claimed were the victims, "safe and sound" or "all grown up."

Halbig spent the summer of 2015 repeatedly visiting Newtown, pursuing macabre theories about the children's identities.

One drizzly afternoon Halbig and an Infowars crew turned up at St. Rose of Lima. Halbig went to the rectory while the crew set up in the parking lot of St. Rose's elementary school. A staffer told him Father Bob wasn't available.

School was dismissed for the day, and the Infowars crew began filming the children streaming out. The school's safety officer called the police. The Infowars crew left before the police arrived. But an officer spotted Halbig, whom the entire force recognized by then, lumbering to his car. He asked him whether he had gotten permission to film children. Halbig dodged, saying he had been there to meet with the monsignor. He got into his blue Honda rental car and bolted. The police let him go; they had no evidence of any crime.

Father Bob told me the crew was pursuing Halbig's twisted belief that a red-haired girl killed at Sandy Hook was still alive and attending St. Rose school. The monsignor knew the child. He had said her funeral mass.

"The families, they were terribly victimized by these people," Father Bob told me. "It's just—it's craziness."

Soon after, the Bridgeport Diocese sent Halbig a letter at home telling him to stay away from the church compound or risk arrest. Halbig dramatized his encounter as more evidence of a cover-up. "Monsignor Weiss conducted nine of the funerals of children. *Do you know he refused to see me?*" he told a conspiracy podcaster. "I was surrounded by six Newtown police cars!"

Father Bob received threatening calls, emails, and letters for years, referencing Halbig. Parishioners guard the St. Rose elementary school, preschool, and cemetery, where several victims are buried.

HALBIG RETURNED TO CONNECTICUT a month later to pursue a macabre claim that a girl who died in classroom 10 was alive and had taken the identity of a girl who resembled her. This child had survived the shooting and lived in Newtown. Halbig knew her name from news reports. The thought of it made Vanessa choke up: "This is a real little girl who now is probably going to experience some problems in her life because of what Halbig put up on the internet."

The living child, whose name I know, was nine years old and in fourth-grade gym class when the shooting began. In the firehouse she reunited with her father and her mom, who had run from her car parked a quarter mile away, so frantic she couldn't recall later whether she had shut off the ignition.

The girl who survived sang in the Sandy Hook Elementary School choir. After the shooting she was among twenty-six choir members invited to the Super Bowl in New Orleans on February 3, 2013, to sing "America the Beautiful."

IN THE SUMMER OF 2015 Halbig; Infowars' Dan Bidondi, the squat former wrestler turned Infowars cameraman; and a few other hoaxers attended yet another public records hearing in Hartford. Halbig had begun to exhaust his legal options. The Freedom of Information Commission rejected his appeal.

Halbig and Bidondi made quite a pair. Infowars paid Bidondi, a doltish post-prime athlete with a thick Rhode Island accent, a couple hundred dollars to show up at official scenes of tragedy or controversy and ask obnoxious questions. Ever eager to please Jones, Bidondi was happy to comply. On April 15, 2013, he earned Boston's eternal ire by asking the first question at Governor Deval Patrick's press conference after the Boston Marathon bombing.

"Why were the loudspeakers telling people in the audience to be calm moments before the bombs went off?" Bidondi blurted. "Is this another false flag staged attack to take our civil liberties and promote homeland security while sticking their hands down our pants on the streets?"

"No," Patrick replied. "Next question."

Inside the cramped, fluorescent-lit hearing room, Bidondi approached Monte Frank, outside counsel for the Town of Newtown and the Newtown Board of Education. Bidondi stuck out his microphone, asking questions. Frank, silent, ducked beneath Bidondi's outstretched arm, grabbed a trolley with his briefcase, and left the room. Bidondi followed. "There goes corruption, right hee-yah," he called out, alerting the hoaxers who waited just outside.

Halbig stood in the hallway outside the hearing room in a white dress shirt and red tie, his hair plastered to his head with sweat. He had just been striding about the hearing room, seething and calling Frank a liar. The lawyer passed Halbig, bound for the elevator.

Halbig exploded. With the Infowars camera rolling, he named the living Newtown girl and the dead girl repeatedly, vowing to track them down.

"Guess what there, Mr. Frank, what do you think?" He again named the two girls. "You're protecting all of those people? You're protecting them? And you're in the synagogue?"

Frank gave up on the elevator and took the stairs.

Halbig's lawyer, Laura Kay Wilson, tried to calm him down. "I can't help it. This is ridiculous!" he said.

Bidondi waved the camera nearer and directed Wilson into the shot. He held his microphone beneath Halbig's chin.

"That is a disgraceful man. To lie, the way he did?" Halbig said, working himself into a soliloquy.

"If we allow government to control what happens in our schools, if we allow government to create panic and fear and then affect the emotions of adults all across the country by putting on this huge illusion, we're in serious trouble.

"We deserve better, we deserve better than what we saw in here today."

As I watched a video of his remarks, it struck me: if you substituted the word "elections" for "schools," Halbig would have precisely echoed the claims of the conspiracists who stormed the U.S. Capitol six years later.

Halbig had deluged Newtown with demands for records of the choir's travel for years. He wanted the identities of the children who sang, which the school deliberately did not release. Halbig showed up at a school board meeting in 2014,[9] accompanied by other hoaxers.[10] He appeared again before the school board in 2015, when a "Celebration of Excellence" was on the agenda. He apparently expected the Super Bowl singers and their families would attend. They did not.

Thwarted in his effort to confront them, Halbig spoke during the public comment period, growing hysterical as he addressed the board.

"How in the world could you not bring them, those twenty-six children, the chorus teacher, and their families into this boardroom and reward them with excellence?" he exploded, his face and neck mottled red.

The room was packed, but no one spoke or moved. Some stared at the floor, others at the back of his fleshy neck, stony with rage at this deluded, vainglorious pest, and his relentless intrusions on their wounded town.

"You can ignore me! You don't have to look at me! But I tell you what, you didn't bother to bring them into this room? To shake their hands? To hug those children?!" Halbig seemed to be crying.

The board chair interrupted him: "Your three minutes are up."

Halbig's fifteen minutes of fame were running out as well. Monte Frank and Connecticut's Freedom of Information Commission soon after found a legal way to end his abuse of public hearings.

I MET MONTE FRANK in his law office in Bridgeport on a January morning in 2019 to discuss Halbig's public records campaign. It was snowing heavily as I crossed the broad plaza at the foot of the law firm's office tower.

Since the shooting, Frank's responsibilities had included laying the legal and organizational groundwork for the new school. He provided legal counsel to the families, much of it pro bono. As counsel for Newtown and its Board of Education, he managed the official response to Halbig's stream of records requests.

When we met, Frank wore leather hiking boots, still damp, with his khakis. His blue-striped oxford shirt was open at the neck. Bespectacled and ascetic looking, he is a long-distance cyclist and cyclo-cross racer. After the shooting he founded Team 26, a cycling group, which in the first few years trekked to Washington to advocate for gun safety legislation. In later years the team rode to other communities touched by gun violence.

We fetched coffee from a hallway pantry and settled at a long conference table.

"To have somebody like Mr. Halbig calling constantly, asking for things that were distasteful to be asking for, and literally having to go look to see if those documents existed, I think it took a tremendous toll on the town and the town employees," Frank told me, all the more so because Halbig wanted the records in order to hunt down traumatized children and parents. Frank worked to comply with the law while protecting them.

Frank became a target. Two months after Halbig's confrontation with Frank at the hearing, a website appeared named Montefrank.com, "dedicated to the disbarment and imprisonment of Monte E. Frank for treason." Its creators posted photographs of Frank and his colleagues, and their home addresses, phone numbers, and emails.

Frank "will not stop until all Americans are defenseless from tyrants," the website read.

He found Halbig's disregard for Newtown's pain repugnant, but said, "He's a citizen, he's entitled to submit freedom of information requests, and we complied."

Halbig didn't understand, or didn't want to, that the records he was asking for were held by other jurisdictions, or simply didn't exist.

Halbig filed at least six appeals. By early 2016, "Halbig's harassment of the town and town officials and employees was such that we asked the commission to avail itself of a provision in the law that they could decide not to hear his appeals." According to records supplied to me by the Connecticut Freedom of Information Commission, the commission voted four separate times in 2016 and 2017 not to schedule hearings on Halbig's appeals. His vexatious requests no longer served the intent of the public records laws, a means for citizens to hold government accountable. The commission had cut him off.

I asked Frank whether he had ever imagined his duties after the tragedy would include having to counter people who denied it ever happened.

"In some ways I'm not surprised," he said. "My mother is a Holocaust survivor. I'm well aware of the proliferation of Holocaust deniers. I represented a family before the Department of Justice after 9/11 for victims' compensation. I'm well aware of the deniers from 9/11.

"So it doesn't surprise me that you'd have people trying to promote themselves and advance their own interests through a denial theory following Sandy Hook."

The maddening part of all this, the infuriating waste of it, was the fact that the Sandy Hook conspiracy theorists had literally hundreds of thousands of pages of evidence available to them, incontrovertibly proving that the shooting happened. Connecticut state agencies from the Office of the Child Advocate to the state police had released reports and research. The *Hartford Courant* had sued and won access to more state police records about the gunman. The Sandy Hook Advisory Commission had released a two-hundred-page report, plus appendices and videotapes of all its meetings. Lenny had released proof of Noah's life and death. Every scrap of this evidence was available online, cataloged and searchable. But instead of it being a source of answers, the conspiracists found in the data proof of the plot, picking at specious details or ambiguities to bolster their phony claims. Halbig had been given hours of hearings and hundreds of pages of records in response to his questions. But he continued to deny the truth, because he had no receipts for porta-potties.

Monte Frank was right. The Holocaust deniers were the same.

When I first became aware of the Sandy Hook conspiracy theorists, I called Deborah Lipstadt, a Jewish studies professor at Emory University, whom in 2021 President Biden named Special Envoy to Monitor and Combat Antisemitism. In 2000, Lipstadt and Penguin UK, her publisher, had won a libel case brought by David Irving, a British Holocaust denier who mocked death camp survivors as liars, claiming their stories, scars, and even their camp tattoos bolstered a phony narrative concocted for profit.

Irving sued Lipstadt in an English court after she wrote about him in her book *Denying the Holocaust: The Growing Assault on Truth and Memory*. In the book she calls out Irving as one of the most dangerous of Holocaust deniers, a man who twists historical facts to suit his antisemitic ends. Lipstadt's book *History on Trial: My Day in Court with a Holocaust Denier* and a subsequent film, *Denial*, trace the legal battle that followed.

The British system differs from the U.S. one in that it reverses the burden of proof in defamation cases, putting the onus on the defendant to prove that the statements at issue are true. The pompous Irving used the high-profile trial to trumpet his false theories. Lipstadt's defense team spent months immersed in such granular horrors as the physical structure of the gas chambers and how many bodies could fit into a twenty-four-yard-long pit.

Lipstadt and her publisher's victory coincided with the early years of the internet. They posted the historical records and trial documents online. Through the miracle of the World Wide Web, they aimed to deploy this mountain of truths to strike a definitive blow against the Holocaust denial movement. Lipstadt found the trial records particularly damning, exposing Irving "not just as a falsifier of history, but as an irrational and foolish figure," she writes in *History on Trial*.

The plan backfired. The website's records provided valuable assistance to historians, researchers, and journalists. But they also proved a motherlode for neo-Nazis, Holocaust deniers, and the websites that cater to them. They picked through the records for anomalies, or twisted the information to "prove" their false claims. So while the trial branded Irving as a Hitler apologist and crank, the internet breathed unintended new life into his hateful theories.

"We were quite naive," James Libson, Lipstadt's lawyer, told me. "No one could have predicted how social media could be an amplifier of the conspiracy arguments."

"I couldn't do the work I do and write the books I write without the aid of the internet," Lipstadt told me. But it fails to separate facts from opinions and outright lies. "And that's where you get the Alex Joneses."

Frank sat silently as I unspooled Lipstadt's story, his lined, somber face reflected in the gleaming tabletop.

"I know what happened during the Holocaust to my family. The fact that my mother's here is an absolute miracle," Frank told me when I finished. "For people to say that what happened to my family and the six million Jews and others who were slaughtered didn't happen—that motivated me a lot, in terms of how I dealt with Mr. Halbig."

12

HONR's HARD-CORE VOLUNTEERS were keen to up the ante and find new tools for combating Halbig, Fetzer, Tracy, and especially Jones, who amplified and encouraged them. Debunking had helped Lenny assert baseline truth about the tragedy but had failed to stop the lies and abuse.

By 2015 an impressive two hundred volunteers had joined HONR, though only about a quarter of those could be called regulars. "People have their own lives and they can't be as devoted as I am because I can't escape this stuff," Lenny told me. "They eventually burn out because it's all-consuming."

Hoax material still appeared at the top of searches for "Sandy Hook." Then and for years afterward, Google would remove a website from search results only if it displayed a person's bank account or social security number. Smaller search engines like Bing and DuckDuckGo were even more resistant. Hoaxers still ran amok on WordPress.com, one of the internet's largest blogging platforms. Facebook roiled with Sandy Hook conspiracy theories. "This was when YouTube was easily monetized,"

Lenny said, so hoaxers earned money for advertising that YouTube placed on their channels. The juxtaposition was surreal: athletic shoe ads and armed forces recruiting pitches appearing next to grotesque videos denying the murders.

Lenny and the HONR volunteers reported hoax content to the companies, citing specific violations to the platforms' terms of service and rules: invasion of privacy, posting images of children, bans on threats. They either heard nothing or got bloodless autoresponses. "Porn would get attention, but nothing else," Lenny told me. Which is ironic, and hypocritical. The publication of pornography is supported by the First Amendment, enshrined by the courts as a signal test of free speech principles. But here were the social platforms, scurrying to take down porn while trotting out the First Amendment to explain why they didn't remove abusive content. Why? Because despite what they say, the platforms are all about pleasing their advertisers, most of whom don't want their ads adjacent to sexually explicit content.

Lenny wondered whether the content moderators were actually human beings. In Facebook's case, they were poorly paid contractors, traumatized by the livestreamed beatings, suicides, and beheadings they watched every day.

No discernible standards governed what stayed up and what came down. "The platforms were concerned with growth and income," Lenny said. "They were not concerned about limiting their inventory of content."

Lenny thought about historical analogies for the social network free-for-all.

Alexandrea Merrell, who had joined HONR to help with public relations, found an article on the *Detroit News* website about the birth of the auto industry. Titled "1900–1930: The Years of Driving Dangerously,"[1] it traced how Americans adopted the automobile so swiftly that it took years for safeguards like traffic lights, lanes, and rules to catch up.

Fatalities soared, three-quarters of them pedestrians. Most were children, wandering innocently into this teeming, lawless new landscape.

The two discussed it as an apt metaphor for the careening, seemingly ungovernable growth of social media. Lenny sent me some thoughts on it.

> Today technology has outpaced the most basic need for rules of the road and online pedestrians, especially those who make easy targets, are paying the price. The answer to the problem is not hiring more "officers" or moderators. Now, as it was then, the answer is creating regulation, codifying "rules of the road," and educating users as to their rights and responsibilities online.

The social platforms often didn't heed their own rules of the road. But they did take consistent action against content that broke federal law, like child sexual exploitation imagery. What about copyright violations?

The 1998 Digital Millennium Copyright Act protects copyrighted content from unauthorized use on any digital medium, regardless of whether the material is registered with the U.S. Copyright Office.[2] The Act doesn't hold internet service providers liable for unknowingly displaying material that infringes a copyright. But it requires that the material be removed when they receive a complaint, called a DMCA takedown notice. Hosting providers must take down or disable access to websites suspected of copyright infringement "expeditiously," which in practice usually means within a few days. The law provides for appeals by both sides in a complaint.

Hoaxers often copied images of Noah from the memorial page Lenny had built on Google+, then used them in their toxic posts and videos. That's stealing: Lenny had taken those home photos himself, and he owned the rights to them.

Copyright "strikes" would grow into HONR's most efficient hoax-content eraser. In early 2015 Lenny used one to hit back at Alex Jones.

AFTER THE HALBIG DEBATE with Keith Johnson, Jones had kept driving the Sandy Hook conspiracy, settling on the Pozners and Robbie Parker as his favorite targets.

"All I know is I saw Cooper with blue screen out there, green screen," Jones said on his December 27, 2014, show. "I mean, something is being hidden there. And then the parent's laughing, and then one second later doing the actor breathing, to cry. I mean, it's just over-the-top. Over-the-top sick.

"How do you even convince the public something is a total hoax?" Jones blustered two days later, on December 29, 2014.

"It took me about a year with Sandy Hook to come to grips with the fact that the whole thing was fake. I mean, even I couldn't believe it. I knew they jumped on it, used the crisis, hyped it up, but then I did deep research. And my gosh, it just pretty much didn't happen!"

"Deep research." Jones's model involved "taking stuff from completely unverifiable internet dumps and pretending that it came from real reporting or investigation," Dan Friesen, the Infowars researcher, told me.

Like plucking mushrooms from a manure pile, Infowars cribbed most of its content from Gateway Pundit, ZeroHedge, Reddit, and RT, the Russian state TV outlet. Jones's staff posted the material on Infowars' website and fed it to Jones, who would riff on it during his broadcast.

Jones would also grab content from mainstream news if it suited his narrative. So it was that in the opening days of 2015, Infowars hijacked a BBC report on a faraway massacre and spun it into a viral falsehood targeting Noah Pozner.

On December 16, 2014, Taliban gunmen attacked a school in Peshawar, Pakistan. The barbaric massacre killed 145 people, including

132 children. Peshawar erupted in grief and rage. At vigils and demonstrations, mourners displayed photographs of schoolchildren murdered in Peshawar and around the world, often using images found on the internet. A couple of the street murals included a photo of Noah, another child murdered in his school, half a world away.

On January 2, 2015, Infowars posted MYSTERY: SANDY HOOK VICTIM DIES (AGAIN) IN PAKISTAN. Host David Knight discussed the article on Infowars' "Evening News." The screen filled with the image of Noah on a placard in Peshawar. Infowars posted the broadcast to its YouTube channel afterward, as it usually did.

The Infowars article was shared, retweeted, and posted tens of thousands of times. It marked a turning point for Lenny. Jones was deploying the technique Lenny had noted long ago—shaping internet traffic by trumpeting a uniquely false theme—on Noah. "Whenever you saw 'Sandy Hook Victim Dies (Again),' that was Jones," he told me.

Infowars had used Noah's photo in its Pakistan segment. Lenny filed a copyright notice. YouTube yanked the entire day's broadcast from Infowars' channel.

Jones went apoplectic. On February 12, 2015, he devoted most of his show to Lenny and HONR.

Jones's backdrop that night was a photo of a tortured-looking man with duct tape plastered over his mouth, FREE SPEECH scrawled on the tape. Jones, in a light gray pin-striped suit and open-collared white shirt, seemed out of breath. He frothed for two hours between ads for Survival Shield, Super Male Vitality, Super Female Vitality, and Ancient Defense diet supplements. He displayed close-ups of the takedown notice several times, sharing with millions of viewers Lenny's name, email address, and the address where he received his mail.

Jones repeatedly showed Noah's photo, defying Lenny.

"The HONR network, Lenny Pozner, reportedly lost his son there, came in and filed a copyright claim on us showing a BBC News article. You can't do that, for those that don't know how copyright works.

"So we're sorry for everybody's losses, whatever. We're investigating this though, because we live in a system where the media exploits things and twists things."

Jones couldn't let it go.

"I just want to tell this network of people something: I'll have to go to Sandy Hook. I'll have to get involved. You're just stirring up a hornet's nest here."

Jones took a call from Brian in Alabama—one of the hoaxers behind *We Need to Talk About Sandy Hook.*

"I can tell you lots about Lenny. This man is something that you've never seen before. He's got a group of trolls—"

Jones interrupted him: "It must be horrible for folks out there that vehemently think this is staged. So just specifically, what have you gone through?"

"I can't even put up a video showing that he has put up a copyright strike against me without him copyrighting-striking that . . . These people are vile," Brian told Jones. "Lenny, if you're listening, your day is coming, my friend. It is coming."

Jones, excited: "Wow, this sounds like a war is going on. I think they made a major mistake involving us."

Brian agreed. "Go after 'em, Alex. Crush 'em."

Jones, Rob Dew, and his Infowars Evening News host, David Knight, aired a roster of falsehoods about Sandy Hook. They played the Robbie Parker video again. "That's an actor," Dew said. They displayed Lenny's information again, this time with Google satellite maps.

"I'm going to have to probably go on up to Newtown," Jones glowered. "I'm going to have to probably go investigate Florida as well."

Having doxxed and stirred up rage against Lenny among millions of viewers, Jones tried to ensure he wouldn't be punished for it. After the broadcast, his producer emailed an all-caps warning to the staff:

DO NOT UPLOAD ANY VIDEO OF SHOW TO YOUTUBE

It didn't work. Lenny filed so many privacy complaints against Infowars for that broadcast that eventually Infowars could only post an audio version of it. Lenny had achieved a kind of takedown multiplier effect, by making privacy complaints about videos in which Jones spoke about Lenny's DMCA takedowns, increasing the number of strikes against Jones's YouTube channel.

LENNY AND HIS VOLUNTEERS knew his escalation would bring a backlash.

"Oh God, Jones was hell-bent on finding out who we were," Vanessa recalled. "He used a lot of language that was accusatory and scary. If his followers knew who we were, we started seeing their wrath."

Jones's followers posted Lenny's social security number online. They posted personal information for HONR volunteers and began calling their workplaces. Some of the young moms who volunteered for HONR found their photos and names on prostitution websites and on sites trafficking in nonconsensual intimate imagery.

Lenny and HONR were putting real points on the board, and they kept going. Lenny settled into a new morning routine. Wake up. Pay a visit to Noah's pajamas, nestled in the dresser drawer. Make coffee. Fire up his Mac, and search for stolen images of his son.

"He spent every waking hour looking for this stuff," Vanessa recalled.

HONR filed scores of DMCA notices each day. Lenny would open a Google doc, the HONR volunteers would paste in links to the day's offenders, and he'd report the whole list. Lenny posted shortcuts to HONR's website so other victims could report content targeting them.

Lenny used a copyright strike to scrub the website targeting Newtown lawyer Monte Frank that Halbig's trolls had set up. It popped up on another provider, and HONR got it taken down again. Then it showed up on a third site—the hoaxers' last resort, based in the Netherlands, where the rules are more lax. But Lenny eventually succeeded there too: Halbig had used photos of the living Newtown girl who sang at the Super Bowl to promote his sick theory. It violated content rules prohibiting potential harm to minors. The website disappeared for good.

Lenny played Whac-A-Mole with the phony filmmakers behind the documentary *We Need to Talk About Sandy Hook*, scrubbing it from one channel after another. One night he found an angry message on Facebook from "Tom": "I have just about had it with your fraudulent complaints," he wrote. "I will see you brought into court to answer for your actions."

Lenny laughed aloud. He sent back a link to a dorky YouTube take on Clint Eastwood in *Dirty Harry*, with the caption "Go ahead. Make my day." He never heard back.

The blizzard of takedown notices put Lenny and HONR on the tech companies' radar. YouTube eventually opened a back channel for Lenny to report content directly. HONR got GoFundMe and PayPal to nuke Halbig's accounts. Ever the victim, Halbig sent a hyperventilating email to Infowars: *"Please tell Alex that they now have shut down my last place people could donate to our legal funds . . . I am called a FRAUD, LIAR and conducting illegal activities . . . Please help me."*

Lenny had complained about Halbig's fundraising to the Florida attorney general. Halbig posted the complaint on his website, using the personal information in it to doxx Lenny. Lenny sued Halbig for invasion of privacy. Halbig posted the lawsuit and doxxed him again, this time exposing Lenny's home address.

Lenny had to leave his apartment in Boca Raton, near the music theater and the beach, where he had begun to build a more peaceful life.

"It was like I was crashing their party," Lenny told me. "They feel justified in anything they do, but when it turns on them, they flip out."

IN LATE 2015, JIM FETZER compiled three years' worth of nonsense claims into the four-hundred-page book *Nobody Died at Sandy Hook: It Was a FEMA Drill to Promote Gun Control.* Fetzer and Mike Palecek, a former small-town newspaperman in Minnesota turned conspiracy chronicler, were coeditors.

Fetzer, Watt, Mead, Halbig, and a dozen others whom Fetzer called "the best students and scholars of Sandy Hook" contributed chapters bristling with false accusations against the families. Fetzer had used Amazon's CreateSpace to produce the book, which went on sale in mid-October.

Lenny and HONR raised hell with Amazon. The company pulled the title less than a month later, having sold five hundred copies.

A day later, an irate Fetzer posted a PDF version of *Nobody Died at Sandy Hook* online for free. Whac-A-Mole again.

Fetzer asked Alex Jones for help promoting the free book. Rob Dew promised that Infowars would post an article about it, adding, "Let us know if there is a bump in downloads."

On November 25, Infowars posted on its website "SANDY HOOK TRUTH BOOK BANNED BY AMAZON: Online Store Suppresses, Censors Book Presenting Controversial Ideas" and tweeted a link. But within a day the article and the tweet had both disappeared.

Infowars sells diet supplements on Amazon,[3] so maybe the relationship was too profitable to jeopardize for the likes of Fetzer. But the hoaxers went into spasms of speculation. "Jones has fallen completely silent on both the copyright run-in with Pozner and the retracted reportage of *Nobody Died at Sandy Hook*'s Amazon ban. Why is this?" Tracy wrote on his blog.

Alex Jones, he wrote, had been "Pozner'd." Tracy would be next.

LENNY WAS FINDING NEWSPAPER opinion columns a useful tactic for educating Americans about the misdeeds of the conspiracists. On December 10, 2015, he and Veronique marked the coming third anniversary of Noah's death by writing a guest essay in the South Florida *Sun Sentinel*.[4] They pointed out that by then, every mass tragedy drew swarms of deluded deniers and new harassment of victims and their families. Then Lenny and Veronique focused on James Tracy.

"This professor achieved fame among the morbid and deranged precisely because his theories were attached to his academic credentials and his affiliation with FAU," they wrote, adding that Tracy had parlayed his newfound exposure into an audience for his blog and radio show popular "among the degenerates that revel in the pleasure of sadistically torturing victims' families."

> Tracy even sent us a certified letter demanding proof that Noah once lived, that we were his parents, and that we were the rightful owner of his photographic image . . . His blog post was echoed dozens of times on conspiracy websites, including one maintained by Tracy's colleague and frequent collaborator James Fetzer, a Holocaust denier . . .
>
> FAU has a civic responsibility to ensure that it does not contribute to the ongoing persecution of the countless Americans who've lost their loved ones to high-profile acts of violence.

The essay unnerved Tracy. Lenny and Veronique had used a prominent local newspaper to alert FAU leadership and donors that it had a fabulist on its payroll who allied with an antisemite to attack a grieving Jewish family. Panicked, Tracy made a fatal move. He again collaborated with Fetzer, on what Fetzer called "A MASSIVE RETALIATORY STRIKE."

On Fetzer's advice, Tracy told me, he sent a letter to the *Sun Sentinel* calling the Pozners "as phony as the drill itself and profiting handsomely from the fake death of their son."[5]

Belatedly realizing his error, Tracy sent another letter he thought was tamer. It ran on the anniversary of the Sandy Hook shooting.

Tracy wrote that he had initially empathized with the victims' parents. "After several days of reflection, however, my instincts as a media historian and analyst took charge," he wrote. "In reviewing news coverage of the Sandy Hook School massacre, I began to recognize very unusual features in the alleged forensics, the emergency response and the overall way the event was being reported," Tracy wrote, unspooling several of his perceived "unusual features."[6]

Tracy's op-ed was long and self-sabotaging, accusing Lenny and Veronique of having "chosen the low road of playing upon the prejudices of decent, good-hearted yet often poorly-informed Americans."

Tracy touted the *Nobody Died at Sandy Hook* book, saying it is "available online as a free PDF."

He closed with an exhortation to Americans to engage in serious interrogation of the mass media. "If that is an outmoded ideal and a skill that can no longer be practiced or taught to young adults, I stand guilty as charged," he wrote.

Two days later, Florida Atlantic University notified Tracy he was fired. He was given ten days to respond, in accordance with faculty union rules. By January he was gone.

"That was the first of the public beheadings that needed to happen," Lenny told me.

The university said it fired Tracy for failing to file paperwork disclosing his blog and radio show. The firing shook the academic community, long accustomed to protections for free expression that went beyond those in most workplaces.

Kevin Carey, an education policy expert at New America, a Washington think tank, addressed the First Amendment protection Tracy claimed.

"Great universities can afford to shelter a few cranks and fools in order to support genuinely original thinking," Carey wrote in the *Chronicle of Higher Education*.[7] "While Florida Atlantic was right to fire James Tracy, it should be forthright about why: Not because of his offenses against truth and decency, though they are many, but because someone so cruel and possibly deranged has no business being employed to teach undergraduate students. Sometimes, the expansive protections of academic freedom are strengthened by defining where they end."

Tracy sued Florida Atlantic University to get his job back, claiming his sacking violated the university's commitment to academic freedom and his constitutional right to free speech. He lost.

CONSPIRACY THEORISTS AREN'T KNOWN for consistency, but they universally insist that the First Amendment shields them from consequences for spreading harmful lies.

The big platforms' creators also trot out free-speech principles when criticized for failing to rein in abuse. That might sound good, but it's based on a false reading of the First Amendment:

> Congress shall make no law respecting an establishment of religion, or prohibiting the free exercise thereof; or abridging the freedom of

speech, or of the press; or the right of the people peaceably to assemble, and to petition the government for a redress of grievances.

The First Amendment shields Americans' freedom of speech against *government* interference, not Facebook interference. As private businesses, social media companies make their own rules about what users can and can't say on their platforms. But the law, namely Section 230 of the Communications Decency Act of 1996, protects them from liability when they fail to enforce them.

Newspapers, television, and news websites enjoy First Amendment protection for the content they publish and air. But the same body of law holds them responsible for its truth. If they knowingly or recklessly spread falsehoods that defame individuals or businesses, they can be sued in the states, whose definitions of "defamation" differ.

In 2017, Beef Products, a South Dakota meat processor, sued ABC News for defamation after a 2012 ABC investigation targeting the company's use of low-cost, processed beef trimmings, which the industry calls "lean finely textured beef," and detractors call "pink slime." The beef processor claimed the ABC segment and its subsequent reports were rife with errors and mischaracterizations of the product, leading to a consumer reaction that devastated its business. Beef Products sought $1.9 billion in damages, which could have grown to nearly $6 billion under South Dakota law. ABC settled with the company for an undisclosed amount that subsequent reports indicated was well in excess of $177 million.[8]

In *Masson v. New Yorker Magazine, Inc.*, the Supreme Court ruled in 1991 that the First Amendment's free expression clause did not shield from liability a reporter who fabricated quotes attributed to Jeffrey Masson, a public figure. The court ruled that although public figures must prove that statements about or attributed to them must constitute a "gross distortion of the truth" in order to be considered "false," the made-up quotes qualified because they differed in their factual meaning from what Masson had actually said.[9]

In *Hustler Magazine, Inc. v. Falwell*, the late evangelical minister Jerry Falwell sued Larry Flynt's adult magazine for a parody ad depicting Falwell engaged in a drunken, incestuous encounter with his mother in an outhouse. Falwell won a jury verdict in a lower court and was awarded $150,000 in damages. *Hustler* appealed. The Supreme Court ruled unanimously for *Hustler* in 1988, holding that public figures cannot recover damages for the intentional infliction of emotional distress without showing that the false statement was made with "actual malice," meaning it was made with knowledge of, or reckless disregard for, its falsity.[10]

The high court further ruled that "the interest of protecting free speech, under the First Amendment, surpassed the state's interest in protecting public figures from patently offensive speech, so long as such speech could not reasonably be construed to state actual facts about its subject." In other words, the First Amendment protects media that lampoon a public figure by making charges so outrageous that no normal person would believe them.

The "actual malice" standard was established in the landmark 1964 *New York Times Company v. Sullivan*, in which the high court set a higher bar for public officials suing the press for libel. The court ruled, famously, that it is not enough for a public figure targeted by a statement to show that it is false. The person must prove the outlet in question either knew the statement was untrue or didn't care.

The liability that these precedents allow to be imposed on publishers applies to all media—except social media.

Although Facebook, Twitter, Google, and YouTube are fast becoming most Americans' main source of news and information, federal law protects them from being sued for any defamatory content they distribute.[11]

In 1996, Congress recognized the internet as an extraordinary information and educational resource for Americans, and its potential as "a forum for a true diversity of political discourse, unique opportunities for cultural development, and myriad avenues for intellectual activity." So they tried to protect the internet from excessive government regulation, giving it nearly unfettered possibilities for growth.

Section 230 immunizes social platforms from liability by treating them not as publishers but as mere pipelines for user-created content.[12] Here's the relevant part:

> No provider or user of an interactive computer service shall be treated as the publisher or speaker of any information provided by another information content provider.

The law defines "interactive computer service" as "any information service, system, or access software provider that provides or enables computer access by multiple users to a computer server, including specifically a service or system that provides access to the internet." That includes services like Facebook, YouTube, and Twitter.

As Lenny learned, the platforms act on content that breaks federal laws against child sexual abuse and copyright infringement. But they do far too little to remove other types of harmful content or users. Free speech principles, they claim, compel them to surrender platforms connecting one-third of the world to democracy's worst actors. The argument betrays either shocking ignorance of the Constitution, monstrous cynicism, or both.

The companies have spent billions on lobbying and marketing campaigns aimed at preserving their Section 230 immunity. Reminding Congress of 230's original intent, they rhapsodize about their good deeds. Mark Zuckerberg pivots to the social justice movements and marriages fostered by Facebook when asked about its role in enabling

Russian election interference, the neo-Nazis gathering on Facebook groups, and the murders streamed on Facebook Live. While misinformation, menace, and calls to violence circulate among the hundreds of millions of videos viewed each day on Google's YouTube, the company accepted plaudits for building "global community."[13]

Twitter publicly celebrated its role in linking activists during the pro-democracy Arab Spring protests and the nascent Black Lives Matter movement in the early 2010s while racist and misogynist attacks skyrocketed. "We suck at dealing with abuse and trolls on the platform and we've sucked at it for years," former chief executive Dick Costolo wrote in an internal memo in 2015.[14]

A year later, Charlie Warzel, then a senior tech writer for *BuzzFeed*, investigated Twitter's abuse problem in an article titled "'A Honeypot for Assholes': Inside Twitter's 10-Year Failure to Stop Harassment."[15] Several women and people of color had left the platform in 2016, amid torrents of abuse and threats, including Leslie Jones, then at *Saturday Night Live*.

"Fenced in by an abiding commitment to free speech above all else and a unique product that makes moderation difficult and trolling almost effortless, Twitter has, over a chaotic first decade marked by shifting business priorities and institutional confusion, allowed abuse and harassment to continue to grow as a chronic problem and perpetual secondary internal priority," wrote Warzel, who now writes for *The Atlantic* and *Galaxy Brain*, an online newsletter.

At the time Twitter said it was investing in better "tools and enforcement systems" to find and take faster action against abuse. The platforms have made improvements over the past several years, adding automated filters and better reporting mechanisms. But abusers have grown more sophisticated in skirting them.

"This is like an arms race," Warzel told me. "The bad actors become more sophisticated, and the company is always a step behind."

Hany Farid has studied this ecosystem, and it's made him a hardened critic of internet culture and the big platforms. Farid is associate dean and head of the School of Information at the University of California, Berkeley. Working from the chilly patio of his house overlooking the Golden Gate Bridge, Farid is an influential voice for social media regulation. His research focuses on digital forensics, image analysis, and human perception. He has provided expert testimony on "deep fake" video and photography in child sexual abuse material, intellectual property, and fine-art authentication cases.

The source of social media poison are its content algorithms, engineered to keep users online for as long as possible while feeding them advertising and separating them from their personal data. The big platforms are paid by advertisers, and the platforms in turn pay content creators for advertising placed on their sites or channels. The more awful or outrageous the content, the more people click, and the more everybody earns.

The big platforms "will have you believe they are neutral arbiters, but it's a complete and utter lie," Farid told me. "That newsfeed is algorithmically determined for you."

The algorithms are secret, but the results are obvious. If you consume conspiracy theories, your newsfeed and recommendations adapt, so you receive more and more of them. "Seventy percent of videos on YouTube is YouTube telling us what to watch," Farid told me. When dangerously false content goes viral, "YouTube is promoting that stuff."

It's the perfect storm, he said. "You have bad people and trolls and people trying to make money by taking advantage of horrible things that happen around the world, you have social media websites who are not only welcoming and permissive of it but are promoting it, and then you have us, the unsuspecting public, who are intentionally or unintentionally propagating it through the economy of the internet, retweets, and likes."

Farid scoffs at the platforms' free speech piety. He points out that "from the earliest days Facebook and YouTube said 'we will not allow adult porn as part of our terms of service,'" even though the distribution of adult pornography has been repeatedly affirmed by the courts, as a signal test of First Amendment freedoms.

"For Zuckerberg to rhapsodize about banning the actual speech that gave us First Amendment law? The hypocrisy and the irony is deep here. The reason they banned it is that they knew their advertisers did not want to be adjacent to it," Farid said. "They can ban all forms of content without running afoul of the Constitution. But they choose not to do it because they're making so much goddamned money."

"The internet is getting close to the point where it's doing more harm than good," Farid told me. "We keep talking about this dystopian future. I wonder if it's here."

The day after Tracy's op-ed ended his career, Infowars' Paul Joseph Watson fired off an alarmed email to "Buckley," probably Buckley Hamman, Jones's cousin who worked on Infowars' operations side:

> Sent this to Alex. This Sandy Hook stuff is killing us.
>
> It makes us look bad to align with people who harass the parents of dead kids. It's gonna hurt us with Drudge and bringing bigger names into the show.
>
> The event happened 3 years ago, why even risk our reputation for it?

If Jones ever got their message, he gave no sign. His reputation, at risk? A Republican candidate for president of the United States had just praised his reputation as "amazing."

13

There were few more blazing warnings of what Americans could expect from Donald Trump than his enthusiasm for Alex Jones. Their meeting signaled a profound change in American politics, the merging of the fringe with the establishment, and the refashioning of truth from an objective, respected standard to a malleable commodity and partisan weapon.

For a couple of years before Sandy Hook, Alex Jones was a regular guest on RT, the international TV network controlled and paid for by the Russian state. Kelly Jones recalls Russia Today, as it was known, calling at all hours, and Jones hopping-to, eager to build his audience by whatever means. The Kremlin was happy too, keen to air his paranoia about the United States—denying citizens their rights, spying on them, and using the police to hunt them down.

Liz Wahl did some of those interviews. An American former RT anchor, Wahl resigned on air in 2014 in protest of Moscow's invasion of Ukraine. Wahl thought Jones seemed harmless at first, an excitable character darkly speculating about the cabal of billionaires at the World

Economic Forum meeting in Davos. But his theories grew more fantastical and darker. "He didn't believe in U.S. democracy as we follow it," Wahl told me. "He thought it was *all* just a cover for some kind of global elite takeover."

Even RT's producers started pushing back. "They were like, 'We can't put this guy on. He's totally nuts,'" Wahl said. "We had crazy people on all the time," she added, but to be effective, RT needed its guests to be at least somewhat credible. They cut Jones loose.

That was a year before Jones began pushing the Sandy Hook conspiracy on his show. And it was three years before Donald Trump, Republican candidate for president, appeared as an honored guest on Infowars.

"DONALD TRUMP IS OUR GUEST, ladies and gentlemen," Jones said on December 2. "He is a maverick, he's an original. He tells it like it is, doesn't read off a teleprompter. Neither do I. He's self-made. This whole media operation that reaches twenty million people a week worldwide, conservatively, self-made. That's why I'm so excited."

Describing Trump, bankrolled by his father,[1] as "self-made" was fanciful. But Jones's father had made Jones too, so that was something else they had in common.

Trump joined from New York. He sat hunched over a desk in his Trump Tower office, his face, spackled in deep ochre, backlit by the window behind him. His bulk blotted out the expensive view of grand buildings, the spire of one of them appearing to protrude, antenna-like, from the top of his famous coif. He bore an indulgent smile, ready to plug his book *Crippled America*, soak up Jones's praise, and move on.

Jones welcomed him, calling Trump "Donald." He began by affirming Trump's false claim that "radical Muslims" around New York had publicly celebrated the fall of the World Trade Center after the 9/11 attacks.

"I took a lot of heat and I was very strong on it and I held my line," Trump said. Jones bobbed his head, star-struck.

"Hundreds of people were calling up my office. I was the other day in Sarasota, Florida . . . twelve thousand people, which is fantastic. And the people were saying, many of the people from New Jersey, four or five people said, 'Mr. Trump, I saw it myself. I was there . . . I saw it myself, Mr. Trump, I was there.' So many people have called in and on Twitter—@RealDonaldTrump—they're all tweeting. So I know it happened."

Trump hijacked Jones's signature claim, that he had predicted the 9/11 attacks.

"I wrote it in a book, 2000, two years before the World Trade Center came down," Trump said. "I talked about Osama bin Laden: 'You better take him out,' I said. 'He's going to crawl under a rock. You better take him out.' And now people are seeing that, they're saying, 'You know, Trump predicted Osama bin Laden,' which actually is true. And then two years later, a year and a half later, he knocked down the World Trade Center." Trump had not in fact "predicted Osama bin Laden."

Trump meandered through talking points on terrorism, Iran, Iraq, China, Obamacare, trade, his poll numbers, and his crowd sizes.

Jones nodded along for a while, placing an index finger to his lips in a show of rapt attention. But then he started cutting Trump off, pushing for more. Jones had a big, paranoid following to impress. Yapping over the older man's braying monotone, Jones began nudging Trump further out there, to satisfy his listeners.

"Donald Trump, let me say this. My audience, I'd say ninety percent support you, OK? And you definitely have shown your knowledge of geopolitical systems . . . People love you for tough talk. Is it not time for impeachment hearings against Obama? I mean, what do we do politically to really try to prosecute Hillary Clinton?"

Trump demurred. He was competing in the primary; he needed to vanquish a full field of Republicans and then draw some Democrats. The timing wasn't right.

"There's so many things to do, Alex. We will do such a good job. There's so many fronts . . . You know, everybody running against me in terms of even the Republican side and Hillary, certainly they're all controlled by their donors and their special interests and the lobbyists. I'm putting up my own money. I'm funding my own campaign. Nobody's going to control me. I'm going to do what's right for you and for the American people."

Jones interrupted. He praised Trump's brains and patriotism, then said, "Let's get down to brass tacks.

"I routinely talk to the"—he paused for effect—"top generals, Special Forces, Pentagon currently, out of the Pentagon, CIA, as I know you do . . . They really know that we've reached the crossroads where the country is *done*. It's a third-world nation, within a few more years. Forget Donald Trump, in four years, if this happens, we're done. I mean, we're talking about resurrection of the dead here. We could turn it around right now. As you've said properly—you're dead-on, sir. You're right. We could turn it around. All the actuaries, all the numbers show it. But it's got to happen in the next few years or we're done."

This was the blather even RT found too weird.

Trump nodded, brow furrowed. "Right, right," he said.

Jones went on, more animated, gesturing with both hands. "And there are globalists that want to have a world government, a system run by select crony capitalists, using socialism at the grassroots to make people dependent. And I've talked to not just high-level folks that have been in government that are on your team, but separately, high-level people in government currently that say there's an internal war going on and that you're a manifestation of that.

"I don't want to get into anything inside baseball with you, but I already know the inside baseball. I know now from top people that you actually are for real and you understand you're in danger and you understand what you're doing is epic.

"I want to tell you right now," Jones said, making sure Trump wasn't too distracted to catch on. "Can you speak about the war for the soul of this country that's happening right now? And really tell people what's happening?"

Trump replied, "We can turn it around. But I would agree with you if we don't get it right this time, I'm not sure if you go another four or eight years with the insanity and the stupidity of these leaders, I'm not sure you're going to be able to turn it around anymore."

Satisfied, Jones jumped in again. "Donald Trump, the man in the arena. His new book that we're talking about at the moment is exposing the fact that this country is being sabotaged by design.

"I know you're for real. You wouldn't be saying the things you're doing. They're scared of you. The whole system is coming out against you. But promise us that you're not going to drop out at the key moment, keeping all the other Republicans out of view and then Hillary races to the head, or Jeb Bush does. Because, as you know, folks are claiming you're a Clinton operative."

Trump looked momentarily surprised.

"You know, I've never heard that," he said, then recalibrated. "I heard it. Actually a few months ago. But I've hit her harder than anybody, times ten.

"I get along great with Clinton . . . I get along great with everybody, because when I needed them, I didn't want to have an argument."

Jones talked over him, coaching him again. "You're not a loser. You don't get in mindless fights. You move forward with your agenda. But now you see America in trouble and you're, 'Hey, that's all sidelined now. Donald Trump's not working for Donald Trump. He wants to work for America.'"

Trump stuck around through a commercial break and gave Jones extra time. The interview lasted more than a half hour.

"So I wrote a book called *Crippled America*," Trump said, signing off. "I hope your audience goes out and buys it as Christmas gifts and everything else.

"And I just want to finish by saying, your reputation's amazing. I will not let you down. You will be very, very impressed, I hope.

"And I think we'll be speaking a lot, but you'll be looking at me in a year, in a year or two years. Let's give me a little bit of a time to run things. But a year into office you'll be saying, 'Wow, I remember that interview. He said he was going to do it and he did a great job.' You'll be very proud of our country."

Jones waved his hands in the air again. "Well, I'm impressed. I mean, you're saying you're fully committed. You know, there's no future if we don't take this country back. Donald Trump, I hope you can help uncripple America.

"Thank you so much, sir. You will be attacked for coming on. We know you know that," he said, wagging a thumbs-up. "Thank you."

Trump's adviser and pal Roger J. Stone Jr., the political dirty trickster, had pushed the candidate to campaign on Infowars, against most other campaign staffers' advice.

Jones and Trump each saw utility and commonality in the other.

Jones has said in court that his audience slumped by "thirty to forty percent" with a Republican in the White House. Conversely, when conservatives were out of power, he could be "more provocative, more interesting and so it gets more viewers." But Trump, the newly minted Republican, dangled a potentially more profitable paradigm. Trump was all about provocation and attention. He cast America as a nation besieged by "enemies" familiar to Infowars listeners: immigrants, the mainstream media, Hillary, Obama, gun grabbers. He resonated with Jones's existentially aggrieved, suspicious audience.

Trump was new to Jones's world of patriot militias, doomsday preppers, and white nationalists. He didn't know all the code words, but he would catch on soon enough.

14

LENNY WAS OUT WITH SOPHIA and Arielle—he no longer remembers where—in the late afternoon of January 10, 2016, when he got an alert from Google Voice. He had four new voicemails on his cell.

It was a pretty winter Sunday in South Florida: a little bit of sun, in the balmy high seventies. Lenny usually ignored his messages when he was busy with the kids. But these had come in rapid succession—four of them in five minutes from the same number, which he didn't recognize. The girls were standing nearby, safe. But Lenny still carried the trauma of the rapid-fire texts alerting him to an emergency at Sandy Hook Elementary School three years before. These messages might be urgent. He opened his phone to listen to the first one.

You're going to die, motherfucker, nigger, kike, Jew bastard, fag, tranny, cunt.

And what are you going to do? Absolutely nothing. You're a loser. You are going to rot in hell.

DEATH. You're going to die. Death is coming to you real soon, motherfucker. You're going to die.

Lenny didn't listen to the whole thing. "As soon as I heard the word 'die,' I immediately pushed stop," he told me. It was a woman's voice, reedy and cracking with hate and menace. Chills flowed through him. He masked his shock and turned back to the girls.

LENNY AND VERONIQUE ALREADY lived in hiding. Veronique refused to live outside a gated community. Since 2014, when he first engaged with the hoaxers, Lenny had avoided being photographed or appearing at events where he would be publicly identified. He was moving every six months or so because trolls kept posting his home address. He received his mail in a series of post office boxes, including one in Newtown and the one at the address Jones had aired on his show. Once, a week after Lenny had moved, Philip Craigie, a conspiracist who went by the name "Professor Doom," called his cell and read him his new address and social security number. "I moved a few weeks later," Lenny told me, jokingly adding, "New apartments give a thirty-day satisfaction guarantee." A couple of years later Professor Doom would be charged with attempted murder after attacking a man with a knife, stabbing him five times.[1]

The caller probably didn't have to look far for Lenny's cell phone number. Lenny used the Connecticut number as a contact on DMCA takedown notices. It appeared on many of the hoaxers' websites.

LENNY AND THE GIRLS arrived home around 5:00 p.m. that Sunday. Lenny closed himself in his bedroom. He put on headphones, careful that only he would hear the threats.

He listened to the first voicemail. Then the second, sent one minute later, at 3:30 p.m.:

> Did you hide your imaginary son in the attic? Are you still fucking him, you fucking Jew bastard?

The third, a minute later:

> Did you let that nigger in the White House fuck you in the ass, you cheap piece of Jew bastard?

And the final message:

> Jew bastard, look behind you. Death is coming to you real soon.

All four messages were in the same woman's voice. Lenny also found two emails, sent the previous day. One read:

> LOOK BEHIND YOU IT IS DEATH.

At 8:01 p.m., Lenny emailed the same FBI agent in Connecticut who in late 2014 had brushed aside Lenny's reports of death threats, telling him to call Florida law enforcement.

This time Lenny would get his attention. "I received Death Threat Voice Mails on my Connecticut Phone Number," Lenny wrote. He included links to recordings of the voicemails.

The agent responded seven minutes later, at 8:08 p.m. on a Sunday.

"Thanks, Mr. Pozner. I'll immediately forward this to the appropriate squad here."

Lucy Richards was enraged that Professor James Tracy had lost his job because Lenny and Veronique Pozner complained about him. Richards was a fifty-six-year-old former waitress living on disability checks and food stamps who rented a single room in a house in Brandon, Florida. A friend had gotten her a cheap deal on a cell phone, which she had used to call Lenny.

Richards liked to watch Alex Jones's show online, where for three years he had been rehashing Veronique Pozner's long-ago interview with Anderson Cooper, claiming it was staged. She told the FBI agent who visited her that she was a fan of Jim Fetzer's too. Like Fetzer, Richards believed nobody died at Sandy Hook. She found Lenny's phone number wherever she had read about Tracy's firing, maybe on Fetzer's website.

When she talked with the FBI, Richards denied threatening Lenny. She said she was expressing her opinion. When she told Lenny Pozner that he was going to die soon, she said she was merely stating a fact.

Richards surrendered in court and was freed on $25,000 bond, pending trial. She pleaded not guilty to four federal charges of sending threats by interstate communications, each carrying a maximum punishment of five years in federal prison.[2] Richards was jailed after she refused to show up in court for a plea hearing. She eventually pleaded guilty to one of the four counts and was sentenced in 2017 to five months in jail, five months' home confinement, and three years' supervised release.[3]

"Your words were cruel and insensitive," U.S. district judge James Cohn told Richards. Referring to Lenny, he said, "I'm sure he wishes this was false and he could embrace Noah, hear Noah's heartbeat and hear Noah say, 'I love you, Dad.'"

Richards was ordered to seek mental health treatment, keep a log of her computer activity, and stay off Jim Fetzer's and Alex Jones's websites.

Lenny found Richards pathetic and scary. "A lot of this hate comes from people who are mentally ill, but there's no filter for the mentally ill online. They are able to spark the fire of hate and keep it going."

Struggling, angry, and probably lonely, Richards had been given access to the global misinformation chain with a bargain cell phone and a cheap computer, which allowed her to torment Lenny from a remove that had made it difficult for him to gauge the threat. Was she a harmless misfit, or was the killer in the house? For many of these people, internet theorizing was their main source of human connection. For others, the internet was all they had left after their pastime hardened into compulsion. That was Doug Maguire.

BY THE END OF 2016 Lenny no longer felt like a faint voice calling for help outside the iron gates of the big platforms. Takedown orders and the public shaming he and Veronique had done in their op-eds meant that when he contacted the companies, usually a human called back. As word of his tenaciousness spread, Lenny received pleas for help from people whose lives had been upended by online hate. Doug Maguire was among the first, and his experiences on both sides of the conspiracy divide were among the most astounding.

Maguire grew up in the mid-1980s in Hyannis, Massachusetts, a working-class town not to be confused with Hyannis Port, home of the Kennedy compound. Maguire and five buddies dreamed of escape to California, and kept a saxophone case in Maguire's bedroom they called the "Cali fund." Whenever they got change back from a dollar or could heist a few quarters from a vending machine, it went into the sax case, to finance their plans for fame in the Golden State.

In the end, only Maguire went. Work, love, and family plans kept his friends home, but for Maguire, California meant freedom from a family in which "we all grew up a little weird," he told me. His father had taken off when Maguire was still in utero, and his mother "had a wild streak." Maguire was twenty-three in 1996 when he packed his Modulus bass guitar and 16 mm film camera into his red 1990 Volkswagen GTI and took off for Hollywood, with no solid plan.

During our first conversation, Maguire unspooled the lowest, most humiliating episodes of his life as if recounting a movie, his candor and lack of exculpation more typical of a child than a man in his late forties. "People say I'm very naive," he told me, and I soon understood why.

Maguire had taken a few film and video classes at Emerson College in Boston. He wanted to write screenplays and, ultimately, direct. Lean and fit, with pale eyes, Roman nose, and plentiful brown hair, he got "sidetracked," he said, by work as an extra, then as a stand-in for stunts. Working with the likes of Brad Pitt and Matt Damon, soon "I thought I was bigger than I was." In the early 2000s he was fired from a set for refusing to remove a Bluetooth headset he was wearing. "Little things like that made me eccentric and difficult, and the jobs stopped coming," he told me.

He bounced around a bit, making a Nike commercial, and starting a graphic T-shirt business called PenalTees. He got married in 2010 to Katherine, a legal secretary, and they had a son, Vincente, in 2012. In between, Maguire wrote and directed a 2011 film called *Bank Roll*, a convoluted tale about a bank robbery that received some indie awards and a couple of truly terrible online reviews.

Maguire stayed home caring for Vincente while Katherine worked. "I had a lot of time," he said, and he spent most of it online. "I kept getting suggested YouTube videos researching Bigfoot," Maguire recalled. The first mass shooting conspiracy video he watched was about the Aurora gunman. After that more kept coming, YouTube's algorithms sending him further down the rabbit hole.

He watched *We Need to Talk About Sandy Hook*. The film's production values were better than most. "It seemed like they were almost there, and he needed somebody like me to kick it up a notch," he told me. Maguire believed the shooting happened, just not the way the media reported it. Characteristically, he imagined roles in a cover-up for a host of Hollywood villains: the mob, a family on the lam, corrupt government officials.

Maguire reached out to Independent Media Solidarity, whose members included Peter Klein, who on his unintelligible LinkedIn page described himself as "a rare combination of both geek and savvy"; Tony Mead, house mover and Sandy Hook Hoax troll; Jonathan Reich, arrested for phone harassment related to Sandy Hook; and Professor Doom, the violent felon who called Lenny and read his home address and social security number to him.

Maguire impressed them with his brief Hollywood résumé and helped them refine their droning, 167-minute film. Then he helped them fan out and get it onto as many video platforms as possible, evading Lenny and HONR, who were using copyright notices to get it taken down as fast as they could post it.

"I was calling Lenny out" for his takedown orders, Maguire told me, and searching for information about the Pozners the hoaxers could exploit.

"If you mess with Lenny Pozner from The HONR Network and get under the man's skin, you win," Maguire wrote on his website, calling a reaction by Lenny a "Badge of HONR." He lampooned an interview Lenny did with the BBC, replaying excerpts accompanied by the theme from the seventies TV show *Charlie's Angels*.

When Maguire mocked a YouTube video from HONR in the comments section, he received a response from the group:

> Doug Maguire will never work in Hollywood ever again.

That worried Maguire, who had burned so many bridges that he wondered the same thing. Who *was* this guy Lenny?

Maguire had a page on IMDb, a website and database of film, television, and actor content, where registered members could edit the site and write reviews. In exchange for a credit line, he helped the hoaxers create a page for *We Need to Talk About Sandy Hook.* To boost the film higher in search results for "Sandy Hook," Maguire added "archival footage" screen credits for some of the characters in the "plot," like Dannel Malloy, Barack Obama, Anderson Cooper, Lenny, Robbie Parker, and Scarlett Lewis.

"IMDb is really strong in SEO," Maguire said—meaning "search engine optimization," the process of ensuring that a website or piece of content surfaces in the greatest number of relevant internet searches. "That's how I figured we could make the movie big. When people would search for these parents, like 'Who is the son and wife of Lenny Pozner?' it was like, 'Click here to watch the movie.'"

We Need to Talk About Sandy Hook was no longer one among thousands of amateur conspiracy videos. "I legitimized it," Maguire told me. He used graphic arts skills honed in his T-shirt business to create an eerie-looking movie poster, with the filmmakers depicted in silhouette, and the gunman's emaciated, haunted face staring out from its center.

"This movie got really popular," Maguire said. "Alex Jones got ahold of it because of IMDb." Jones brought the filmmakers onto his show in 2014, before the shooting's second anniversary.

Maguire deluded himself that conspiracy videos might be a path to filmmaking relevancy. He made "Two Bombs in Backpacks," a Weird Al Yankovic–style YouTube video claiming that the 2013 Boston Marathon bombing was a false flag. He browsed Google+ and the r/dankmemes subreddit, a niche forum crawling, back then, with images of splayed torsos and severed genitals. Someone sent him footage of a man falling to his death from the top of a Ferris wheel. The memes reminded him of episodes from his childhood. "We would go down to catch bullfrogs, and my friends would bring tennis rackets and smack them into the middle of the pond. It used to make me scream," he said. "I was catching frogs too, but I just can't take it to that level of cruelty." But he couldn't look away either.

His relationship with his wife soured. When he mentioned that one of his new online friends thought the Holocaust was a hoax, "she thought I was dangerous," Maguire told me. Maguire started mixing it up with his wife's brother, who lived next door. His wife took her brother's side and kicked him out of the house she and her brother jointly owned. When the brother texted him, threatening to toss out his belongings, Maguire freaked out, calling Katherine fifteen times in a row. She finally picked up, and "I said if any of my property is ruined, I'm going to dump a gallon of piss on her brother's car and throw rocks through [her] window." Katherine called the police. The lawyers in the firm where she worked helped her secure a restraining order. The couple divorced, and Maguire lost custody of Vincente.

It was 2016. Jobless, Maguire lived in his car, collecting $1,000 a month in disability, money that didn't go far in Los Angeles. But he didn't give up conspiracy theories. Maguire spent his days trying to find work and stay afloat. At night he communed on Facebook with Sandy Hook Hoax and Independent Media Solidarity. Maguire messaged with them for hours, his laptop on his knees in the driver's seat of his broken-down VW parked in the lot of the Burbank Metrolink station, eating convenience-store food and stuffing cigarette butts into the ice melt of a Big Gulp.

He worked hard on a video for Independent Media Solidarity, exposing an evangelical Christian YouTuber as a fraud. The hoaxers killed the video, refusing to out another YouTuber. Angry, he berated them, calling them sick obsessives who "can't stop talking about Sandy Hook." They told him they wouldn't be needing his help anymore.

After several days things died down a bit, Maguire thought.

One of the Sandy Hook hoaxers, who went by Mollygirl (I have changed the screen name), reached out to Maguire. She seemed warm and sympathetic, and she and Maguire developed a virtual friendship. She told him she was a real estate entrepreneur. "Doug I will have an awesome attorney, private investigator and apartment for you in the next couple weeks. Your ex wife's family apparently does not comprehend the consequences of what they did to you, and they are going to learn what I do to people who hurt children and make false allegations," she wrote to him on Facebook. She signed off with a string of Russian-language boasts like "We are anonymous! We are legion! Expect us!"

In fact Molly lived with her toddler in her parents' basement in New York. The child's father was serving a long prison sentence. Molly, still in her twenties, was in and out of recovery.

Molly offered to help Maguire clean up his Facebook account as part of his effort to regain custody of Vincente, by then four years old. Maguire gave her his Facebook password. She tried the same one for Instagram and got in. She got into his Yahoo email account and set up a "backdoor," allowing her discreet access if ever she were locked out. Molly professed love for Maguire, "which made me really uncomfortable," he told me. "But by then she had all my social media accounts, and I couldn't just back out."

Molly launched an attack on Maguire, revenge for his bucking the Sandy Hook filmmakers. She contacted the domain registrar and hosting company GoDaddy and talked them into transferring all Maguire's work-related domain names to herself. She deleted Maguire's work videos, photos, and artwork from his Facebook, Flickr, DeviantArt, and

Instagram accounts. Cruelly, she sent him a video of herself erasing every image of Vincente from Maguire's Facebook albums, deleting them one by one as striptease music played. She and other hoaxers created simulated porn using images of Maguire's face and posted them on Google+. She fabricated a suicide note from him and posted it on his LinkedIn page.

During one of their online conversations, Molly texted Maguire a photo of her child in the bathtub, then called police to say he had images of sexual abuse on his phone. Maguire had deleted the photo without downloading it and explained the situation when the police turned up. Molly had told them where he was.

Maguire had lost his virtual and real lives in the space of a year, his chances of seeing his son destroyed by a troll he barely knew. Desolate, he wondered whether he had it in him to kill himself. He decided he couldn't do that to his son.

Vincente was born in early 2012. After the shooting in Newtown, Maguire went on Google during his son's nap. He searched on "Sandy Hook" and Google autofilled the word "hoax."

While sitting in his car, not having seen Vincente in nearly a year, "it just kind of hit me," Maguire said. "This is all karma. I started understanding better what it is to be a dad with loss. And here was Lenny, his son dead, taking on a world of hoaxers."

Maguire resolved to contact Lenny. Again, it was not difficult. That year, 2016, Wolfgang Halbig had posted Lenny's TransUnion report online. It contained one hundred pages of contact and financial information for Lenny and most of his family members. Maguire messaged Lenny and in time, a note came back:

"You have five minutes of my time."

They spoke for a half hour. Maguire explained about the YouTube videos he couldn't stop watching, how he'd descended to a dark online place with walls so steep he couldn't climb out. He told Lenny about the loss of his son, about Mollygirl and the hope she offered. He asked Lenny's forgiveness for spreading lies about the tragedy that killed his boy. Finally, ashamed, he asked for help. Lenny agreed to try.

Maguire wept telling me this. "Whatever I put him through, he ended up dusting me off when I was at my worst," he told me.

When I asked Lenny why, he was typically reticent. He never reveals more about his online acquaintances than they reveal about themselves. That's policy, Vanessa told me, and Maguire's predicament reinforced its wisdom.

"Doug has a good heart. I immediately sensed that he was gentle," Lenny told me. "It's easy to get a sense for him because he's really pure when you talk to him, you know?" I did.

Maguire had been snarky and stupid, but he hadn't threatened Lenny or mentioned the children's names online. Lenny knew plenty about Mollygirl already. "She was sick in every sense of the word," he told me. She had lifted videos of Noah from Lenny's Google+ page and made a video of her own from them, set to porn music.

I wondered why anyone would trust a random like Molly for help in a child custody case. Lenny knew: "She actually used to work as a Child Protective Services field investigator."

The HONR volunteers received messages from people who lived in her town, describing run-ins with her neighbors and calls to police.

"But on the internet everyone is equal," Lenny told me. "You can be in filthy underwear and sitting in the basement, and it's no different than a PhD," he said with a laugh, since some of his more persistent assailants are PhDs.

Lenny told Maguire to go to the police. But when Maguire, homeless and distraught, his hair uncut and greasy, entered the Burbank station to report he was being stalked online, they shrugged and sent him on his way.

"Every time he wanted to file a police report, they told him none of this was a crime," Lenny told me. "It has to be a crime. But they were not interested in him. He handed me the phone in the precinct, and the police were like, 'Do you know this guy? Do you have any idea who this guy is?'

"They weren't interested. They didn't understand. That was the way cybercrime was interpreted."

HONR tracked at least ten bogus Facebook accounts to Mollygirl. Lenny registered a new blog in Maguire's name, for him to build a healthier virtual home. Maguire uses it to record journal entries for his son.

Maguire was still living in his car in late 2016, "eating 7-Eleven hot dogs, Doritos, Powerade, and two bananas every day for a year," he told me. But he had found a new mission. Jammed into the driver's seat, his laptop warming his legs, he built a website exposing pseudonymous trolls.

One of his first targets was Mollygirl.

"I was fighting back. And at the same time trying to find a girlfriend on Bumble or Tinder," he said. He told them he was a writer who goes after online trolls, a hero.

Maguire still lives in California, in the home of a girlfriend who is helping him in his effort to see his son again. He works on the packaging line for a cannabis company and wrote the screenplay for a film that garnered modest buzz on the indie circuit in 2020. He managed to hang on to his Modulus bass, vintage now.

Lenny leveraged his newfound contacts at the social media companies to get Maguire's social accounts and websites back, one by one. It took four years.

15

ON A CRISP AND BRILLIANT October morning four years after Emilie's death, Robbie and Alissa Parker rallied Madeline and Samantha for a road trip. They packed overnight bags with dressy clothes and bathing suits, loaded them into their Honda minivan, and set off from their small farm in the cedars of Brush Prairie, Washington, for Seattle, two and a half hours away.

Alissa and Robbie had worked to overcome feelings of anger and lost innocence that lingered after Emilie's murder. By autumn 2016 they had begun to reclaim some measure of peace. Alissa had nearly finished her book, *An Unseen Angel*, about her post-shooting journey and her faith that "evil didn't win that day."[1]

After the battle with the United Way, the Parkers had used some of the money given to each victim's family to create a foundation in their daughter's name. The Emilie Parker Art Connection offers art therapy for traumatized and neglected children.

"Especially when you lose somebody young, there's this life that you knew that they had, all this potential," Robbie told me. "You kind of just want to have that seen."

That venture had prompted their trip to Seattle that day. Art with Heart, another nonprofit that uses creative expression to help heal children, had invited the Parkers to its fundraising gala in the city. Robbie and Alissa would be part of its presentation that evening. The program was meant to be light and celebratory, showcasing the year's artistic achievements, so Madeline and Samantha also were invited. The girls chattered with anticipation as they motored north, past mossy wetlands and regiments of conifers. They looked forward to staying in a hotel, to dressing up and being, as Emilie would have said, "fancy."

They arrived around 4:00 p.m. in Seattle. The girls wanted time to swim in the hotel pool before they all changed clothes and got ready for the dinner. Robbie dropped Alissa and the girls off in front of the Hyatt at Olive 8, a boutique hotel on the corner of Olive Way and Eighth Avenue. To avoid the pricey valet fee, he parked the car in a public lot about two blocks away, locked it, and set off up Eighth Avenue to the hotel.

On Eighth Avenue, he crossed Pike Street and was walking toward Pine when he passed a man headed in the opposite direction. "Just a normal dude," Robbie recalled, in his forties, dressed in khakis, a collared shirt, and sport coat. The man paused in faint recognition.

"I recognized that look, so when he started to say something I stopped," Robbie told me. "In Connecticut that would happen all the time. People would stop and were always very compassionate," he said. Sometimes a bit too much so: in Newtown, trips to the grocery or hardware store sometimes turned into gantlets of effusive, total strangers. The Parkers' daughters wondered back then why people were always hugging them.

In Washington, Robbie and Alissa enjoyed more anonymity. So it was surprising but not unwelcome when the man asked Robbie the familiar question: Did he lose a child at Sandy Hook?

"Yeah, that was my daughter," Robbie said, extending his hand.

The man ignored it. "How do you fucking live with yourself, you fucking piece of shit?" he hissed.

The man trailed him for blocks, "jabbering in my ear," Robbie recalled.

You fucking liar, Sandy Hook never happened; how much money did you get from the government, you evil son of a bitch?

"Absolute venom," Robbie told me. "He was absolutely disgusted with the person that he believed I was."

Heart banging, Robbie abruptly changed direction and walked faster, back the way he had come. He strode farther and farther from the hotel, determined to put distance between the harasser and his waiting family.

The man kept pace with Robbie, continuing his stream of curses. For the first time, he used Emilie's name. Though he has tried, Robbie cannot recall the words. Only his revulsion at hearing his daughter's name in this monster's mouth.

Robbie spun around. "How do you live with *yourself*?" he shouted.

"I didn't care that he was swearing at me, but when he started using Emilie's name, that tripped me," Robbie recalled.

He faced off with the man. They shouted back and forth, their voices growing louder. Robbie vaguely noticed passersby staring. Some stepped from the sidewalk into the street, giving the two men a wide berth in case they came to blows.

"For years I had been silent, because who was I gonna defend myself to online?" Robbie told me. But in that moment, "I'm defending my family and myself and I'm defending Emilie."

He cannot remember exactly what he said, only that he said plenty.

"You always hear things about how when somebody gets so enraged and so emotionally flooded, and the next thing they're saying, 'Honestly I dunno what happened. I woke up with blood on my hands,' like they blacked out. I understand that now."

The man, over six feet tall, looked down into Robbie's face as if he might strike him. Then he abruptly stalked off.

Robbie walked for ten, fifteen minutes more, making sure he was truly gone and not following him. His body shook.

"This was the only time anybody actually personally engaged with me," he recalled. He struggled to get himself together. He needed to act like nothing was wrong when he reached the hotel room.

Alissa opened the door. The girls stood behind her, blond hair tumbled, beaming expectantly in their tropical-print bathing suits.

"Alissa took one look at my face and said, 'What happened?'" She urged him into the bathroom just inside the room's entrance, shutting the door behind them.

"I started bawling," Robbie said.

THE PARKERS' MOVE to Washington State after the shooting had seemed almost preordained after Emilie's death. Robbie and Alissa had traveled there for a friend's wedding and realized how much they had missed it, and the innocent, happy time they had spent there with the girls before their move to Newtown. A neonatal ICU job had opened up at a hospital outside Seattle, and a professor of Robbie's had put in a word for him. In Brush Prairie the Parkers could choose how much to engage in the causes and battles that followed the tragedy. They could speak about Emilie's death when they chose to, not because others demanded it.

Still, Robbie once described his and Alissa's decision to move away from Newtown as going "rogue," and they had struggled with it.[2] They missed the support and safety they felt in the company of the other victims' relatives. But the Parkers had never fully shed their newcomer status in Newtown. They hadn't met the Sandy Hook relatives before tragedy threw them all together, a heartbreakingly large number of grieving people, each with individual needs and perspectives.

"Every family kind of took on their own thing," Robbie told me.

Alissa and Robbie had become convinced that the shooting might have been prevented, or at least the death toll may have been lower, had more basic safety features been in place at Sandy Hook. Alissa and Michele Gay, whose daughter Josephine had died with Emilie, tapped security professionals willing to contribute their expertise and created Safe and Sound Schools, a nonprofit advocating for practical, relatively inexpensive ways to better secure American school buildings.

Reserved and intensely private, Alissa found herself traveling around the country, reliving the day she lost Emilie and urging audiences to "'let my hindsight be your foresight,'" Robbie recalled.

Alissa had supported Robbie through college and his early career, and he liked being able to hold down the fort while she traveled. "I became really reclusive and I didn't explain a lot of stuff. I know that there were Sandy Hook families that went down to Washington and lobbied and stuff, but I wasn't one of them."

Robbie didn't want the public to assume where he stood on gun policy simply by dint of his daughter being murdered. Alissa found more purpose in what she and Michele were doing. "People are angry when things like this happen, and they want to point to the nearest thing to blame. If they can solve that one thing, they feel better. They feel like they can go back to their normal lives," she told an interviewer for *The Atlantic*.[3]

The couple came from a close-knit world of family, childhood friends, and shared faith. Tragedy had torn away their privacy, and they needed it back.

They had also learned, Robbie told me, "Anytime we put ourselves out there, we're going to get harassed."

The resigned sadness of his words struck me. Deserving of Americans' protection, they had been conditioned to expect abuse instead.

The confrontation with the man in Seattle unsettled Robbie. He and Alissa had always comforted themselves with the knowledge that very few people act on their online obsessions. Now one of them had jumped the virtual divide.

AFTER EMILIE'S DEATH but before they moved to Washington, the Parkers had found letters in their Sandy Hook mailbox. "They said they know where I live, telling me stuff like they're waiting with a baseball bat to meet me," Robbie recalled. Such mail was another reason they were keen to move. But one of the letter writers lived a half hour from Brush Prairie, where they were moving.

Kevin Purfield lived in Portland, Oregon. In early 2013, Purfield had called a friend of Robbie and Alissa's, then sent the couple two long, rambling letters denying Emilie was dead. But the Parkers resisted taking action. They were reeling and exhausted, and as Robbie had told CNN when he declined to be interviewed, they did not want to dignify such claims with a response.

Robbie and Alissa didn't know that while Purfield was contacting them, he was simultaneously stalking survivors of the Aurora theater shooting, which had happened five months earlier, in July 2012. Police in turn did not know that Purfield was contacting the Parkers, because they hadn't reported it.

Purfield, forty-five, with a history of mental health problems, had by his own admission called, emailed, and posted on the social media accounts of relatives of at least eleven of twelve people murdered in Aurora, most of them many times.[4] He told them that the Aurora shooting never happened and that their loved ones were still alive. When challenged, he grew abusive.

On his blog, Facebook page, and YouTube channel, Purfield talked about harassing the family members. He said he believed that Aurora and Sandy Hook had never happened and that the FBI had orchestrated the 1995 Oklahoma City bombing.

Purfield recorded scores of YouTube videos about his efforts. They betrayed his obsessive reading of news reports about the shootings, and a chilling knowledge of their details, including the nature of the victims' injuries. A large man with a broad face, and at times a graying, well-kept beard, Purfield spoke slowly, breathing deeply and looking at the ceiling between sentences.

In one video Purfield described calling the brother of a young woman murdered in Aurora. He named both the woman and her brother.

"I remember asking him if he saw his sister's body," Purfield said. He detailed leaving voicemail messages for a police officer described in news reports as having tried to save the woman's life. He questioned how the officer could have done so "if [the victim's] head was blown off."

"Everyone ignored me," Purfield said.

No, they didn't. The brother of the young woman who died had alerted police. He had pleaded with Purfield to stop, but Purfield left at least five more voicemail messages. Police in Aurora asked Portland law enforcement to help track down Purfield using his cell and other records. In April 2013 police apprehended him not far from his home.

Purfield pleaded guilty to one count each of misdemeanor telephone harassment and stalking.[5] In June 2013 he was sentenced to a year's probation and ordered into a mental health treatment program.

A couple of months later, Purfield disappeared. In September and early October 2013, Portland law enforcement and county officials in Multnomah County, Oregon, received more than thirty phone calls from a man they believed to be Purfield. The caller said bombs had been planted in the Multnomah County Justice Center, a complex with courtrooms, a police precinct, and the county jail where Purfield had been held on the Aurora-related charges.

Police and the FBI tracked Purfield to Long Beach, California, where they approached him on October 10. Purfield ran, and police had to use stun guns to subdue him.[6] In December, Purfield pleaded guilty to two counts of conveying hoax bomb threats. He was held in county facilities for more than a year, while medication restored him to competence. In March 2015, a U.S. district court judge in Portland sentenced him to time served and a year's probation.[7]

PURFIELD HAD NEVER CONTACTED Robbie and Alissa at their home in Brush Prairie. He was arrested before they moved to Washington, and while he was hospitalized and on probation, the Parkers hadn't heard from him.

But Robbie's confrontation with the man in Seattle suggested Purfield wasn't a one-off. Four years after the shooting, the episode was a brutal reminder that he remained a kind of Exhibit A in the Sandy Hook deniers' bogus case. They had discounted everything about Robbie's media appearance the night after Emilie's death except his reflexive smile as he came face-to-face with the cameras. Nothing he said afterward, no emotion he betrayed, mattered. They had decided he was a liar before he had uttered a word.

Alex Jones had referenced that moment so often that "parents" and "laughing" had become hoaxer code for Robbie. Robbie could guess when Jones had mentioned him on Infowars because online attacks spiked. "To those people who have that belief system, Alex Jones was like a prophet," Robbie told me. "Just like spiritual people like to read the Bible, if you believe in a conspiracy theory and you think I willingly sacrificed my daughter, you have a place to go to confirm what you believe."

That fleeting moment swept from Infowars outward, replayed in posts and tweets, podcasts and videos. Lenny and HONR had helped take down what they could, but the tsunami of material was overwhelming. Robbie had repeatedly asked the platforms to take material down. "The absolute silence we received from them suggested that it wasn't going to hurt their image enough, and it wasn't going to hurt their pocketbooks," he said. "What a huge gap in reality and feeling, that they could put that aside."

The Parkers had failed to outrun the harassment.

"I was randomly walking down the street in a random city three thousand miles away from Newtown. For him to see me, recognize me, put me in the right context, I can't imagine how many videos or how many things he had read and seen about me," Robbie told me.

"I can't say directly who Alex Jones's audience is, but I have seen the effects of people who absorb that content and what they do. And that's a direct impact."

16

THE DIRECT IMPACT ROBBIE DESCRIBED would be felt in many corners of the United States in that election year, as Donald Trump's campaign against Hillary Clinton unleashed a fervid, radical conspiracism among Trump's most ardent supporters. Unlike previous Republican presidential candidates, Trump embraced and echoed their paranoia, emboldening bullies everywhere to take action against their perceived enemies.

In the summer of 2016, a preposterous theory involving top Clinton operatives and a child sex ring germinated in the fetid depths of the internet and began to spread. A team of journalists from *Rolling Stone*, the Investigative Fund, and the Center for Investigative Reporting's Reveal later found[1] that in early July on 4chan, someone with the handle FBIAnon posted on an anonymous message board falsely touting dark secrets about the Clinton Foundation.

"Bill and Hillary love foreign donors so much," FBIAnon wrote. "They get paid in children as well as money."

That and similar bizarre posts constituted the early traces of what grew into Pizzagate, a nutty web of total nonsense claiming that Hillary

Clinton and top Democrats operated a child sex slavery ring from the basement of Comet Ping Pong, a pizzeria in Washington, D.C.

Although the Sandy Hook families had already suffered for years, Pizzagate jolted people awake to the real-world consequences of "fake news." The term, defined by Hunt Allcott of New York University and Matthew Gentzkow of Stanford University, referred to "news articles that are intentionally and verifiably false, and could mislead readers."[2] At least that was its original meaning, before President Trump repurposed it to discredit reports critical of him, and authoritarians around the world followed suit.

In their article "Social Media and Fake News in the 2016 Election," Allcott and Gentzkow framed the problem. "Social media platforms such as Facebook have a dramatically different structure than previous media technologies. Content can be relayed among users with no significant third party filtering, fact-checking, or editorial judgment," they warned.

Febrile, far-fetched conspiracy theories have swirled around Bill and Hillary Clinton for three decades. During Bill Clinton's presidency, conspiracists had to use the U.S. mail to send a list of "100 people murdered by Bill and Hillary." But in 2016, thanks to the internet, "an individual user with no track record or reputation can in some cases reach as many readers as Fox News, CNN, or the *New York Times*," Allcott and Gentzkow wrote.

Pizzagate was stoked by twin events that damaged Clinton in the closing days of the 2016 campaign.[3]

In early October, Wikileaks released thousands of pages of emails stolen from John Podesta, Hillary Clinton's 2016 campaign manager, released, interestingly enough, on the same day as the *Access Hollywood* video, in which Trump was recorded lewdly boasting about groping women.

Some of the emails embarrassed Clinton and the emails' senders, detailing Clinton's paid speeches to special interests, staff backbiting, and craven efforts to curry favor.[4] Many of the emails, though, concerned gruntwork like ordering takeout food. Podesta and his lobbyist brother, Tony, sometimes ate at Comet. The pizzeria is a Washington institution, a warehouse-style restaurant where families played ping-pong on the tables in the back while awaiting their food. The owner, James Alefantis, was friends with Tony Podesta and through him knew his brother. Among the leaked emails was one from eight years prior, a September 27, 2008, invitation from Alefantis to John Podesta, inviting him to an Obama fundraiser that a group of young lawyers hosted at Comet.

"Would you be willing to stop by around 8 o'clock or so and make a little speech. They (and I) would be thrilled to have you of course," Alefantis wrote.[5]

On October 28, then-FBI director James Comey said in a letter to Congress that he was reopening an investigation into Clinton emails sent from a private server in her home while she was secretary of state. That investigation had been closed, but then the FBI found some of Clinton's emails on a computer belonging to Anthony Weiner, the estranged husband of Clinton aide Huma Abedin. Weiner, a former New York congressman, was being investigated for allegedly exchanging sexually explicit text messages with a fifteen-year-old girl. (He was later convicted and jailed.)[6]

The development resonated among online conspiracists. Wild rumors began to bubble on 4chan and Reddit that Clinton and her top lieutenants ritually abused and trafficked children imprisoned in the basement beneath Comet, which doesn't have a basement.

References to pizza and its toppings, the theory went, were Clinton campaign code words denoting child sex trafficking. In this imaginary lexicon, "cheese pizza," for example, meant child pornography.

Pizzagate is the direct predecessor of QAnon, the false worldview that several years later caught fire among some Americans. Both delusions rest on a pastiche of ancient tropes. Secret satanic meetings, the ravaging, selling, and killing of innocents by bloodthirsty elites—Pizzagate and QAnon contained elements of blood libel, the hateful, centuries-old falsehood that Jews murder Christian children as part of religious rituals.

Days before the November 8 vote, Pizzagate popped up on the radar of Will Sommer, a reporter then working for the *Washington City Paper*, who had been covering the far right during the 2016 election and had been to Comet many times.

On Reddit, Sommer found photos of kids playing ping-pong at Comet, and commentary claiming the children were captives of Democratic predators. Someone had written: "What the actual fuck . . . Absolutely disgusted. No wonder Donald thinks the Clintons are animals."[7]

Sommer called Alefantis to ask, "'Hey, are you getting any weird responses to what's going on online?'" Alefantis laughed, ridiculing the atrocious spelling in the online posts and the pure absurdity of it. He thought the whole thing would fade after the election, when partisan tensions would ease.

The hoax surfaced on Facebook in late October, then spread to Twitter, helped on its viral way by online clout chasers followed by influential far-right outlets. Infowars was late to the game, but once Jones saw how much attention Pizzagate was getting, he jumped in with both feet.

"When I think about all the children Hillary Clinton has personally murdered and, and, chopped up and, and, and raped, I have zero fear standing up against her," Jones, quaking with menace, said on Infowars. "I just can't hold back the truth anymore. Hillary Clinton is one of the most vicious serial killers the planet's ever seen."

Jones urged his followers to travel to Washington and investigate the satanic abuse of children at a family-friendly pizzeria.

At the time, YouTube told the *Washington Post* that Jones's video had been viewed more than 427,000 times.[8]

JONES'S GROTESQUE CLAIMS drew the attention of Edgar Maddison Welch, a twenty-eight-year-old father of two young daughters, from Salisbury, North Carolina. Welch, who went by his middle name, Maddison, was estranged from his wife and lived in a home near his parents.

Slender, with a striking, angular face, shaggy blond hair, and a short beard, Welch's peripatetic life had included stints as a volunteer firefighter and local actor, an abortive pursuit of a college degree, and training to become an emergency responder. An evangelical Christian, he traveled on a mission trip to Haiti after the 2010 earthquake.

In October 2016, Welch was driving to work the night shift at a Food Lion warehouse when he hit a thirteen-year-old boy riding a bicycle. The child survived, but Welch's parents, who had lost another son, Maddison's brother, in a car crash, said Maddison's personality changed after the accident. He went from energetic and daring, a dad who played sports and video games alongside his girls, to morose and brooding.

Welch had just gotten internet service and spent hours watching conspiracy-themed YouTube videos on his cell phone. Pizzagate's false claims about tortured and wounded children struck a deep chord. Welch believed it all and searched for Comet's address.

On the morning of Sunday, December 4, 2016, Welch left his sleeping girlfriend at home with his girls and set off on the 350-mile trip to Washington. With him in the Toyota Prius that morning he carried an AR-15 rifle, a shotgun, a .38 revolver, and a folding knife. Forest and farmland spinning past the car windows as he sped to Washington, Welch, wearing a dark-knit beanie and a purple hoodie, chunky silver rings adorning the hand gripping the wheel, propped his cell phone on the dashboard and recorded what he seemed to believe would be a final message to his daughters.

"I can't let you grow up in a world that's so corrupt by evil, without at least standing up for you, for other children just like you," he said, struggling with his emotions. "Like I've always told you, we have a duty to protect people who can't protect themselves, to do for people who can't do for themselves. That's what I'm trying to do. I hope you understand that one day. I love all of y'all."

EXHAUSTED, JAMES ALEFANTIS HAD slept in a little that Sunday. He and his staff had been dealing with months of harassment. At one point, Comet staffers counted five #Pizzagate posts per minute on Twitter.[9] They were getting a total of fifty violent threats a day on social media and by phone. One caller told Alefantis he was coming to shoot him in the face and watch him bleed out on the restaurant's floor. It was a brilliant December day, and Alefantis had stopped at a holiday bazaar on his way to Comet when he got a call from Lisa Larson, the manager on duty. "She's sobbing, saying 'a guy's come in with a gun,'" Alefantis told me. "So I go racing to the restaurant and it was just this horrible, terrifying scene, where the whole block is shut down and the police are storming the building."

Welch had set off from North Carolina intending to "self-investigate" Pizzagate.[10] But once he reached Washington a different impulse took hold. Carrying the handgun and knife, brandishing the AR-15 assault-style rifle, Welch walked through Comet's front door, through the dining room, past the ping-pong tables, and to the back of the restaurant to begin his search.

Larson and some of the waitstaff had spotted him as he came in. They moved from table to table, quietly urging patrons to leave by the front door. Several refused, wanting to show support, until the staffers told them the guy had a gun.

As Welch passed the open kitchen midway through the long, narrow restaurant, with its dome-like brick pizza oven, Comet's salad cook on duty, a Howard University student, looked up and saw the rifle. He quietly left his station and followed Welch to the last room in the back, a high-ceilinged space hung with stage lights, where Comet hosted local bands and birthday parties. While the waitstaff emptied the front dining room, he hustled patrons from the ping-pong area out the rear door. When it was empty, he returned to the kitchen to guide any staff who hadn't already fled out the same way.

Left alone in the empty restaurant, Welch discovered its only locked door, leading to a makeshift storage closet near the restaurant's side entrance, whose walls stopped a couple of feet short of the ceiling. He tried to pick the lock with a dinner knife, and when that didn't work fired the rifle more than once, trying to shoot it off, the bullet entering the closet, which could have had someone sheltering inside. A staffer carrying a tub of pizza dough unknowingly entered through the side door, surprising the gunman. Welch swung his weapon around and pointed it at the head of the shocked young man, who dropped the vat and ran.

Welch scrambled atop a storage unit and peered over the closet's red-painted plywood walls, finding no one, and no basement entrance. Welch abandoned his search. He laid the rifle, handgun, and knife on beer kegs and a table across from the closet, walked back through the dining room and out the front door, surrendering to police, who made him lie starfished in the center of Connecticut Avenue, one of Washington's busiest thoroughfares.

Business owners and workers along a two-block strip of Connecticut Avenue were locked down inside their establishments for an hour. Alefantis joined his staff, who had been evacuated to a firehouse a block away. They waited for several hours while police cleared the scene and interviewed Comet staff, one by one.

It was late at night by the time Alefantis was allowed to reenter Comet. He stood in the center of the dining room. "The pizzas were in the kitchen window, food was half-eaten, beers half-full, but all the people were gone," Alefantis said. "At that moment I imagined what it could have looked like—instead of it being an empty room, it being a bloody, horrific scene."

The next day, customers called constantly, wanting to return to show their support. Comet's young staff urged Alefantis to reopen right away. "In my gut, I was like, 'I don't want to do this. I'm scared and this doesn't seem worth it,'" Alefantis told me.

They said they were scared, too. But they felt a responsibility to the community and an obligation to the truth, Alefantis said they told him.

The restaurant reopened two days after the gunman's visit, on Tuesday, December 6. A private security guard stood outside the entrance, and Metropolitan Police squad cars sat in front and in back. Scores of news cameras lined the sidewalk in front, and the queue of customers snaked down the street. "I'm at the front door saying, 'Welcome, c'mon in,' and I'm thinking, 'This is so fucked up and I can't trust anyone,'" Alefantis recalled. He spotted one of his dearest friends in the line, smiling, her three little girls in tow.

He teared up at the memory of it. "I'm thinking, 'Please don't come in here, and don't bring your children in here, because it is not safe.' And it turns out when we spoke months and months later, she was thinking the exact same thing," he said. But she said her girls wanted to go. Comet was their favorite place.

"There was this element of extreme disappointment. And the incentive to not ruin everything we've worked for," Alefantis told me. "One of my missions in Comet was that it would be this place where people go there every week, and when you grew up, you'd say, 'There's this pizza place I would go in Washington where I lived, and we would play ping-pong. And my dad was always so happy.' Sort of joyous family moments, you know?" I nodded, struck by the beauty of the thought.

"If we don't do this, we're going to ruin this for these kids. The social media companies can't take this away. These crazy people can't take it away."

WELCH TOLD MY *TIMES* colleague Adam Goldman from jail that "the intel on this wasn't 100 percent"—a pretty dramatic understatement.

Alefantis testified in court to the terror and emotional damage wrought by Welch and the online lies that spurred him to action. He said he hoped that "one day in a more truthful time we will remember this day as an aberration," a twisted period in history when "lies were seen as real and our social fabric had frayed."

Prosecutors argued for a significant sentence for Welch, "to deter other people from pursuing vigilante justice based only on their YouTube feed."

U.S. District Court judge Ketanji Brown Jackson sentenced Welch to four years in prison. She said sentencing had proved challenging because no one had committed a violent crime so damaging to so many people, prompted purely by internet delusion. At the time, anyway.

In a written statement to the court, Welch said, "I felt very passionate about the possibility of human suffering, especially the suffering of a child, and was prompted to act out without taking the time to consider the repercussions of my actions."

The apology troubled the judge. She couldn't tell whether Welch was sorry for his actions or sorry he didn't find any children to save.

"The fear is now that even though no one was physically harmed in this case, other people who are worried about other issues will take up arms with the intent of sacrificing lives in order to achieve what they believe is a just result," she told Welch in court.

"And as I'm sure you know, that kind of system of justice is utterly incompatible with our constitutional scheme, and with the rule of law."

Her fears were well founded.

ALEFANTIS'S LAWYERS SENT JONES a letter demanding that he retract multiple statements he made on Infowars between late November and early December 2016, spreading the Pizzagate theory and telling his audience, "It's up to you to research it for yourself," comments they said inspired Welch to bring his high-powered rifle into Comet.

After Alefantis's lawyers made it clear they were serious, Jones delivered a careful, legalistic statement on his March 24, 2017, broadcast.

"To my knowledge today, neither Mr. Alefantis, nor his restaurant Comet Ping Pong, were involved in any human trafficking, as was part of the theories about Pizzagate that were being written about in many media outlets and which we commented upon," he said.

"In our commentary about what had become known as Pizzagate, I made comments about Mr. Alefantis that in hindsight I regret, and for which I apologize to him. We were participating in a discussion that was being written about by scores of media outlets, in one of the most hotly contested and disputed political environments our country has ever seen. We relied on third-party accounts of alleged activities and conduct at the restaurant. We also relied on accounts of reporters who are no longer with us. This was an ever-evolving story, which had a huge amount of commentary about it across many media outlets." Infowars also removed videos of Jones's Pizzagate broadcasts, which had drawn millions of viewers.

"To my knowledge today . . ."
"We were participating in a discussion . . ."
"This was an ever-evolving story . . ."

Pizzagate was a viral lie, not an "ever-evolving story," or a "discussion."

Jones's shouting, weeping, on-air peddling of the lie, his exhortations of "God help us, we're in the hands of pure evil," contained a call to action that brought a deluded man with a rifle into a family restaurant, convinced he was saving children.

For Alefantis and his team, the fear, if not the memory, of Welch's visit would eventually ease. Facebook, Instagram, and YouTube began closing accounts and removing content using the word "Pizzagate." But the Pizzagaters hadn't gone away. They were devising other ways, new code words to discuss the imagined plot.

In late January 2019, Ryan Jaselskis, a twenty-two-year-old, Los Angeles–based model, actor, and wellness coach with a history of mental illness, entered Comet's back room during the dinner hour, doused a curtain used as a stage backdrop with lighter fluid, and eventually set it alight. Staffers put out the fire, and no one was hurt. Alefantis's new security cameras recorded the attack, and Jaselskis was captured a few days later after he assaulted a police officer near the Washington Monument. Law enforcement found a Pizzagate-themed video on his parents' YouTube account, posted hours before the Comet arson. In April 2020, Jaselskis was sentenced to four years in federal prison.[11]

Burning along the same social media fuse, and sparking on new platforms, Pizzagate begat QAnon, a new, more virulent mass delusion. QAnon, some of whose adherents see Trump as an avenging hero in a child-trafficking scheme led by Democratic politicians and Hollywood liberals, first appeared on 4chan around 2017, grew steadily, then surged during the coronavirus pandemic.

In 2020, *Times* technology columnist Kevin Roose described lurking[12] in QAnon Facebook groups and watching them "swell to hundreds of thousands of members," spreading misinformation about the coronavirus along with the claim that Hillary Clinton and liberals drink the blood of children. The FBI began the 2020 election cycle by warning that QAnon posed a potential domestic terror threat. The social media platforms cracked down, but the hoaxers adapted, using hashtags like #SaveTheChildren as camouflage.

In August 2021, a forty-year-old QAnon believer shot his ten-month-old daughter and two-year-old son in their chests with a spear-fishing gun. He told investigators that by killing them he was saving the world from monsters. QAnon followers have allegedly murdered a New York mafia boss, vandalized a church, and run for office by the score. One, Marjorie Taylor Greene, was elected a U.S. representative from Georgia. Greene has also shared posts on Facebook calling Sandy Hook a hoax, claims she has since renounced.

Will Sommer, now a reporter at the *Daily Beast*, calls QAnon a "re-branding" of Pizzagate.

"It's fair to peg the Comet arson to QAnon," he told me. "These kids are walking the same path tread by their Pizzagate forebears, and it all starts anew."

I VISITED JAMES ALEFANTIS on a summerlike afternoon in late October 2020, exactly four years after he first heard the word "Pizza-gate." I wanted to learn whether he still believed that what happened was an aberration, a localized seizure induced by a short circuit in our political consciousness.

Alefantis dug two bottles of mineral water from a refrigerator near the red door of the storage closet, where Welch had expected to liberate a huddle of terrified children but instead found a messy tangle of coats, mop handles, and computer cables. We walked out the rear door, through which Comet's panicked kitchen staff had fled, and where the dishwasher had surprised the gunman. Welch had just been released, and Alefantis thought he was still living in North Carolina. "I don't think *he's* coming back to Washington," Alefantis told me in a tone suggesting he hoped he was right.

We sat in the parking lot behind Comet, which Alefantis and his staff had transformed during the coronavirus pandemic into an outdoor dining area, with a bar and firepit, picnic tables for families, and an arbor space for quieter groups. Alefantis told me he had chosen not to live his life in fear. But at one point during our conversation, two men sat down nearby and Alefantis, eyeing them, asked that we move out of earshot.

Alefantis was in his mid-forties, with an open, almost innocent way and a clever wit. He was dressed like his young colleagues, in artist's black and sneakers, and he spoke like them, ending sentences with an upward lilt. A lifelong Washingtonian with many powerful friends, he retained an outsider's awe at the business he'd built over the past fifteen years, including three restaurants and a nonprofit art gallery called Transformer.

Alefantis and I were speaking in the wake of a new crackdown on QAnon by Facebook, Twitter, and YouTube. For the first time since 2016, he could go a day or two without any threatening messages. Still, every few weeks another of Alefantis's friends would call him, panicked. Pizzagaters had unearthed their connection to him and flooded their social media accounts with threats. Alefantis would carefully walk each through the drill: "Take down your Instagram, take those photographs off, go private—this is all we can do. You're probably safe, but here's how you report to the FBI. Here's how you report to the police. Follow all the protocols."

I asked whether he thought it would ever fully go away. He shook his head.

"Recently I've thought about changing my name," he said. He'd just opened the third restaurant. "I told the *Washington Post* food critic not to mention me in the review. Don't write my name. It's going to get out there, but at least when you google the new restaurant, it's not the first thing that comes up, and my staff isn't getting attacked."

Alefantis's lawyers went on to represent the family of Seth Rich, a young Democratic National Committee staffer gunned down in July 2016 in what police believe was a botched robbery attempt. Far-right conspiracists wrongly linked Rich to the theft of thousands of DNC emails published by Wikileaks during the campaign, speculating that Hillary Clinton ordered his killing in retribution. Police, Rich's family, and federal investigators repeatedly debunked those false claims. But the theories persisted, spread by a coterie of Trump confidants, including Roger Stone and Sean Hannity, the Fox News commentator and Trump whisperer. Rich's parents sued Fox News. The Second U.S. Circuit Court of Appeals in Manhattan ruled that Fox's exploitation of their son's murder amounted to a campaign of emotional torture. In late 2020 the case was settled.[13] Neither side disclosed the financial terms.

I asked Alefantis whether it helped to enlist lawyers in his case.

"Maybe for Jones," Alefantis said. "I do think he deserves to be taken down. He's very, very dangerous, and people like him are very dangerous.

"But the way the internet works, it's constantly repopulating. I could pay lawyers or whatever to take everything down. But that doesn't mean anyone's going to stop sharing it."

Alefantis paused, flipping his iPhone, encased in vivid green plastic, over and over in his hands. He is a chef and restaurateur, not "the face of fake news," he told me. He didn't know much about how social media works until it nearly destroyed his life.

"These people are in business. Their business is to sell you products. And if there's no product to sell, they're going to go out of business. And that's it. It's as simple as that. So they have no interest in policing at all. Pizzagate was a huge seller."

We had been talking for a couple of hours. The first families in Comet's weeknight dinner crowd had drifted into the parking lot, where the firepit blazed. The sun no longer lit our table, and the breeze had picked up. We stood and passed through the rear door. Alefantis again showed me where Welch had fired, climbed, and left his weapons. We walked to the room crowded with chairs and stage lights in the back, where Jaselskis had set the stage curtain aflame. The restaurant was full of families with little kids.

We passed through the front door and onto Connecticut Avenue. The wind, stiffer now, stirred burnished gold mums in pots along the street. I gathered my sweater around myself. Alefantis walked with me a few paces south.

I asked Alefantis what it had cost, the new panic button behind the host station, a full-time security guard for a year, the cameras, the security assessment by Gavin de Becker, author of *The Gift of Fear*, who trains people to trust their gut when it comes to other people's evil intentions.

"It's hard to tell," Alefantis said, adding in his head. "The security guards alone were one hundred and fifty thousand dollars, at least. And the security company things, seventy-five thousand dollars. Another seventy-five thousand dollars . . . fifty thousand." He stopped.

"It's cost a lot. But we're still here. There was a part of me that wanted to win. The winning was about truth over fiction," he said.

He turned from watching the street to look at me. "Those things for me are not Zuckerberg talking points," he said. "I believe them. I live them."

Five days after our talk, Alefantis sent me an urgent-sounding email. "There is a very important detail about Alex Jones and how he relates to Comet," he wrote. "Can I call you? . . . I haven't talked about it before."

We spoke an hour later. Alefantis had recalled a critical episode a week after Welch's arrest. He'd been working late and returned exhausted to his duplex in Washington's Logan Circle neighborhood. Collapsing on the couch, he brought up the newly filed complaint, *United States of America v. Edgar Maddison Welch,* on his phone.

As he read, Alefantis learned that two days before walking through Comet's door, Welch sought his friends' help "raiding a pedo ring, possibly sacraficing [*sic*] the lives of a few for the lives of many. Standing up against a corrupt system that kidnaps, tortures and rapes babies and children in our own backyard . . . defending the next generation of kids, our kids, from ever having to experience this kind of evil themselves. I'm sorry bro, but I'm tired of turning the channel and hoping someone does something and being thankful it's not my family. One day it will be our families. The world is too afraid to act and I'm too stubborn not to."

On page 4 of the complaint, Alefantis read text messages gathered by the FBI from Welch's cell phone. He learned that on December 1, 2016, three days before Welch traveled to Washington, he sent a text message to his girlfriend, telling her he'd been researching Pizzagate and it was making him sick.

Welch had watched Pizzagate videos for hours that day, YouTube algorithms sending him one after another. He visited Comet's website.

That night after 8:00 p.m., Welch texted a friend a link to a YouTube video, writing, "Watch PIZZAGATE: The Bigger Picture on YouTube."

Alefantis searched for the video. It was an Infowars broadcast.

"I'm like, 'What the fuck?'" Alefantis told me. "Like, holy shit, this fucking gunman came here because of Jones!

"I called the attorneys and said, 'Do whatever you need to get Jones.'"

He held for a beat, letting me take it in. "So there you go. I hope that helps."

17

BY LATE SUMMER 2016 the Clinton campaign faced a decision similar to what James Alefantis and Seth Rich's family had struggled with, as falsehoods sparked by Clinton's candidacy caught fire across the internet. Clinton had met some of the Sandy Hook relatives and knew about the torment fed by Alex Jones. Erica Lafferty, the daughter of Dawn Lafferty Hochsprung, the slain Sandy Hook principal, had appeared in an ad for Clinton's campaign.

Should the Democratic presidential nominee fight back against the internet-borne hate fostered by Trump? Or would that just feed the trolls? How much of this was harmful but idle rhetoric, and how much posed a real threat?

In hindsight, the Republican National Convention in July provided plenty of clues.

The convention in Cleveland marked a debut of sorts for Alex Jones, a former political outsider whose blossoming bromance with Trump surprised his own staff. Jones's popularity had surged in the years since Sandy Hook. His frenzied defense of gun ownership after the shooting had drawn mainstream media notice. Independent audits suggested that average monthly page views for Jones's two main websites had more than doubled between 2013 and 2016,[1] to roughly fifty million views per month.[2] Infowars' YouTube videos had together been viewed more than one billion times.[3] His radio show, aired on 129 terrestrial radio stations and simulcast on Infowars' website, drew a daily listenership of about five million.[4]

To kick off the convention on July 18, Jones sponsored the America First Unity Rally in a park on the Cuyahoga River downtown. For five hours, alt-right cranks, trolls, and conspiracists came to the stage to hail their new standard-bearer, Donald J. Trump.

"Hillary for prison!" Jones yelled into the microphone. Standing onstage in a light breeze, the river glistening behind him, he looked like a country club Republican in khakis, a navy blue sport coat, a crisp white shirt, and dark Ray-Bans. Infowars-branded HILLARY FOR PRISON T-shirts adorned many in the crowd.

"Donald Trump is surging," Jones exulted. "If you think the awakening we've seen so far is big, this planet and the globalists haven't seen anything yet!"[5]

I was covering the convention as the national politics writer for the *New York Times* editorial board. People in the park that day shared absurd beliefs that Hillary Clinton would eliminate the Constitution, that most American slave owners had been Black, and that America is "the last free nation, because Europe has been taken over by Muslims."

I asked one attendee, a retired air traffic controller named Dennis Ragle, where he got his news. Mostly from Facebook, he told me. I asked him what he thought Trump meant when he said President Obama's "body language" when he speaks about police shootings suggests "there's something going on."

"I'm not much of a conspiracy theorist," Ragle said. "But there could very well be something going on."

Republican politicians to that point had generally avoided people who held the views trumpeted from the stage that day. In a town hall on the presidential campaign trail in 2008, John McCain had famously chastised Republican voters in Minnesota who declared themselves "scared" of a Barack Obama presidency, including an older woman who said she distrusted Obama, calling him an "Arab." "No ma'am, no ma'am, no ma'am, no ma'am. He's a decent family man, citizen, that I just happen to have disagreements with," he said, gently taking the microphone from her.

That was then. Trump, who ventured into politics in 2011 by amplifying the racist "birther" lie about Obama's citizenship, announced his campaign with a slur against Mexicans, and ran on dark fears of globalist takeover, had ridden paranoia and nativism to the nomination. At the start of the 2016 campaign, Trump had been an outlier candidate in a crowded field. Needing an edge, he pulled conspiracy-minded, anti-establishment people suspicious of politics off the sidelines and into his fold.

Trump's appeal had little to do with traditional Republican values. "He was able to convince people of things they already wanted to believe," the conspiracy scholar Joe Uscinski told me.

In this, Trump was very much like Alex Jones.

Like a shark that must keep swimming to breathe, Jones would go belly-up without attention. At the convention he gorged on it, strutting through the crowds trailed by a half-dozen bodyguards, a cameraman, and fans. Jones had teamed up for the occasion with Roger Stone, erstwhile Trump campaign adviser who had connected Jones with Trump. Stone was exiled by the establishment GOP in the mid-1990s, when, while working as an unpaid adviser to Bob Dole's presidential campaign, it emerged that Stone and his wife were cruising for sex partners with explicit ads. Ever since, Stone had pushed fringe candidates and hatched wild plots, a jackal circling the campfire's light, sniffing for an opening.

Stone and Jones, committed conspiracists, had met in 2013 at a gathering in Dallas marking the fiftieth anniversary of John F. Kennedy's assassination. They made a strange pair in Cleveland: the manic Texan stuffed like sausage into his shirt, given to spontaneous fury and tears, and the anachronistic dandy in white suit and two-tone wingtips, his murine visage stretched into a rectangular, joyless smile, his white hairline a stark, straight-edge row, like crops planted atop the parched clay of his forehead.

In scenes reminiscent of high school, when Jones preferred being beaten up to being ignored, he looked for trouble at the convention that he could parlay into viral videos. He dove into a group of peaceful communist demonstrators and had to be rescued by police, and almost came to blows with liberal commentator Cenk Uygur when Jones and Stone crashed his show. One of Jones's sidekicks, Joe Biggs, a hard-drinking, profane Army veteran, got into a fight with a flag-burning demonstrator, drawing a lawsuit. A few years later, Biggs would join the Proud Boys extremist group and help lead the Capitol insurrection on January 6, 2021.

"Every day it seemed like there was some type of stunt," said Josh Owens, a former Infowars video editor who filmed Jones's convention antics and uploaded them to YouTube. "Basically he saw the RNC as his playground."

On the evening of Thursday, July 21, Jones stood on the convention's main floor, a VIP invitee to Trump's nomination acceptance speech. Owens stood next to Jones as the nominee strode onto the gilt-bedecked convention stage, "TRUMP" projected in gargantuan letters behind him.

Trump's acceptance speech sounded at times like an Infowars broadcast. "Our convention occurs at a moment of crisis for our nation. The attacks on our police, and the terrorism in our cities, threaten our very way of life," Trump said,[6] later adding: "Americanism, not globalism, will be our credo."

Owens looked over at Jones. "He was crying, genuine tears streaming down his face," he told me.

"Trump was what Jones had been working for, for years."

DONALD TRUMP AND ALEX JONES echoed each other all that summer, sowing suspicion of Islam, undocumented immigrants, and the Clintons.

"There is only one core issue in the immigration debate and it is this: the well-being of the American people," Trump said in one of his speeches. "We have no idea who these people are, where they come from. I always say Trojan horse. Watch what's going to happen, folks. It's not going to be pretty."

Jones praised Trump's words as "sure to win him support from those who've been conned by the lying media into thinking he's some evil demon creature, when the truth is he's a man with a heart of gold."

Some among Trump's core campaign staff hadn't known who Jones was until Trump went on his show in late 2015. Trump seemed oblivious to Jones's attacks on the Sandy Hook families, but in any event it would not have benefited him to criticize. Instead, that summer Trump invited "Angel Moms," women whose children were killed by undocumented immigrants, onstage at a rally to bolster his wildly false claim that "thousands" of Americans were killed by undocumented immigrants every year.[7]

"Mr. Trump is the only one who can protect your children from being slaughtered like my son was," one of the moms, Agnes Gibboney, said at a rally in Texas.[8]

Study after study demonstrates that the violent crime rate is actually lower among undocumented immigrants than among native-born Americans. Only 8 percent of Americans ranked immigration as "the most important problem facing this country today," according to a Gallup poll taken at the time.[9] But that thin slice of the electorate included Trump's staunchest supporters, many of whom viewed immigrants as an existential threat.

Trump's reality-bending claims and combative appeals to bigotry alarmed Americans of both parties. On August 25, 2016, in Reno, Nevada, Hillary Clinton scrapped plans to talk about her economic agenda and delivered what became known as her "alt-right speech."

Clinton spoke from a stage at Truckee Meadows Community College, backed by a grouping of American flags.

"He is taking hate groups mainstream and helping a radical fringe take over the Republican Party," Clinton said of Trump.

"A man with a long history of racial discrimination, who trafficks in dark conspiracy theories drawn from the pages of supermarket tabloids and the far, dark reaches of the internet, should never run our government or command our military," Clinton said, as the audience burst into applause. "Ask yourself, if he doesn't respect all Americans, how can he serve all Americans?

"This is someone who retweets white supremacists online," she added. Clinton reviewed a litany of falsehoods amplified by the Republican nominee.

"You remember he said that thousands of American Muslims in New Jersey cheered the 9/11 attacks. They didn't.

"He suggested that Ted Cruz's father was involved in the Kennedy assassination," she said, as some people laughed. "Now perhaps in Trump's mind, because Mr. Cruz was a Cuban immigrant, he must have had something to do with it. And there is absolutely, of course, no evidence of that.

"Just recently, Trump claimed that President Obama founded ISIS. And he has repeated that over and over again."

Trump had said both Obama and Clinton founded ISIS—but Jones had said it first.[10]

Clinton toggled between ridiculing Trump and sounding the alarm. Trump's claims were so wild, so previously inadmissible, it seemed impossible that anyone would believe them, or elect someone who did.

"His latest paranoid fever dream is about my health. All I can say is, Donald, dream on," she said, as the crowd laughed and whooped and started a chant of "Hillary, Hillary."

Then Clinton homed in on "the radio host Alex Jones, who claims that 9/11 and the Oklahoma City bombings were inside jobs.

"He even said—and this, really, is just so disgusting." Clinton's voice dropped in revulsion, making sure that her audience grasped Jones's most loathsome claim.

"He even said the victims of the Sandy Hook massacre were child actors, and no one was actually killed there," she said, as a high-pitched exclamation rose from a few people. "I don't know what happens in somebody's mind, or how dark their heart must be, to say things like that.

"But Trump didn't challenge those lies. He actually went on Jones's show and said: 'Your reputation is amazing. I will not let you down.'

"This, from the man who wants to be president of the United States."

THE "RADICAL FRINGE" ATE IT UP.[11] Beneficiaries of the attention economy, it thrilled them that the Democratic Party standard-bearer had given them the time of day. A couple of them live-tweeted Clinton's speech, hoping she would name them individually. White-nationalist-friendly website VDare launched a fundraising drive, saying, "Hillary wants to ignite a witch hunt against the alt-right because she knows we are finally starting to make an impact on the public's thinking about immigration." White nationalist Richard Spencer, credited with coining the term "alt-right," gave yet another round of coy interviews, objecting to the suggestion that racist groups like his National Policy Institute promote violence.

Critics on the left thought Clinton either should have ignored the issue—"Don't feed the trolls," again—or called out more of them by name. Nobody knew how to counter this truth-bashing, divisive force in mainstream politics, and the swath of the Republican electorate that unabashedly embraced it. The term "alt-right" began to disappear from political discourse, slammed by anti-racism groups and media critics[12] as too antiseptic for describing the hatemongers and neo-Nazis gathering under its umbrella. That too seemed a sign that nobody knew what to do.

Alex Jones loved being singled out. He aired a "special report" after Clinton's speech, basking in her recognition of him. "Alex Jones? Little old me is one of the main opposition points against these monsters?" he crooned.

Jones replayed Clinton's speech during his broadcast, providing live commentary.

At the part where Clinton spoke about Sandy Hook, Jones grew excited.

"Wow! How dark my heart must be? My black heart? Baby! It's big, juicy, and red, and you know it. This is ecstasy, actually. We've really got her here," Jones said.

"I didn't say there were child actors! That's lies and twisted disinformation. Some people say it's all child actors. That's not what I'm saying.

"She's such a liar."

Unsurprisingly, it was Jones who was lying.

Jones had of course said that the Sandy Hook victims and their families were actors, sometimes several times in a broadcast. Here's what he told a caller on his radio show on January 13, 2015:

"Yeah, so, Sandy Hook is a synthetic, completely fake with actors, in my view, manufactured. I couldn't believe it at first. I knew they had actors there, clearly, but I thought they killed some real kids. And it just shows how bold they are, that they clearly used actors."

Pointing that out wouldn't stop Jones. Infowars was drawing more traffic to its website at the time than the renowned fact-checking site PolitiFact.[13] Jones wasn't wooing people interested in truth.

OFF CAMERA, JONES BROODED about what a Trump victory might do to his business. Trump seemed like a boon, but there was that Republican-in-the-White-House audience slump to consider. Jones had been hearing from libertarian listeners and some friends, telling him his shilling for Trump had dinged his image as an establishment skeptic. But once Trump clinched the nomination, Alex Jones and the mainstream GOP were inextricably linked.

As the weeks ticked down to Election Day, Jones grew more erratic.

"He would come in and say, 'Everyone better find a new job, because once Trump is president we're not gonna be making money anymore,'" Josh Owens, Infowars' young video editor, told me. "He was struggling to navigate what the future would be like for him, in the mainstream, attached to the president of the United States. It was new territory."

Jones was hard on his staff, and this inner turmoil didn't make things any easier. "We used to call Jones the energy vampire," Owens said. Jones would summon staff to his house in the middle of the night, to film an alcohol-fueled rant that had just popped into his head. They'd hustle over, but Jones would have gone out again, and they'd have to wait for hours for him to return.

Buckley Hamman, Jones's cousin and production manager, warned young staffers not to trigger Jones. "If Alex sees you screwing up over and over, he starts to dislike you. And when he starts to dislike you, you don't have a future here, I'm telling you right now," Hamman said in a staff training meeting. "If something's going wrong, it's not working, just be very calm," he added. "Keep him calm, or he'll pick up on that and you'll have a bad day. He's like a dog that smells fear."

Infowars employed about seventy-five people then, some inside the 7,500-square-foot headquarters building and its seven studios, and the rest in the products warehouse across the back parking lot. Very few staffers had offices, including Jones, who liked to wander around. A hive of desks snaked around the control room and studios, expanding that year into former warehouse space. Lighting was low, and most walls and ceilings were darkly paneled or painted black. Jars and bottles of Infowars "neutraceuticals" sat on most desks, since Jones insisted his staff try his products.

The place had the aura of a college dorm on move-out day. A reception area inside the front door was furnished, if you could call it that, with a mismatched assemblage of office chairs and faux-wood Formica-topped tables, littered with carryout boxes and coffee and soft-drink cups. Posters and banners with libertarian and pro-gun slogans like *Molon Labe*—Greek for "come and take them," a challenge popular among some gun owners—blared from the walls, and from the T-shirts of several employees. Several staffers wore sidearms, but that was not unusual for Texas.

The office air-conditioning was set to sixty degrees year-round. Jones tended to overheat and hated looking sweaty on camera. Some staffers stashed space heaters under their desks and worked in jackets and wool beanies. In that summer of 2016, Smashing Pumpkins frontman Billy Corgan, a Jones pal, flew in for a twenty-eight-hour "Operation Sleeping Giant" Infowars donation drive, a cash grab patterned on Ron Paul's "moneybomb" fundraisers. Corgan spent almost as much time grousing about the office temperature as he did about Hillary Clinton.

Jones often wandered about the office shirtless. A former employee shared a photo of him, his bare, woolly torso parked behind a conference table, shoveling down carryout. He'd wash it down with beer or vodka, wiping his mouth on the back of a thick hand.

Josh Owens is a stocky man with shaggy hair and beard and heavy-lidded eyes that make him look thoughtful, or sad. He first heard about Alex Jones as a teenager, talking with a buddy about the scene in Stanley Kubrick's *Dr. Strangelove* in which Sterling Hayden, as the deranged Brigadier General Jack D. Ripper, tells Peter Sellers's Group Captain Lionel Mandrake that water fluoridation "is the most monstrously conceived and dangerous communist plot we've ever had to face." Owens's friend recommended Owens check out Alex Jones, who made similar claims.

Owens was a twenty-three-year-old film student at the prestigious Savannah College of Art and Design in 2012 when his entry in an Infowars video contest won him a job offer. Owens dropped out of school and drove to Austin. "I went there as a directionless, mostly ignorant kid, thinking, 'Why make a movie when you can go live in a movie?'" he told me. He didn't much care about Jones's views. Jones's theatrical gift was the draw.

Owens wrote a regret-filled memoir about his time at Infowars for the *New York Times Magazine* in 2019.[14] He told me *Times* editors had urged him to describe Infowars' editorial process, and how he had struggled to explain it didn't exist. "It's like they have an assumption and they never ask questions. It's just assertions."

Jones's obsession with Robbie Parker's press conference likely had similar roots, Owens told me. "Jones comes into the office in the morning, walks by Rob Dew's desk, and Rob says, 'Hey, come over to the computer and look at this.' It's whatever Rob Dew says and what Jones thinks at the moment."

Still, Jones treated him kindly. On Owens's first day in Austin, Jones took him to see *Iron Man 3*. Through his pal Dave Mustaine, he got Owens a backstage pass for a Megadeth concert. Owens met Jones's friends Mike Judge, creator of the animated TV series *Beavis and Butt-head*, and fellow 9/11 conspiracists Charlie and Martin Sheen.

Owens recalled the stunt that earned him Jones's trust. Not long after Owens started at Infowars, Anthony Gucciardi, Jones's new partner in the supplements business, and Weldon Henson, Infowars' products manager, flew to Utah to visit a manufacturer that bottled nascent iodine drops under Infowars' label.

Gucciardi was a twenty-one-year-old college dropout writing articles about natural healing online when he linked up with Jones in September 2013 to broker new supplements lines. Nascent iodine was one of Infowars' biggest early sellers, pushed by Jones to counter radiation in the atmosphere and "support healthy hormone levels." Gucciardi earned $2 for every bottle delivered to Infowars' warehouse, raking in $300,000 in the last three months of 2013 alone, he estimated in court testimony. After Jones's marriage fell apart in 2013, he moved next door to Gucciardi in the Ashton, a tony building in downtown Austin, paying $13,000 a month for a penthouse.

Owens was supposed to film the iodine plant for an Infowars ad, but its managers wouldn't allow them inside. Jones called with another manic idea: infiltrate the National Security Agency's Utah Data Center. The sprawling secure facility, code-named Bumblehive, stored data collected by the American intelligence community. Its secrecy and function made it a prime target of speculation by conspiracists and ordinary Americans worried about unlawful collection of their personal information.

The trio drove 140 miles west to the center's base in the U.S. Army's Camp Williams. Ignoring the No Trespassing signs, they made it to an employee parking lot. Owens had begun filming when security guards surrounded them. Told to erase the video, "I lied to them and uploaded the footage online," he told me. "Jones got a lot of mileage off that, and that's the moment he kind of started to like and respect me."

A big part of Owens's job was making Jones look better on camera. "He was always like, 'Hold the camera higher,'" to hide Jones's double chin. Owens remembers editing "before" and "after" photos of a shirtless Jones flexing for an ad for Infowars' Tangy Tangerine weight-loss supplement, which Jones claimed helped him lose thirty-seven pounds in two months.

Gingerly, Owens told Jones, "'I dunno if these photos are a great indicator that anything's changed,'" Owens recalled. "And he was like, 'I don't care.' In his mind his pants were fitting a little looser." HBO's John Oliver would later resurface those photos, ridiculing Jones for looking exactly the same except that in the "after" photo his skin was redder.[15]

"I tried to do everything he wanted and keep him happy," Owens told me. "So I did a lot of stuff I will regret forever."

Rob Jacobson worked in Infowars' control room. He had worked with Jones for a decade and said that he had never seen him shill for any politician like he did for Donald Trump.

"He actually did a video of himself voting in the Republican primary, for the first time in seventeen years or something," Jacobson told me.

Jones for years had disparaged his talk show rivals Glenn Beck and Rush Limbaugh, calling Limbaugh a corporatist "whore" for serving up the race-baiting, anti-feminist commentary his establishment Republican masters wouldn't utter in public.[16]

While Jones slammed Hollywood culture as the epitome of evil on his show, he chased celebrities "like a little girl screaming at the Beatles. It was an inside joke," Jacobson told me. Jacobson, who is Jewish, left Infowars in 2017 and filed a discrimination claim with the Equal Employment Opportunity Commission. Jacobson says the case ended inconclusively, and Jones's lawyer said he won.

"He was jazzed by Donald Trump and Roger Stone," Jacobson told me. "He was thinking, 'I'm stepping into the next dimension here. I'm rolling with Donald Trump and Roger Stone, and we're heading to the White House.'"

Jones punched walls and file cabinets, threw a microphone down, and knocked a video monitor off Rob Dew's desk, breaking it, Jacobson told me. He yelled at employees, including his father, David. "He was definitely a victim at times of Jones's rage," Owens told me. "He seemed to take it. Jones is who he is partly because of nurture."

Owens relayed an incident that he documented in the *Times Magazine* article. Jones had wandered, shirtless, into the communal space Owens shared with Dew and two other employees and poured himself a drink from a bottle of Grey Goose vodka he kept in a storage cabinet.

"Hit me," Jones told an employee sitting near Owens. It was a familiar game, which staffers knew never ended well. "*Hit me*," Jones said, repeating the command until his resigned target delivered a half-hearted punch to his tricep.

"*Harder!*" Jones demanded. The staffer slugged him again.

On Jones's turn, he threw his bulk into the punch. Back and forth it went, Jones growing wilder, until he slammed his beefy fist into the man's arm with such force that his flesh gave way. Owens watched "blood dripping down the guy's arm," he told me. A week or so later, the man showed Owens his still-healing wound.

Few dared protest publicly.

Jones had named his parent company Free Speech Systems LLC, but free speech applied only to him. Infowars employees were required to sign a lengthy nondisclosure agreement restricting them from disclosing any information or materials related to Infowars or Jones's family, "from this date to infinity."

Owens had signed Jones's NDA and sought legal advice before writing the *Times* article. He told me his attorney said: "As a lawyer, I'm saying you can't do it. But as a person, I'm saying you have to."

ELECTION DAY, NOVEMBER 8, 2016, found Owens in the passenger seat of Jones's new Dodge Hellcat. Jones was driving them to a polling station in an Austin strip mall to make a video claiming that he had been turned away in a Clinton-led voter suppression conspiracy. They both knew cameras were prohibited in polling places. It didn't matter: "Alex Jones Denied Right to Vote" would be the first of many phony election-rigging claims Jones had made on Trump's behalf.

Jones drank vodka from a paper cup while Owens streamed the drive on Facebook Live. After they were tossed from the polling place, Owens gratefully accepted a ride back to Infowars HQ from another staffer, who wasn't drinking.

As the vote counting began that night, Roger Stone joined Jones on the Infowars broadcast desk in Austin. They dressed in somber-toned suit jackets and blue shirts, Stone's with a deep blue necktie tied in a double-dimple knot.

"It was a moment of enormous tension. I'm working the phones very aggressively, working my contacts," Stone told PBS's *Frontline*, which revisited election night inside Infowars in a documentary, *United States of Conspiracy*.[17] In fact, Stone was intensely disliked by many in Trump's roiling inner circle, but he usually wormed his way back in with extreme demonstrations of loyalty.

The mood was sedate at first, staffers pacing themselves for a long night. Then Trump started winning key states, and "There was this loud chatter, like, 'I can't believe this,'" Jacobson recalled. Infowars staffers gathered around the on-set television. Infowars' audience swarmed online, watching Trump's emerging upset play out on Jones's face.

"The metrics were off the charts about how many people were tuning in to Infowars. Numbers that were comparable to the networks," Morgan Pehme, a creator of the 2017 documentary *Get Me Roger Stone*, who was in the Infowars studio that night, told *Frontline*.

At about 1:30 a.m. Austin time, Jones and Stone tapped champagne flutes. Trump had clinched victory. "Love you," Jones told the older man, who frowned at the fratty, keg-party display. Choked up, Jones gestured at his staff with his glass: "Love you guys, love all of you."

"There was a combination of elation and confusion," Pehme said. "They realized that, 'oh my God, we just played a role in making the president of the United States.'"

Infowars staffers howled and ran through the studios. They climbed onto the Infowars set, unfurling maroon T-shirts that read TRUMP IS MY PRESIDENT, refilling the two hosts' glasses. Joe Biggs, wearing a Trump T-shirt and hat, bellied up to the desk, puffing a stogie.

They watched Trump come to the stage in New York to declare victory.

"Here goes Donald Trump," Jones said, his voice sounding muffled by the sides of his cheeks. He stared into the bubbles inside his glass and heaved a growling sigh. "I'll tell ya, he's chargin' into a goblin's nest."

Stone murmured, "This is amazing."

"As long as he dudn't, long as he dudn't kiss a goblin," Jones said, hunched over the desktop microphone, his champagne flute listing toward his left ear.

"I don't think there's any danger of that," Stone told him.

"No, he defeated the goblins. He did it, HAH HAHA!" Jones said, throwing his head back in theatrical mirth, then reaching over to toast Stone again.

As the evening ended, Jones sat alone at the broadcast desk, locking onto the audience for a messianic, 3:00 a.m. soliloquy.

"We are bound from heaven. And if we don't deliver this plan and free humanity, we will be bound to the ninth circle of hell," he intoned, holding up a clenched right fist, his face working furiously. A tear escaped his left eye, streaking his cheek.

"I'm bound to this truth, and I will never stop delivering. I've said it. I want to run that course.

"I already know my entire life purpose has been completed. I will continue on," he said, rising from the anchor chair and unbuttoning his suit jacket, preparing for a dramatic exit.

"Now I realize—I've won."

18

NEWTOWN'S LEADERS FEARED that President Trump's conspiratorial supporters would cross the weakening membrane between the online and real worlds, as the Comet gunman had in Washington.

Town and state officials still received hate mail, calls, threats, and visits from people "investigating" Sandy Hook. If anything, the Sandy Hook hoaxers' newfound proximity to power had emboldened them. When Hillary Clinton mentioned Alex Jones in connection with the Sandy Hook hoax, Wolfgang Halbig had complained that he, not Jones, deserved the credit.

Newtown leaders wanted to write to the new president, asking him to disavow Jones and the hoaxers plaguing the town.

Newtown had voted only narrowly for Hillary Clinton, who bested Trump by fewer than 300 votes, with a total of 7,448.[1] Newtown officials haggled for weeks over the wording of the letter, laboring to eliminate any language suggesting partisanship. They circulated it among several town boards for review. Board of Education member Daniel Cruson Jr. worried that sending the letter could "feed the hoaxers" or attract unwanted media attention after officials had "tried very hard to convince the media to leave us alone."

In a meeting, Dan Gaston, Newtown's Board of Finance vice chair, sought to allay concerns that sending the letter was a political act. "We're not asking to pass a law, we're asking for support. We received support from President Obama, now we're asking for support from President Trump," he said, according to the *Newtown Bee*.[2] He pointed out that the abuse had been going on for four years at that point.

"If you're being bullied you don't sit there and take it," he said.

The families approved the letter draft that the Board of Education would send to the White House on February 20, 2017. It had taken a month to reach consensus on a single-page letter to the president. It was a surreal endeavor: a small town scarred by a mass murder, begging the President of the United States to tell America that they didn't make it up.

Newtown's plea read:

> *As a town, we continue to mourn, question why and try to find the kindness and goodness that we believe is out there.*
>
> *One of the significant roadblocks to our future is the continued rumors and viciousness spread by many people outside of our homes who believe that our tragedy was a staged government event that never happened, that the children and educators we lost never existed.*
>
> *We are asking you, as the new President, to help in putting a stop to these horrific lies and demonization of the adults who so bravely protected our children and died doing so. One perpetrator of these lies is radio host Alex Jones. He continues to spread hate and*

lies towards our town, towards the people and organizations who came to help us through those dark days. Jones repeatedly tells his listeners and viewers that he has your ears and your respect. He brags about how you called him after your victory in November. He continues to hurt the memories of those lost, the ability of those left behind to heal.

We are asking you to acknowledge the tragedy from 12/14/12 and to denounce anyone spreading lies and conspiracy theories about the tragedy on that December morning . . . We are hopeful that with your help these lies will end. Please clearly and unequivocally:

—Recognize that 20 children and 6 adults were murdered at Sandy Hook Elementary school on December 14, 2012; and
—Denounce any and all who spread lies that the tragedy was a hoax; and
—Remove your support from anyone who continues to insist that the tragedy was staged or not real.

We know that you have many and un-ending responsibilities as the President, but we are hopeful that with your support our town can continue to find ways to move forward and try to heal.

Respectfully,
Newtown Board of Education
Keith Alexander, Chair

They sent the letter and waited. Trump never answered.

THOUGH PRESIDENT TRUMP WASN'T about to rein him in, Jones's new notoriety was bringing a slow reckoning. As Lenny had discovered, no amount of shaming or debunking would dissuade a fixture of the conspiratorial-industrial complex like Jones. But two other tactics showed promise: legal action and cutting off Jones's social media oxygen supply.

James Alefantis's legal threat forced a stilted apology from Jones. But Alefantis's lawyers also demanded that Jones remove his Pizzagate videos, including the one the Comet gunman sent to his friends, from YouTube and other platforms. This was serious: Infowars posted its broadcasts to social media, where they garnered tens of millions of views. In 2017 Jones and his staffers boasted a total of eighteen YouTube channels. The biggest among them had racked up 1.2 billion views.[3] On YouTube three "strikes"—meaning violations of its community guidelines—within ninety days would get a channel permanently removed.[4]

Jones could not afford to get crosswise with the social media giants. Tormenting the parents of murder victims and endangering the patrons of a restaurant didn't seem to bother him much. But getting booted from social media would cost him money.

Infowars' internal emails demonstrate over and over how Lenny's copyright notices against Jones and his moves against the other hoaxers prompted Infowars to recalibrate to avoid sanctions. After Jones slammed and doxxed Lenny in response to Lenny's copyright strike on Infowars' broadcast on Noah in Pakistan, the memo had gone out warning Jones's staffers not to upload the broadcast to YouTube. Infowars nuked its article about Amazon yanking Fetzer's *Nobody Died at Sandy Hook* book within a day, likely to protect its deal to sell Infowars supplements on the e-commerce behemoth.

On the day Lenny and Veronique's op-ed got James Tracy fired, Rob Dew, who with Jones was the Sandy Hook lie's leading internal booster, covered the news on Infowars without saying Tracy's name or Noah's. "I'm not even gonna mention the son's name, the son that they're referring to, because we've gotten letters from this son's dad in the past," Dew said on the show.

Infowars' online "article" about Tracy's firing was even more careful. It relied on mainstream news sources, did not repeat Tracy's false claims, and included a disclaimer:

> Editor's note: Infowars offers its support and prayers for all the victims of gun violence. This article isn't an endorsement of Professor Tracy's claims about Sandy Hook.

It almost looked like real journalism.

When Jones was a fringe phenomenon, he got away with plenty. But his association with Trump brought him a vast new audience and a lot more scrutiny.

A MONTH AFTER JONES'S on-air apology to Alefantis, he landed in more legal trouble. Chobani operates a yogurt plant that is the largest employer in Twin Falls, Idaho, a city of forty-nine thousand, south of Boise. Former refugees make up nearly one-third of the workforce, in keeping with the philosophy of Chobani's founder, Hamdi Ulukaya, who immigrated to the United States from Turkey: "The minute a refugee has a job, that's the minute they stop being a refugee."[5]

Plans to resettle Syrian refugees in Twin Falls had already stirred Islamophobia, when in mid-2016, two refugee boys, seven and ten years old, were implicated in an incident of sexual misconduct involving a five-year-old American girl in an apartment building in Twin Falls. Details were initially sketchy, in part because the incident involved underage children. Online anti-immigration groups filled the void with wild, xenophobic claims, falsely casting the incident as a "gang rape" by Muslims of a child "at knifepoint."

The online attack against Chobani and Ulukaya was led by Breitbart

News,[6] where Stephen Bannon was a top editor and Trump campaign official. Drudge blared the knifepoint-rape claims at the top of its home page. Russian online miscreants also played a role.

On April 11, 2017, Infowars broadcast reports falsely linking Chobani's refugee employees to crime and to an increase in tuberculosis cases, posting the bogus report to Jones's YouTube channel with the headline MSM COVERS FOR GLOBALIST REFUGEE IMPORT PROGRAM AFTER CHILD RAPE CASE. Infowars blasted links to its Chobani coverage to its 600,000 Twitter followers, with the headline IDAHO YOGURT MAKER CAUGHT IMPORTING MIGRANT RAPISTS.

"People he brought in and force-fed on America have been implicated, indicted, and pled guilty," Jones falsely claimed on Infowars, inspiring calls for a Chobani boycott.

On April 24, Chobani slapped Jones with a lawsuit for defamation and unfair competition in a federal court in Idaho.

"I will win or I will die!" Jones shouted on his show, launching into a purple-faced, bigoted rant. "We had barely covered this story, but don't you worry—I'm gonna be covering it now, Mister Hah-MEED Ulooo-kai-YAH," he said, mispronouncing the Chobani owner's name.

Jones waved the lawsuit around like he did Lenny's YouTube take-down order so his viewers could see the contacts for Chobani's lawyers. And he begged for money: "I want to see a record surge in funding . . . I want to grow in the face of the enemy.

"There's a total Islamic takeover taking place behind the scenes! They got Muslims following me around. You wouldn't believe the crap they're pulling, man—*it's a damn cult!* And I'm ready to take them on. And I need your prayers."

Closing his eyes, tilting his face heavenward, "Christ, please help us win this," he said.

Three weeks later Jones folded.

At the end of his May 17 broadcast, Jones, all bravado gone, said, "During the week of April 10, 2017, certain statements were made on the InfoWars Twitter feed and YouTube channel regarding Chobani LLC that I now understand to be wrong.

"The tweets and video have now been retracted and will not be reposted. On behalf of Infowars, I regret that we mischaracterized Chobani, its employees and the people of Twin Falls, Idaho, the way we did."[7]

A MONTH AFTER THE CONFLICT with Chobani, Jones got into fresh trouble over Sandy Hook.

In June 2017, Megyn Kelly, the former Fox News superstar newly hired by NBC, flew to Austin to interview Jones for her new show, *Sunday Night with Megyn Kelly.*

Jones's staffers had warned him against the interview. Jones was already struggling to regulate himself under scrutiny, narrowly avoiding two potentially devastating lawsuits.

On Fox, Kelly had been the rare prime-time personality to call BS on conservative spin and mendacity. Her confrontational questioning of Trump's misogyny during a 2016 presidential debate had surprised and angered him, earning her new fans among feminists, not exactly Jones's target audience.

But Jones, swooning over Kelly's celebrity and keen to stoke his own, said yes. It was a disastrous decision that plunged him into deep legal jeopardy with the Sandy Hook families, for whom another backhanded apology would never suffice.

On June 11, 2017, NBC aired an announcement that Kelly, its star hire, would profile Jones on her show the following weekend.

The Sandy Hook relatives learned of NBC's plan from a cringey promotional blurb in which Jones sidestepped Kelly's question about the

Sandy Hook conspiracy theory. Her half-smiling protest, "That's a dodge . . . That doesn't excuse what you did and said about Newtown, you know it," came off as flippant, given the misery caused by Jones's claims. Just as bad, the interview was scheduled for Father's Day.

Sandy Hook victims' relatives blasted Kelly and NBC on social media and in newspaper op-eds. Seven victims' families signed on to a certified letter sent to NBC brass by Koskoff, Koskoff & Bieder, a Connecticut law firm representing several Sandy Hook families who were suing Remington Arms Company, maker of the rifle used in the shooting. "We urge you to consider the ethical and legal ramifications of broadcasting this interview to millions of Americans," it read. "It should be clear to NBC that airing the interview will cause serious emotional distress to dozens of Sandy Hook families. NBC—and NBC alone—has the power to prevent that harm."

New York mayor Bill de Blasio, gun control advocates, even the *National Enquirer* jumped on, calling Kelly the "Most Hated Mom in America."

The furor caught Kelly and the network flat-footed. Kelly thought grilling the Infowars impresario was important. She knew the Republican landscape, and Jones's putative coziness with Trump, his entry into mainstream Republican politics, disturbed her. Kelly knew several of the Sandy Hook relatives and understood the pain and threats Jones had inflicted by harping on their loss.

"I find Alex Jones's suggestion that Sandy Hook was 'a hoax' as personally revolting as every other rational person does," she said in a statement at the time. "It left me, and many other Americans, asking the very question that prompted this interview: how does Jones, who traffics [*sic*] in these outrageous conspiracy theories, have the respect of the president of the United States and a growing audience of millions?

"Our goal in sitting down with him was to shine a light—as

journalists are supposed to do—on this influential figure, and yes—to discuss the considerable falsehoods he has promoted with near impunity."

Kelly declined to speak with me on the record. In retrospect, her defense stands up. But in the politically overheated environment that followed Trump's victory, it failed to quell the uproar.

WITH DAYS TO GO before the broadcast, NBC producers scrambled to address the legitimate complaint that they had not given the Sandy Hook families equal time.

Kelly's producers issued repeated public reminders that those criticizing the show had not yet seen it, insisting it was much tougher than the promo suggested. Then they set out to make sure they were right, by seeking a Sandy Hook parent to include in the broadcast.

Meanwhile Jones had convinced himself that the interview was a conspiracy to place him in the worst possible light. He demanded it be pulled, then tried to compromise Kelly by releasing a friendly-looking photo of them together and a recording of her pitch to him, in which she promised the interview would not be a "hit piece." It wasn't exactly a gotcha moment, when you consider that no reporter seeking an interview would ever say the opposite.

Lenny didn't agree with family members who wanted NBC to pull the Jones profile. Jones had been braying about his First Amendment rights for years. Killing the show would hand him a free-speech cudgel, while helping him quash an interview he had realized too late was a mistake.

The broadcast had become a hot potato, and Sandy Hook relatives NBC contacted refused to appear. Lenny avoided most television interviews because he didn't want the conspiracists to recognize him. Instead he wrote a guest essay in the *Hartford Courant* supporting the broadcast.[8]

"With the aid of media platforms such as alternative talk radio, YouTube, Google, Facebook and Twitter, scores more are being reached and indoctrinated into the cult of delusional lunacy every day," Lenny wrote. Ignoring people like Jones "feeds the movement rather than diminishing it," he argued.

"The very fact that Jones has some semblance of influence over our president's thinking speaks to my position that we should challenge his warped and pernicious views out in the open public forum . . . Maybe then people will see the monster that he truly is."

Neil Heslin, Jesse Lewis's dad, felt the same way. When Kelly's producer called him, he agreed to go on the show.

"She was loyal to the families and nothing but respectful and supportive," he told me of Kelly. Though they inhabited vastly different worlds, Kelly and Neil had grown up in the same region of upstate New York, and there was a toughness to Kelly that Neil admired. Most of all, Neil thought it was important to confront Jones.

Shortly before his interview with Kelly aired, Jones released a self-serving, one-minute video with a Father's Day message to the Sandy Hook victims' families.[9] Jones sat at his broadcast desk in a navy suit jacket and white shirt, the video screen behind him displaying a streaming American flag. Naturally he opened with a reference to himself.

"I woke up this morning on Father's Day, and I was holding my young infant daughter in my arms, looking into her eyes," he said, clasping his hands, eyes heavenward.

"Sitting out on the back porch, hearing the birds sing, and it just brought tears to my eyes, thinking about all the parents that have lost their children on Father's Day or Mother's Day, who have to then have to think about that. Parents should never have to bury their own children. And that's why on Father's Day I want to reach out to the parents of the slain children at the horrible tragedy in Newtown, Connecticut, and give you my sincere condolences," he said.

"I'd also like to reach out to any of the parents who lost a child at Newtown to invite them to contact me to open a dialogue, because I think it's really essential we do that, instead of letting the MSM misrepresent things and really try to drive this nation apart."[10]

If ever a moment called for an abject apology, the day of Jones's prime-time interview was it. But Jones didn't do it.

On the broadcast Kelly opened her conversation with Neil with comments from Charlie Sykes, a conservative radio commentator who had broken with Trump. Sykes expressed his shock that the president was echoing themes advanced by the likes of Jones.

"He has injected this sort of toxic paranoia into the mainstream of conservative thought in a way that would have been inconceivable a couple of decades ago," Sykes said of Jones. "We're talking about somebody who traffics in some of the sickest, most offensive types of theories."

At the top of that list, Kelly said, was "Jones's outrageous statement that the slaughter of innocent children and teachers at Sandy Hook Elementary School, one of the darkest chapters in American history, was a hoax."[11]

Neil wore his brown tweed jacket, his shirt open. His face in close-up looked weary, his close beard streaked with gray, his eyes sad.

"I lost my son. I buried my son. I held my son with a bullet hole through his head," he said.

Kelly narrated: "Neil Heslin's son Jesse, just six years old, was murdered along with nineteen of his classmates and six adults on December 14, 2012, in Newtown, Connecticut."

Neil again recalled that morning. "I dropped him off at 9:04, that's when we dropped him off at school, with his book bag. Hours later I was picking him up in a body bag."

The show cut to footage of Jones talking about Sandy Hook on his show, saying, "The whole thing was fake."

Kelly confronted Jones, pressing as he tried to slither away. "Listeners and other people are covering this. I didn't create that story . . . I watched the footage, and it looks like a drill," he waffled.

Kelly narrated: "We asked Jones numerous times what he now believes, and he never completely disavowed his previous statements . . . The families say that Jones's words have caused lasting pain, and they fear the harassment will continue."

The program cut back to Neil, telling Kelly, "You know, it's disrespectful to me. In fact, I did lose my son. And the twenty-six other families lost somebody. And I take that very personal."

Kelly told him: "You know, this piece is going to air on Father's Day. What is your message to him?"

"I think he's blessed to have his children to spend the day with, to speak to," Neil said. He sighed heavily.

"I don't have that."

KELLY CALLED NEIL in tears afterward, to thank him. "They bashed her really bad," Neil told me. "I made it clear I was not part of it, I did not support it, and it didn't reflect on all the families."

Most reviewers found Kelly sufficiently tough on Jones. "Of course she should have gone ahead and done that story," Jim Rutenberg, who had been the *New York Times*' media columnist at the time, told me. "Jones was the living embodiment of the conspiracy wing of the Trump coalition."

If anything, Kelly could have plunged more deeply into Jones's relationship with Trump, which Jones liked to brag about but which was downplayed by the White House.

Reflecting on the episode years later, Rutenberg thought Kelly had become ensnared by what was then a new phenomenon. "Social media was training everyone to see only the reality they wanted to see," he said.

That mindset had bled from social media to television, where Fox affirmed its conservative viewers' worldview, and NBC, particularly its MSNBC cable outlet, catered to the left. As well, Rutenberg said, "There was this extra sensitivity in the NBC leadership over Trump," because Trump's star turn on NBC's *The Apprentice* had helped pave his way to the White House.

"Megyn Kelly was a trailblazer" for casting a prime-time spotlight on Jones, Lenny said. "She put a bright light on Alex Jones. Nobody remembers that now. They think Alex Jones was always somebody."

Jones had become nearly impossible to ignore, his link with the White House constantly discussed, including by him. His higher profile encouraged young wannabes on his staff, like Owen Shroyer, who joined Infowars during the 2016 campaign, to imitate Jones's worst abuses in an effort to ride his coattails.

This was terrible for the Sandy Hook families targeted by the conspiracists. They noticed that when anyone protested, Jones doubled down, like the president he so admired, stopping only when his targets threatened to drop a lawsuit on him And sometimes not even then..

Alex Jones would soon turn his wrath on Neil. That proved to be a poor choice.

19

A WEEK AFTER NEIL HESLIN discussed his final moments with Jesse in the Megyn Kelly interview, Owen Shroyer, the twentysomething Infowars host, dismissed Neil's recollection as a lie.

"He's claiming that he held his son and saw the bullet hole in his head. That is his claim," Shroyer told the Infowars listeners on June 25, 2017. "That is not possible."

Shroyer, a baby-faced Jones imitator, had joined Infowars less than a year before. In case the boss was watching, Shroyer prefaced the segment about Neil by saying, "I don't even know if Alex knows about this . . . Alex, if you're listening and you want to—if you just want to know what's going on . . ."

IN THE NEARLY FIVE YEARS since Jesse's death, Infowars had not targeted Neil by name. But other conspiracists had attacked him, and he refused to endure more abuse in silence.

Neil is a straightforward man who churns through grief like men from his background do: silently, with occasional outbursts. He didn't have it in him to understand the gunman as a victim of bullying lost to mental illness, as Scarlett Lewis, Jesse's mother, could. He had derived no comfort, as the Parkers had, from meeting the gunman's father.

After Jesse's murder, Neil joined some other parents in their efforts to curb gun violence. The work felt important to him, a way to honor Jesse and the others they had lost. Neil had owned guns and hunted for sport as a young man, so he thought he could bring a different perspective to the debate. But advocacy plunged him into an unfamiliar world.

After Obama's visit to Newtown for the vigil, Neil flew with other families to Washington on Air Force One, to lobby Congress for change. Scarlett had not joined the effort. Neil, thrilled to be on the president's plane, called her from the air and got voicemail.

Neil returned to Washington in February 2013 to testify before Congress, his plane ticket and hotel paid for by one of the nonprofits that had sprung up in Newtown.

Neil carried with him a bulky gilt-framed photograph of him and Jesse, which Neil had had made as a gift for his mother when Jesse was six months old. Walking back and forth through the hallways of the Capitol in the clothes he had worn to Jesse's funeral, the portrait seemed to weigh almost as much as Jesse did at that age, but maybe Neil was just tired. Jesse had been a sturdy boy, but Neil had carried him effortlessly on his shoulders.

On February 27, 2013, Neil carried the portrait to the Senate Judiciary Committee's elegant, paneled hearing room to testify in favor of a proposed assault-style weapons ban.

Neil stated his name, and "I'm Jesse's dad," then broke down. Senator Dianne Feinstein, who was chairing the hearing, pressed a tissue to her face.

"It's hard for me to be here today talking about my deceased son. I have to, I'm his voice," Neil said. "I'm here to speak up for my son."

Appearing before Congress and crying in public were two things Neil had never done in his life. He lacked the patience for politesse, for always saying "gun safety" instead of "gun control," for the sensitivities he provoked when he spoke about "committing" mentally ill people like the gunman. Neil believed that these things would have saved Jesse, so why not say them?

Neil toured the Capitol and the White House, and saw historic desks used by historical figures. Wandering the city the morning after his testimony, he stopped before the sleek white facade of the Newseum, now-defunct temple to journalism and democracy. Glass-front cases outside displayed newspaper front pages from around the nation. A photo of Neil holding Jesse's portrait, his face contorted in pain, appeared on most every one.

The gun legislation opponents and conspiracy theorists who overlapped on the internet viewed Neil, a conservative former gun owner, as a traitor and a fraud.

The hoaxers posted court records of Neil's DUIs, from a troubled time he'd gone through a decade earlier. While lobbying in the Capitol, he missed a court date to account for the bad checks he'd passed the year before Jesse's death, for $102.35 in tire repairs, home heating oil, and construction materials to keep his contracting business going. "One crooked gun control hero," his anonymous critics wrote online. "Sandy Hook dad has sordid past."

A reporter from the Danbury paper called Neil and asked if he thought his past legal troubles hurt the Sandy Hook families' cause. "I never gave it much thought," Neil said.[1]

In the spring of 2014, a couple of guys from an appliance store delivered a new refrigerator to Neil's house. As he came through the door, the driver recognized Neil and offered condolences. They stepped into the dining room to talk. The other deliveryman walked past and overheard the words "Sandy Hook."

"Oh yeah, that's that thing where they're trying to take the guns away," he chimed in as he passed by. Neil looked at the man standing beside him in the dining room.

"He didn't mean anything by it," the embarrassed driver said. "But don't even bother—you're not going anywhere with it. You're not changing his mind," he told Neil, who didn't say anything.

"I don't know where it would have went if he knew who I was," Neil told me. "He wasn't peddling it. He wasn't a hoaxer or a nutjob . . . It was drilled into him, like all these other people."

Neil let it go.

One night while Neil was out, someone drove by and fired rounds at his house. He called the police, but there wasn't much to go on. He got a big dog, then two more. By 2017, Neil had long since stopped testifying. Then Owen Shroyer suggested to millions of Infowars fans that Neil was a fraud.

ATTACKING NEIL'S CREDIBILITY ON-AIR was Shroyer's way of sucking up to the boss. Jones had been launching broadsides against Megyn Kelly and NBC, saying he had been tricked, lied to, conspired against. Jones's grotesque efforts to capitalize on Sandy Hook denial had never been criticized by a parent before a national television audience. Jones had come off horribly, and he knew it. What made the attack on Neil even more offensive was how casual it was, Shroyer's facile assumption that he could drag a grieving father into Jones's celebrity beef about his bad press.

Infowars posted the broadcast to YouTube, as it usually did. It drew not one but a couple takedown notices, including for violating YouTube's community standards. Infowars had been nailed with many such notices by 2017, some of them filed by Lenny and HONR. But these were apparently filed anonymously, depriving Jones of a target.

His staff cautioned Jones to let it go. But Jones had seldom in his life heeded warnings of any kind. When provoked he saw red and charged, heedless of the costs.

Owen Shroyer had followed a path to Infowars that originated in conventional radio, then veered off through the muck of partisan social media.

Josh Owens, the young Infowars cameraman who abandoned film school to join Infowars in 2012, had met Shroyer during the turbulent summer of 2014, when the fatal shooting of Michael Brown by a Ferguson, Missouri, police officer sparked days of demonstrations that grew into a national movement. The two men were both in their mid-twenties and looking to make a name for themselves.

Owens had traveled to Ferguson from Austin, along with Joe Biggs and other Infowars colleagues. Owen Shroyer was the entertaining local yokel. He lived in St. Louis at the time. He was born there, had attended a local Catholic high school, and graduated from the University of Missouri–St. Louis, where he had studied media and psychology. When he met Josh Owens and the rest of the Infowars team on the street in Ferguson, Shroyer had been working as a journeyman reporter hosting a 10:00 p.m.-to-midnight talk-radio shift on St. Louis's KXFN. Shroyer's listenership doubled as a result of his Ferguson reporting, and he boosted it further by posing on an I-64 overpass holding posterboards on which he'd written TYRANNY IS HERE #FERGUSON.

Shroyer seemed to be on the side of the protesters back then. They "don't deserve tear gas and guns pointed at them," he told the *St. Louis Post-Dispatch*[2] as a photographer took his picture. "The issue is a tyrannical government, an oppressive system that we all need to buck."

Like Jones, Shroyer had honed that libertarian message by 2016, emerging as a die-hard and confrontational Trump supporter, with a provocative social media presence. Shroyer and Owens met again in spring 2016, when the Infowars crew returned to Missouri to cover a Trump rally in St. Louis. Trump supporters had lined up for blocks to enter the evening rally, yelling back and forth with demonstrators carrying signs calling the Republican candidate a racist. Weaving through crowds on a sunny street outside the rally with the Infowars team that afternoon, Owens bumped into Shroyer, who was in the middle of a crowd of anti-Trump protesters. Shroyer's compatriots filmed him as he debated with them. At the end of the confrontation, Shroyer savvily recorded a shout-out to Infowars and posted the eighteen-minute video of the heated argument to YouTube. The pro-Trump internet targeted a couple of the more awkward social justice activists in the video for ridicule, and the video drew millions of views seemingly overnight, Owens told me. "Jones noticed that, and I encouraged Rob Dew to reach out to Shroyer and bring him in," he said.

Jones plucked Shroyer from the chorus line of newbie, far-right online provocateurs fighting for attention that election year. In August 2016, Jones interviewed Shroyer during his show, seeking to introduce him to the Infowars audience and drive more traffic to the videos Shroyer had made.

"Thank you, Alex. You were my original inspiration to get into current events and politics, and it's just great to be here," Shroyer said. Jones stayed uncharacteristically quiet, soaking up the praise. "I'm motivated to have people experience the same awakening process that I experienced. That there's things out there that matter more than the Kardashians, more than their local sports teams, more than Drake's latest album drop. There are things that actually affect our future, and our lineage."

Jones by all accounts has few friends outside those on the Infowars payroll. His demeanor toward Shroyer that day was a combination of alpha-male bravado and solicitousness toward someone he viewed as a rising Young Turk. Shroyer just wanted to do what Alex did.

In a bullying pas de deux, Jones and Shroyer took turns urging the two men in the video to appear on Infowars.

Jones shifted to race, Shroyer nodding as Jones said, "They're teaching college courses called 'How to Eradicate Whiteness.'

"Can you feel the spirit of liberty rising?! Arrrrrgggghhhh!" Jones said, clenching his fists, throwing his head back, and roaring. Shroyer roared too, making his fists like Alex's.

Jones asked Shroyer to talk a little bit about himself.

"My dream was to be in sports media, and I was well on track. I was on that path. I was working in sports media. But what happened was for the first time in my life after the Boston Marathon bombing—and I mean this, *for the first time in my life*—I watched a news report on Fox or CNN. I paid attention to anything to do with news or current events for the first time in my life," Shroyer said.

Jones interrupted, revisiting the false flag claim he had spread after the Boston bombing. "The whole thing is a cover-up, psy-op, drills, weird stuff involved. I mean, really, really crazy."

"Just crazy stuff happening," Shroyer agreed. "So it just led me down the rabbit hole. I started questioning everything from 9/11 to vaccines to the lines that everybody sees in the sky, but never questions. And it all just led me to the rabbit hole. And I became a big fan of yours."

Jones responded to the ego-stroking by unspooling a torrent of falsehoods about forced government vaccinations.

"Okay, so you just laid out a bunch of facts," Shroyer responded. "And considering that, I looked at myself in the mirror and I said, 'How can I dedicate myself to sports media? How can I sit here every day and talk ad nauseum and follow sports when I've got real issues?' And it's not just issues, Alex. This is attacking our livelihood. This is attacking our future!"

Shroyer off camera was a nice guy, Owens recalled. He was easy to talk with, a friend who during a rough patch in Owens's personal life let him crash at his place. Shroyer was one of the few people Owens sought out in April 2017 when he decided he couldn't work at Infowars another day.

"Things had gotten so crazy and hectic, so much was going on, that I was at a little bit of a breaking point," Owens told me. Owens told his friend Shroyer that having seen how the sausage was made, he realized Jones wasn't questioning government power as much as he was selling snake oil. Owens couldn't stand the fakery, the creative editing in service to Jones's mendacity, any longer. He was tired of making things up, of targeting innocent people, stirring racism and fear to sell diet supplements and survivalist gear nobody really needed. He hated himself for being taken in by it all.

Owens didn't give notice. He waited until Jones had left the studio that day and then walked out.

SHROYER DIFFERED FROM OWENS in that he wanted Jones's popularity but didn't seem to care how he got there. Shroyer was more condescending and smirking, slower on his verbal feet than Jones. But he labored to imitate him, down to Jones's disturbing bruxism and facial tics. Out in the field, Shroyer began shouting instead of debating, disrupting events, and barking about his First Amendment rights when authorities moved him off or arrested him. Lanky, with a peachy complexion and a patchy beard, Shroyer was in all aspects a pale shadow of his boss.

To *Knowledge Fight* podcast founder and Infowars student Dan Friesen, and to Lenny, Shroyer personified the weak, mutant next generation of American conspiracism. By the time he hired Shroyer, Jones had for years been fanning hatred while moving the merchandise that made him rich. But Jones had pursued his obsessions in obscurity for twenty years before then, studying forebears like his hero Gary Allen. For better or worse—and by 2017, overwhelmingly for the worse—Jones was part of a long tradition of American political conspiracism. But for Shroyer, "It's all attention and reaction, as opposed to being rooted in a philosophy like the strident anti-communists that were Alex's heroes," Friesen told me.

Back in the 1990s, when Jones began airing his views from the unused baby's room in his and Kelly's Austin house, conspiracy broadcasters "were boring men in public access TV studios dissecting the all-seeing eye on the back of a dollar bill," Jon Ronson, the author and filmmaker who accompanied Alex to Bohemian Grove, told me. "People were yearning for somebody who would be funny and eloquent." In that subterranean world of truth distortionists, "Alex was a star.

"But it got darker. The money and the power got to him," Ronson said. "Hounding parents of shooting victims, the Islamophobia—I might be wrong, but I don't know of occasions when Alex was that nefarious or malevolent in the nineties."

Shroyer's behavior spoke to Jones's devotion to click-driven commerce and politics, human consequences be damned. Social media minted hundreds of bullies like Owen Shroyer every day, across the political spectrum. Doubting Neil's recollection of Jesse seemed pretty much the same to Shroyer as mocking a couple of awkward anti-Trumpers, as long as it got a lot of views.

"So, folks, now here's another story," Shroyer said on Infowars that day in June 2017. He was dressed in a dark suit and white shirt, just like Jones had taken to wearing since Trump ushered him into the big time.

"Alex, if you're listening," he began, alerting Jones to a story on far-right website *Zero Hedge*, "Megyn Kelly Fails to Fact Check Sandy Hook Father's Contradictory Claim in Alex Jones Hit Piece."

Infowars often borrowed from *Zero Hedge*, a libertarian-ish, far-right financial blog that sprang up in 2009, founded by a Bulgarian former hedge fund trader whom regulators had banned from the industry. *Zero*

Hedge had branched into conspiracy-themed content, all of it appearing anonymously under the pen name "Tyler Durden," a psychopathic character from *Fight Club.*

"This broke, I think it broke today, I don't know what time," Shroyer said, working to imbue the *Zero Hedge* concoction with real-news urgency.

"Neil Heslin, a father of one of the victims, during the interview described what happened the day of the shooting. And, basically, what he said—the statement he made—fact-checkers on this have said cannot be accurate.

"He's claiming that he held his son and saw the bullet hole in his head. That is his claim. Now, according to a timeline of events and a coroner's testimony, that is not possible.

"This is only going to fuel the conspiracy theory," Shroyer said, pressing his lips together in faux consternation like Jones did. "And here's the thing too. You would remember if you held your dead kid in your hands with a bullet hole. That's not something that you would just misspeak on.

"So let's roll the clip."

The screen filled with the image of Neil in his funeral suit, his face sagging in sad recollection. "I lost my son. I buried my son. I held my son with a bullet hole through his head," he told Kelly. "I dropped him off at 9:04, that's when we dropped him off at school, with his book bag. Hours later I was picking him up in a body bag."

Shroyer returned. "Okay, so making a pretty extreme claim, that would be a very thing vivid in your memory, holding his dead child," he said with the jumbled delivery of the amateur sportscaster he was. He aired a network news clip from the day after the shooting, of H. Wayne Carver, the Connecticut medical examiner, explaining that the parents were offered the choice of identifying their children from photographs. To Shroyer this suggested Neil owed Infowars an explanation.

"Okay, so just another question that people are now going to be asking about Sandy Hook, the conspiracy theorists on the internet out there that have a lot of questions that are yet to get answered. I mean, you can say whatever you want about the event. That's just a fact. So there's another one. Will there be a clarification from Heslin or Megyn Kelly?

"I wouldn't hold your breath. Ha ha! So now they're fueling the conspiracy theory claims. Unbelievable."

"That punk, that Al Jones puppet," Neil said, when he found out. Jones and his lackeys had a gripe with Megyn Kelly, but Neil hadn't done anything to them. He couldn't understand the reason for their disbelief and disrespect.

"We could try to figure that out all day long," Neil mused. "Why does Al Jones do what Al Jones does?"

AFTER COMPLAINTS PROMPTED YOUTUBE to yank Infowars' video of that day's broadcast, Jones flew into a fury, adding to Infowars' problems.

"On every platform, the Democrats and the liberals have organized into groups that go around making false copyright claims and false community claims," Jones warned on July 20, 2017. He sat before one of Infowars' new neon backdrops, its logo and a map of the world rendered in Las Vegas–style green and purple lights. He looked puffy that late July afternoon, with deep bags beneath his eyes. He had lost a major battle in his war with Kelly Jones over custody of their children earlier that year, and had been living on his own, partying hard with Gucciardi and other guys on his staff.

"YouTube is announcing that they're looking at shutting down and basically kicking us off YouTube, for people complaining that I've reported on Sandy Hook and had Wolfgang Halbig, a former school safety administrator, on for a debate about whether the official story was true or not. Then the media misrepresents what I say, saying that I say it never happened when I've looked at both sides!"

Jones repeatedly wiped sweat from his upper lip and chin, creating a circle beard of pink irritation.

"It's my right to say it. I can question big PR events like Sandy Hook, where there are major anomalies, like I'm saying none of the parents were allowed to see their kids that day in school. And they have people on NBC saying they held their kid, dead, in the school!

"I've already got the law firms in D.C. and others ready. They know I'm going to sue whoever files a fake copyright claim on me again. I am going to sue you. I CANNOT WAIT!

"We're gonna expose you bullies! You understand? Next person, you're sued! So line up. You wanna get into a big fat lawsuit with me? Whoever you are, I don't care who you are! You make up crap, you lie about us, you try to take my free speech and gag me, and take my speech so you can have your way with my family and my children—IT AIN'T HAPPENING ANYMORE!"

He hit the desk with an open hand, then with his fist, pounding.

"Wake up, everybody! We're in a fight against the globalists! They're trying to put our president in prison, they're funding radical Islamicists and terrorists, saying our president's a Russian agent because he doesn't wanna fund al-Qaeda. If they're able to shut us down, they're going to shut everybody else down!"

Jones hunched over, yelling, his face purple, as the camera pulled back for a break.

"THIS IS A TOTAL WAR, PEOPLE!"

Then Jones dropped his fury like flipping a switch, and did a product ad. He did that a lot, making such instant shifts to shameless hucksterism that I wondered if it was intentional self-parody. But Jones didn't seem to possess that level of self-awareness. He played the segment with

Neil again and repeated his doubts about Neil's story, daring whoever had complained to do it again. He repeated the other headline he'd gotten into trouble for, SANDY HOOK VICTIM DIES (AGAIN) IN PAKISTAN, that Lenny had had taken down.

"They're using Sandy Hook, and they're using the victims and their families as a way to get rid of free speech in America. That's the plan," he said.

When Jones rails about his constitutional right to speech, the truer interpretation is that he constitutionally can't stop speaking. His Alamo-like belligerence had cost him plenty. But it solidified his public persona, while his personal life was anything but solid.

IF JONES SEEMED MORE unhinged than usual, and he did, it was likely related to his family life. Earlier that same year, in April 2017, Alex and Kelly Jones had appeared before a jury in Austin for a trial over custody of their three children. Their son, fourteen, and two daughters, ages twelve and nine, had been living almost exclusively with Alex since the couple's 2015 divorce, and Kelly was seeking to make her home their primary residence, with at least joint custody. Alex Jones's Trump-era notoriety drew national attention to the case.

The case cost nearly $3 million before the couple set foot in a courtroom, including $2.4 million that Kelly had spent on lawyers, therapy, and professional assessments over the nearly two years between their divorce decree and the ten-day custody trial.

In pretrial agreements, Alex Jones's big-ticket legal team had painstakingly placed discussion of most of Jones's on-air activities off-limits.

At home with his kids, Jones is a totally different man than he is on Infowars, his lawyers insisted in an astonishing gambit.[3] "He's playing a character," his lawyer Randall Wilhite told Texas district judge Orlinda Naranjo. "He is a performance artist."

Jones is simply driving "a message" on his show, another of his lawyers, David Minton, told the jury.[4] "He does it with humor. He has done it with bombasity. He has done it with sarcasm. He has done it with wit."

He has done it by bullying the survivors of tragedy too, but Jones's lawyers got away with omitting that. If Jones could have behaved himself in court and avoided evoking the Infowars persona his lawyers had so painstakingly kept off the record, he could have helped his case. He couldn't do it.

Jones mugged for the media on his way into court. The judge admonished him against making "bodily comments," but he muttered and grimaced almost constantly during the trial, leveling menacing stares, eyes narrowed, at his ex-wife, about whom he testified, "She doesn't have any good qualities . . . any good is sandwiched with bad." He took personal, public aim at Kelly Jones's lawyers, telling one of them, "I just don't trust you, man." He interrupted the judge while she was telling him to stop interrupting. And he testified that he means what he says.[5]

Charlie Warzel covered the saga for *BuzzFeed*.[6] He described a moment when Kelly's lawyer pursued an excruciating line of questions about the couple's sexual history.

Watching Jones in the witness box, "I thought he was gonna have an aneurysm. He was truly about to explode," Warzel told me. "And then he did."

"HAVE YOU NO DECENCY?!" Jones erupted.

"In a more formalized court setting he would have been taken out by a bailiff," Warzel told me. The jury voted 10–12 against him. Kelly Jones won joint custody; the children's main residence would be with her.[7]

Jones's wild behavior is itself algorithmically addicting; his audience keeps tuning in to see what he'll do next. Most of what Jones says is false; he lies and embellishes even when the truth would serve him better. But his screaming inability to get out of his own way is real, a personal

Infowar that rivets his fans like watching WWE personalities pound themselves into oblivion.

JONES DUG IN and kept digging as he went after Neil that July afternoon, boiling over with self-righteous rage.

"I'm going to air this again," Jones said of Shroyer's segment questioning Neil.

"We're not backing down. We're not giving up. We're getting more affiliates across the United States because America is done being intimidated. America is done bowing."

A half hour later: "Let's play the censored report," Jones said. "Quite frankly, the father needs to clarify. NBC needs to clarify. Because the coroner said no, the parents weren't allowed to touch the kids or see the kids.

"You can't ban free speech of people that are asking questions. And for us to simply look at the Megyn Kelly public event where someone sat down and was interviewed and to *politely* discuss it, if you ban that, you ban free speech in total.

"Very, very dangerous. Here it is."

"All right," Jones said after he played the episode again. "That's the full clip that's been censored on YouTube, that's hateful and evil they say, and that we're harassing people with."

Jones could not let it go.

"Bottom line, there was massive PR around this. This was used to blame the American people to say gun owners were at blame for this and that we had killed these children. That's why America rejected it and said it was fake!

"We have not seen a clarification!" he yelled, as if Neil owed him one.

20

Lenny and Neil had talked about suing Jones for years. They just hadn't found the right lawyer. Lenny had wanted to haul Jones into court in early 2013, when Jones suggested Veronique had faked her CNN interview reminiscing about Noah five days after they buried him, pointing as proof to glitches in a video copy made by his own staff.

Lenny had wanted to sue Jones again in early 2015, when Infowars created a falsehood that to this day still circulates, claiming Noah had been murdered "again" in a Taliban-led massacre in Pakistan, a lie based on an image of a grieving woman in Peshawar displaying Noah's photo in solidarity. He wanted to sue after Jones attacked Lenny on his show for daring to have that video removed from YouTube, saying of Noah, "You can say the name of any*thing* you want as long as it's free speech." He had wanted to sue Jones when Lucy Richards, Infowars superfan, threatened his life.

But Lenny viewed suing conspiracists as a nuclear option: draining, potentially expensive, and unpredictable. He knew. He had tried it once with Wolfgang Halbig, and it hadn't worked out very well.

BY THE END OF 2015, after three years of Halbig's torment of his and the other families, Lenny said, "I had had it."

Both men were living in South Florida at the time. Lenny figured suing Halbig was easier than trying to visit him had been, when Halbig responded to his offer to show him records of Noah's life with an email from Watt, telling him to exhume Noah and "prove to the world you lost your son."

Lenny sued Halbig for invasion of privacy after Halbig used Lenny's complaint to the Florida attorney general about Halbig's fundraising to doxx him.

"Suing Halbig is symbolic," Lenny had told Reeves Wiedeman of *New York* magazine. "If I can show that if you go after a victim, a victim is gonna sue you, that's real."[1]

But after filing the suit, "I found out that Halbig was much more of a fighter than I expected him to be," Lenny told me. Halbig turned the lawsuit into a cause. He created a legal defense fund and stepped up his records requests, saying he needed the material for his defense. And he used his access to records to destroy what was left of Lenny and Veronique's privacy.

Among the emails Halbig forwarded to Infowars that have surfaced as part of court proceedings is one he sent to his lawyer at the time, Caleb Payne.

Under the subject line "Thought we are getting records?" Halbig wrote:

> Caleb:
>
> URGENT:
> Nothing on Veronica [*sic*] Pozner background check yet?
> How about the Traxware Inc corporate Tax filings?
> How about checking out the fake social security number used by Pozner?

> Please schedule a hearing with the courts on Your Motion to Tax cost . . .
>
> Please contact the courts to schedule a hearing they do not deserve to escape this frivolous and harassing lawsuit, Please also start on the Counter Suit to be filed.
>
> Please help me it is my life and I want it back
>
> Wolfgang

In late 2016, Halbig posted Lenny's TransUnion background report.[2] It contained one hundred pages of phone numbers, addresses, and credit and financial information for Lenny and his family members.

Lenny didn't attend the initial hearing in his case against Halbig. He and Veronique had moved a half-dozen times by 2017, to stay ahead of the constant doxxing, and Lenny refused to be photographed or filmed, unless his face was obscured.

Tony Mead joined Halbig in the courtroom, as did a conspiracy theorist who filmed the hearing and posted it on a YouTube channel that referred to Lenny as "that Jew."[3] When he saw the hearing video, Lenny realized the man who made it lived in his apartment building. He had seen him walking his dog.

Faced with sitting for a deposition he feared would be posted online, or testifying in a courtroom alongside hoaxers and potential stalkers, Lenny dropped his lawsuit against Halbig. It had cost him nearly $30,000, money he didn't have. Midway through the suit, journalist Mark Hill interviewed Lenny and pointed out that HONR's online fundraiser "is sitting at 235 dollars."[4]

Halbig, by contrast, had raised nearly $100,000.

MEANWHILE, LENNY AND HONR had remained focused on content removal.

"That was very linear, very two-dimensional. That, too, was frustrating, but miraculously, I was getting results," Lenny told me.

Lenny's reputation continued to grow. His op-eds and features about him appeared in mainstream media. Survivors of mass tragedy and victims' families contacted HONR, pleading for help in getting the big platforms to enforce their own policies and to remove dangerous or harmful content. Because the platforms seldom acted unless publicly shamed, celebrities or big names in tech usually received quick attention to complaints of abuse.

"I can't tell you how many times I've forwarded things, and things do happen," tech reporter Kara Swisher said during an online discussion of social media harassment. "But that's ridiculous. I'm not a help desk for these companies."

Lenny became part of that informal clout system, using his influence on behalf of desperate people like Doug Maguire and Andy Parker, whose daughter Alison Parker, a news reporter at WDBJ in Roanoke, Virginia, and her cameraman, Adam Ward, were gunned down during a live interview. The gunman was wearing a GoPro camera and uploaded the footage to social media before taking his own life.[5] The video spread wildly, often accompanied by comments that the Parkers had faked their daughter's death.

Lenny expanded HONR's work on behalf of other victims after a few HONR volunteers reached out to Alexandrea Merrell, a woman in New York whose story they heard on a radio broadcast. Merrell had endured a thirteen-year campaign of cyberstalking by a person whose identity she never learned. After her ordeal, Merrell had turned her public relations firm into a crisis management firm for people whose lives and businesses had been upended by online abuse. Lenny's volunteers were concerned about how his nonstop effort to defend Noah online was affecting him and thought

Merrell could help him. The two spoke every day, and Merrell became the HONR network's director of public relations, helping to publicize HONR's work and deploy its volunteers to assist a growing pool of victims.

By 2017, Lenny and the HONR volunteers had gotten two thousand pieces of content removed from YouTube for community guideline violations alone, plus tens of thousands more based on copyright violations. Of those two thousand individual web addresses HONR had reported and YouTube had removed, some were single videos and many others were entire channels with one hundred videos or more.

Lenny sent me lists of the content HONR had reported. There were dozens of pages, each line a sky-blue link to lies about Sandy Hook that Lenny had gotten erased from the internet.

Removing that much content changes what YouTube recommends to users and it flags as suspect. So two thousand reported videos could lead to exponentially more of them being taken down, he said.

"So the impact is hard to quantify because it's huge," Lenny added. "That impact had to do with the squeakiest wheel. Which is me. And that was the main reason why people hated me."

The conspiracy theorists reacted to HONR's efforts as conspiracy theorists do: they imagined plots within plots. One of them sent up a flare to Jones, using Infowars' anonymous "whistleblowers" channel.

"Just an FYI to Alex. Members of the HONR network (Lenny Pozner/Sandy Hook) have stated that they are going after Alex . . . The speed at which they act and the reach of their hand leads be [*sic*] to believe DHS or DOJ involvement . . . I just want to let you guys know that if you get any harassment by these people to take it seriously. They are organized, connected and ruthless."

Alex Jones epitomized Lenny and the other families' challenge in confronting the online conspiracists. When YouTube removed Infowars' content, Jones pivoted to his broadcast, using the "attack" on him to amplify the original lies, plus inflame his audience. And then that tirade, too, wound up on YouTube.

Like the algorithms that spread their content, the conspiracy theorists continually adapted, migrating their websites from one hosting company to another, creating new video channels, squeezing amoeba-like into new, dark spaces. Jones ruled over them all, his association with President Trump and his money-spinning supplements business having transformed him into an icon of the post-truth era. On his show that year, Jones called himself and his listeners "the operating system of Trump."

"I'm making it safe for everybody else to speak out just like Trump's doing, on a much bigger scale," he boasted.

JONES'S ON-AIR EXPLOSION AFTER YouTube removed Shroyer's broadcast brought matters to a head. The idea of suing Jones had begun to percolate among the families. But Lenny and Neil resolved to go it alone so as not to lose time. Jones's broadcasts had referenced both of them by name, giving them a firmer basis for a defamation claim.

Neil wanted revenge. He would tear Jesse's name from Jones's mouth if he could. Lenny, characteristically, took a more dispassionate view.

"Jones is just really good at what he does. He has this very large megaphone, this great power, and he should have shown responsibility with this great power. And so he had to be slapped," Lenny told me. "Everyone has this hatred for him, and I get that. But I really don't give a fuck that way."

That reminded me of what Lenny had said when I asked if he had wanted to meet the gunman's father. No, he told me, a meeting like that held nothing for him. The gunman had taken enough from them. Neither he nor his family deserved further consideration.

"He's a sociopath and a narcissist," Lenny said of Jones. "You could never make him feel bad about this, so why bother? It's a waste of energy."

They wanted Free Speech Systems, the empire Jones had built on lies, to pay for his recklessness. But to succeed they needed a skilled attorney who also understood Alex Jones's twisted world.

It was in the aftermath of another horrific school shooting that Lenny and Neil found that lawyer.

On Valentine's Day in 2018, Nikolas Cruz, a troubled nineteen-year-old who had been expelled from Marjory Stoneman Douglas High School in Parkland, Florida, opened fire on his former classmates and teachers. Seventeen of them were killed, and an equal number were wounded. Students recorded the shots from Cruz's AR-15 rifle, the bloodied bodies, and the chaos on their cell phones, and soon channeled their grief into furious protest. The Parkland kids galvanized young people and gun violence survivors to demonstrate in more than eight hundred cities around the world on March 24, 2018, under the banner "March for Our Lives."[6] Nearly two hundred thousand marched at the main event in Washington, demanding Congress act to end the gun violence killing scores of them each year.

March for Our Lives grew into a durable national movement. Not since Sandy Hook had a mass shooting created such a groundswell against gun violence. But unlike after Sandy Hook, the hunting of Parkland's survivors by conspiracy theorists came as no surprise.

Bogus false flag claims had grown into a feature of the country's post-shooting ritual, as predictable and swift as "thoughts and prayers" statements from members of Congress. In 2018, conspiracists shifted from blaming former president Obama to blaming socialists and Obama holdouts in the federal "deep state" for the bogus gun-grab plot.

Infowars usually piggybacked onto whatever claim had drawn the most interest online, siphoning off some of the traffic to its own social media accounts. After Parkland, Infowars amplified and embellished a false claim that bubbled up from 4chan, wrongly linking twenty-four-year-old Marcel Fontaine to the Parkland shooting.

Fontaine, a shy man with a severe stutter, lived in Boston and had never visited Florida. After the Parkland shooting, someone had lifted a photo of Fontaine from one of his social media accounts and posted it on a message board on 4chan, adding the caption "Shooter is a Commie." The red T-shirt Fontaine wore in the photo apparently made him the target for the fake charge. It was emblazoned with a hammer and sickle, and a cartoon spoofing the Communist "party": Lenin, Mao, and historic Communist figures were caricatured drinking from red Solo cups; Karl Marx wore a lampshade on his head.

Rushing to capitalize on the spreading falsehood, Infowars minion Kit Daniels posted the photo of Fontaine under headlines like REPORTED FLORIDA SHOOTER DRESSED AS COMMUNIST, SUPPORTED ISIS. Infowars' website boasted thirty million views a month by then, and the threats targeting Fontaine began surging almost immediately.

Mark Bankston, an internet-savvy Houston personal injury lawyer active in far-left circles, watched the Fontaine attacks gain speed. In him Fontaine found a sympathetic and like-minded advocate.

In early April, Bankston sued Jones and Infowars for defamation on Fontaine's behalf, seeking a judgment of $1 million.[7] Bankston filed the suit in Travis County, Texas, where the Austin-based Infowars operates.

Lenny read the lawsuit, which contained this passage: "Mr. Jones' recklessly opportunistic career is littered with the fallout from his willful pattern of malicious defamation, most notably a series of high profile incidents over the past few years. Mr. Jones garnered significant attention for his slander against the victims of the Sandy Hook massacre, claiming he has seen 'evidence' that could lead people to believe 'that nobody died there.'"

Before Bankston filed the Fontaine suit, Lenny and Neil's efforts to find a lawyer had been met with frustration. A couple of the attorneys Lenny and Veronique contacted wanted the participation of a big group of relatives, many of whom still seemed lukewarm on the idea at the time. Several lawyers Lenny spoke with didn't seem to get it, Lenny said. One said he and Neil didn't have a case because "it's okay to defame dead people."

Neil liked the idea of suing Jones in Texas. "What better place to serve it to him than in his home state and with a home jury?" he told me.

Neil called Farrar & Ball, the Houston firm where Bankston was a partner.

Bankston had recalled the news photographs of Neil testifying in the Senate shortly after the shooting, holding his portrait of himself with Jesse.

Talking with Bankston, Neil mentioned Lenny. Bankston thought Lenny had a case too, "but he needs to hurry, because the statute of limitations here is about to run out."

Neil replied, "I'll give him your number."

Defamation is a matter of state law. In Texas, people who believe a published statement has defamed them have one year from the time of publication to file a lawsuit. Jones had cast doubt on Neil's recollection of his last moments with Jesse in late July 2017; Neil had three months remaining on the clock. But Jones's most recent, and potentially damning, airing of his false accusation that Veronique had faked her CNN interview was broadcast on April 22, 2017. Bankston filed both lawsuits on April 16, 2018, making the deadline in the Pozner case by less than a week.

Five years after the murders of their sons, Lenny, Veronique, and Neil became the first Sandy Hook parents to take Jones to court to try to hold him accountable for years of denying their loss.

"On April 22, 2017, Mr. Jones broadcast a video entitled 'Sandy Hook Vampires Exposed.' The information presented was not new. It contained a continuation and elaboration of the same stale attacks Mr. Jones has made about the honesty and identity of the Sandy Hook parents for years," read the lawsuit Bankston filed on Lenny and Veronique's behalf.[8] Veronique's name was by then Veronique De La Rosa. She had remarried in 2015 and taken her husband's last name.

The lawsuit went on to say: "During the April 22, 2017 broadcast, Mr. Jones discussed a CNN interview with Plaintiff Veronique De La Rosa and Anderson Cooper, stating: 'So here are these holier than thou people, when we question CNN, who is supposedly at the site of Sandy Hook, and they got in one shot leaves blowing, and the flowers that are around it, and you see the leaves blowing, and they go [gestures]. They glitch. They're recycling a green-screen behind them.' The gist of this statement by Mr. Jones is that Mrs. De La Rosa's interview was faked and did not occur at the Edmond Town Hall in Newtown . . .

"The comments made by InfoWars in 2017 did not occur in isolation. Rather, the statements were a continuation and elaboration of a years-long campaign to falsely attack the honesty of the Sandy Hook parents, casting them as participants in a ghastly conspiracy and cover-up," the complaint read.

Neil's lawsuit[9] against Jones, Owen Shroyer, and Infowars reads: "This case arises out of accusations by InfoWars in the summer of 2017 that Plaintiff was lying about whether he actually held his son's body and observed a bullet hole in his head. This heartless and vile act of defamation re-ignited the Sandy Hook 'false flag' conspiracy and tore open the emotional wounds that Plaintiff has tried so desperately to heal . . . The underlying point or gist of Shroyer's report is that Plaintiff's version 'is not possible' and 'cannot be accurate,' and that Plaintiff was lying about the circumstances of his son's tragic death for a nefarious and criminal purpose . . .

"This conspiracy theory, which has been pushed by InfoWars and Mr. Jones since the day of the shooting, alleges that the Sandy Hook massacre did not happen, or that it was staged by the government and concealed using actors, and that the parents of the victims are participants in a horrifying cover-up."

In both cases, Bankston argued, Jones and his cohorts "acted with actual malice. Defendants' defamatory statements were knowingly false or made with reckless disregard for the truth or falsity of the statements at the time the statements were made."

Both lawsuits sought damages in excess of one million dollars. Six months later, Scarlett Lewis filed suit against Jones too.

The parents sought jury trials. They wanted Alex Jones to account for himself in public. In a post-truth culture where lies careen through cyberspace, smashing lives and reputations, they wanted Americans' verdict on whether our Constitution protects purveyors of these lies.

21

JONES OF COURSE WENT BANANAS. On April 17, the day after Lenny and Neil filed their suits across town in Austin, Infowars aired "Alex Jones' Full Statement on Frivolous Sandy Hook Lawsuit."

Jones's "statement" was essentially a four-hour jeremiad, interspersed with ads for ProstaGuard, Real Red Pill, and Bodease supplements, Superblue fluoride-free toothpaste, and Alexapure water-filtration systems.

"The Deep State is going into hyperdrive war against President Trump and basically all of his supporters in an attempt to bully and silence free speech," Jones began, grinding his teeth more than usual.

"It's George Soros financing the lawsuits against me. Two more filed yesterday." He heaved a theatrical sigh.

"Talked to two top law firms. They read the lawsuits, and they said, 'These are the most frivolous things we've ever seen. They are not even based in reality.'

"What they say we said against the Sandy Hook families we did not even say."

Jones re-aired his false flag claims about the shootings in Parkland and Las Vegas. He repeated most of his false claims about Sandy Hook, complaining he'd been unfairly edited, taken out of context, just asking questions, playing devil's advocate.

"It's just the mainstream media and the whole combine, using children and using hurt families to try to get them on board, to not just go after the Second Amendment but also the First Amendment."

The families weren't going after the First Amendment, they were going after Alex Jones. His insistence that the Constitution freed him to torment them would finally be tested, and in his home state.

By 2018 I was working as a writer in the *New York Times* Washington bureau, having left the *Times* editorial board to return to news reporting. A day after Lenny, Veronique, and Neil sued Jones, I told Elisabeth Bumiller, our bureau chief, that I found their story a heartbreaking example of the human costs of online abuse, and their lawsuits a signal test of free speech protection for damaging falsehoods spread by people from Jones to the president. She agreed, and I flew to Houston to meet the families' lawyers.

The law firm of Farrar & Ball was located on the sixteenth floor of a steely tower in downtown Houston, behind heavy, mission-style wooden doors that contrasted with the building's glassy sheen.

I met Mark Bankston in his office, NASA swag littering his credenza and a Rage Against the Machine poster hanging on the wall.

Bankston was thirty-nine in 2018, with a short beard, thick brown hair, and a smooth brow he furrowed theatrically when arguing in court. He and his wife, an environmental prosecutor in the Harris County District Attorney's Office, have one child, Ben, who that year was the same age as the first graders killed at Sandy Hook. My first mention of the massacre caused the lawyer to tear up.

"Ben's birthday is Noah's birthday," he told me. "The day of his birth was the first birthday that Noah wasn't around.

"There's something about Noah that reminds me of my own son," he said. "Sweet and mischievous at once, like he came out of the womb with a map of Europe and plans to invade."

Bankston was a third-year lawyer on December 14, 2012. He was vaguely aware of the theories spreading after Sandy Hook, but it wasn't until Ben was born that the damage of it sunk in.

As a native Texan, Bankston had known about Alex Jones since he was a teenager, checking in from time to time on Jones's evolution from a local flavor on Austin Community Access TV "who let you call in and talk about anything," to 9/11 truther, to supplements pitchman with a darker agenda.

"As weird as he is, he is a very smart person, a very good analytical thinker. So when you see what he says now, it just does not square with who that person is," Bankston said. "He absolutely knows that his viewership just eats this up with a spoon, at the same time they're eating his supplements up with a spoon. That's what this is all about.

"I don't think he believes a word of it," Bankston said. "And that really, really bugs me."

Infowars' theme, "There's a war on for your mind," seemed emblematic of an emerging ethos in which lies in service to an agenda are no longer lies but, as Kellyanne Conway, senior adviser to President Trump, damnably put it, "alternative facts."

Bankston frowned and fiddled with one of the plastic gewgaws on his desk.

"He knew about this, coming up in the late nineties. That's what's so terrifying about it to me. He's one hundred percent right. We're in the middle of a war over who controls information, who's the arbiter of truth. And if you destroy the arbiters of truth, anybody can be an arbiter of truth."

Bankston grew up in the Houston suburb of Richmond. His

upbringing by a father who was a criminal defense lawyer and a journalist mother felt like preparation for taking on Jones. "The collection of facts and the reporting of them to the public, there was something sacred about that. I grew up with that and really felt that very much."

His father, Donald Bankston, an associate judge in Fort Bend County, was appointed to the bench after a lifetime of private practice, during which he went after dirty cops and overreaching prosecutors on behalf of his clients. While in law school, Mark sometimes helped his father prepare cases, which afforded him an education in the darker side of human nature.

Mark Bankston graduated from the South Texas College of Law in 2009. He came out of law school not sure whether he wanted to be a defense lawyer like his father. He started working for a small personal injury lawyer whose health began to fail, so Bankston gained plenty of courtroom experience. In 2010 that got him in the door at Farrar & Ball, a firm whose founding partners were both under forty.

Kyle Farrar and Wesley Todd Ball graduated from Baylor University's law school in 2002 at the top of their class. Three years later, tired of partner politics and defendant work, they hung out their own shingle.

After a slow start, they won a series of cases on behalf of the victims of construction accidents, vehicle tip-overs and the like, building a modest national reputation.

In Florida, they met lawyers from a personal injury firm that had built a substantial defective-products reputation. They were nearing retirement age and struggling to manage a heavy caseload. Farrar, Ball, and their two associates at the time took on the overload. "All these tire cases in particular, and these boys just lit them up," Bankston said proudly. The young firm made a small fortune going after tire manufacturers for blowouts and failures that courts ruled were the cause of catastrophic injuries and deaths. Their success led to other defective-products cases.

"It was so much fun to have a firm with five guys under forty, just going after corporations," Bankston told me. "As we got incredibly successful at it, it allowed us to free up a huge part of our practice to say, 'What do we want to do? What do we want this law firm to be about? What kind of values do we want to go after?'"

Bankston introduced me to Bill Ogden, the associate working on the Sandy Hook cases, a thirty-year-old father to a four-year-old, Jack. Ogden, now a partner, worked at Farrar & Ball as a summer associate while still attending South Texas College of Law and joined the firm upon graduation. Well over six feet tall, with wavy, black Beethoven hair, Ogden has a sharply observant sense of humor, amusing his colleagues with telling details and anecdotes about opposing lawyers, defendants, and themselves.

It had fallen to Ogden to catalog everything Jones had said about Lenny, Veronique, and Neil, by reviewing hundreds of hours of Infowars broadcasts.

"It legit started messing with my mind," Ogden told me. He knew he needed a break when late one night, after hours immersed in Jones's claims of coming doom, he thought, "What if he's right?"

Ogden's work had also made him an expert on Infowars' product line. His favorite item: Combat One Tactical Bath wipes, which Ogden described as "baby wipes for middle-aged men who serve in a thrown-together militia out in the woods."

Bankston, Ogden, and I poked our heads into Kyle Farrar's corner office. Affable and slight, with sandy stubble on his cheeks and a gap between his two front teeth, Farrar came out from behind his massive antebellum desk, a rural roadside buy. We sat in leather chairs around a table in Farrar's conference room, adorned with the mounted heads of two tusked wild boars that Farrar and Ball had killed with knives in Florida.

The three lawyers loved that Jones had called them "ambulance chasers" and "exploding-tire lawyers" on Infowars. "He did say we were the best ambulance chasers," Farrar said.

"Yes! He called us the 'cream of the crop,'" Bankston added. "I think he thinks of us as Democratic Party apparatchiks." Actually Bankston is a socialist, which in Jones's estimation is even worse.

The partners considered their firm "an incubator," where profits from lucrative consumer safety suits sustained their work on pro bono "hobby cases." Bankston had worked on behalf of Houston's homeless people, threatening to sue the city over its plans to evict them from Tent City, an encampment beneath the U.S. 59 overpass, and sued an industrial cattle operation on behalf of a motorcyclist who hit an escaped steer.

All three lawyers began talking over one another to tell me about their favorite hobby case: suing the owners of a lemur named Keanu that attacked a mail carrier named Reeves.[1] "She got torn up," Farrar said.

"We got justice for a government employee," Bankston said. "Google it." Farrar & Ball secured a $600,000 settlement for Marla Reeves from Keanu's owners' homeowners' insurance.

"Largest lemur-related judgment in Texas history," Ogden said.

Farrar said the national attention around the Sandy Hook lawsuits surprised them.

"Yesterday we were in an elevator, and there was a random guy that we've never seen in our lives, and he's like—pardon my French—'I hope you kick that fucker's ass.' That is crazy, for people to recognize the lawyers."

The firm was inundated with calls from people offering to help pay the families' legal fees, volunteering research and expertise. The firm works on contingency; clients pay one-third of the award if they win and nothing if they lose.

I asked how they would define a victory in the cases. How far were the families prepared to take this?

"A 'Sorry, my bad,' is not gonna solve this," Bankston said. "The families know, and we understand too, that a verdict from a jury of his neighbors and peers is going to be very culturally meaningful, and have more impact on how he has to do business, than a settlement will ever be."

Farrar remarked that in the space of several weeks, four parties had sued Jones for defamation: Lenny and Veronique; Neil; Marcel Fontaine; and a musician and former State Department employee named Brennan Gilmore. In August 2017 while demonstrating against the white nationalist Unite the Right march in Charlottesville, Virginia, Gilmore filmed a neo-Nazi speeding his car into a crowd of counterprotestors, killing Heather Heyer, thirty-two, and wounding a score of others. Infowars and others falsely accused Gilmore of being a government operative, fomenting violence to discredit Trump.

In response to each suit, Jones had only doubled down.

"I'm sort of shocked he doesn't have a lawyer trying to tackle him, going, 'Dude, just stop talking about this. You're not doing yourself any favors, you know?'" Farrar said.

The three lawyers wondered who would defend Jones in Lenny's and Neil's cases. Jones could certainly afford counsel, and had lawyers at his disposal in other cases. But defending him against the Sandy Hook parents presented a different and unsavory challenge.

"I don't think any of what I would consider some of the prestigious defense firms would want their name associated with him," Farrar mused. "Who wants to sign on to that?"

Months later, the Texas lawyers got an answer. Mark Bailen is a Washington, D.C.–based partner at BakerHostetler, one of America's best-known law firms. For reasons even his fellow media lawyers told me they struggle to understand, Bailen quietly represented Jones on several matters during the Trump era.

In early 2017, after James Alefantis discovered that an Alex Jones video had brought the gunman to Comet, Bailen brought Alefantis, his lawyers, and Jones together in BakerHostetler's elegant offices near the White House. While Jones sputtered and raged, convinced Hillary Clinton lurked behind the legal threat, Bailen brokered Jones's stilted public retraction, avoiding a lawsuit.

A month later, when Chobani founder Hamdi Ulukaya sued Jones for falsely accusing him of "importing migrant rapists" to the U.S., Bailen again represented Jones. Jones broadcasted a belated retraction, and Chobani dropped the suit.[2]

BakerHostetler and Bailen were defending Infowars[3] in the Gilmore suit in Charlottesville, too.

It would take the Texas lawyers[4] a while to discover traces of Baker-Hostetler's work in at least one Sandy Hook case. In a court filing in Neil's lawsuit dated August 27, 2018, Jones's father, David Jones, referred to Mark Bailen of BakerHostetler as "one of our lawyers."

The elder Jones's declaration included a letter Bailen sent to Google on Infowars' behalf on August 16, 2018, after Google terminated its content hosting services agreement with Free Speech Systems, LLC, Infowars' parent company.

Not long after that, Bankston found a 2018 Infowars email in a trove of Infowars documents released as part of pre-trial discovery, in which Mark Bailen and Texas lawyer Eric Taube "were deciding on what evidence to secure, what transcripts to get for discovery," in Neil's lawsuit against Jones, Bankston told the judge in an August 2021 court hearing. Separately, I read another 2018 Infowars email released in court proceedings that described Infowars staffers' efforts to gather and send documents to Bailen.

"Bailen is a well-recognized, well-respected First Amendment defamation lawyer," a lawyer who dealt with Bailen in his capacity as an adviser to Jones told me. "I couldn't tell you whether this is something he really believes in passionately, or it's a client he got that pays the bills."

Bailen doesn't like to answer questions about his work for Jones.

In late 2021 Bailen's wife, Jessica Rosenworcel, was the acting chairwoman of the Federal Communications Commission. She was campaigning to be President Biden's nominee as permanent chair of the powerful broadcast and telecommunications regulator. I had not met Bailen or Rosenworcel, a communications lawyer from Connecticut with long experience and a solid record at the commission, but the two are a Washington "it" couple, well-known in D.C.'s overlapping legal and government circles. In a speech to the National Association of Broadcasters in 2018, Rosenworcel urged them to fight back against Trump's efforts to undermine a free press, saying that her own young son had used the term "fake news" at the family dinner table.[5] But the other parent presumably at the table that night was Bailen, whose client Alex Jones claimed to have coined Trump's phrase "enemy of the American people" to discredit the mainstream media.

In an interview for the *Times* about her bid for the FCC job, I asked Rosenworcel about Bailen's work for Jones. They lead "very separate professional lives," she told me, so I should ask Bailen. Rosenworcel added that BakerHostetler had ended its relationship with Jones. Bailen said BakerHostetler had stopped representing Jones in the Gilmore case in Charlottesville in late 2020, a couple of months before Biden's election.

Rosenworcel had the strong support of Senator Richard Blumenthal, Democrat of Connecticut. I asked him whether it mattered that Bailen had worked for Jones. The senator had not known about it. He said it was "irrelevant" to him, adding that he didn't know Bailen. Then one of his staffers called Rosenworcel, saying the senator had been "ambushed" with the question.

I invited Bailen, who had once handled a couple of cases for the *Times*, to lunch for an interview. He agreed, postponed, and asked for questions in writing. He repeatedly contacted the *Times*' legal department, complaining that news of his work for Jones could unfairly hurt his wife's chances for the FCC job, and angry that I had raised it with Blumenthal, who represented most of the Sandy Hook families in the Senate. David McCraw, the *Times*' principal newsroom lawyer, encouraged Bailen to sit for the interview he had agreed to.

A couple of weeks later, Biden nominated Rosenworcel as FCC chairwoman. I wrote an article for the *Times* that included a description of Bailen's unsuccessful efforts to engage *Times* Legal on the couple's behalf.[6]

Bailen had raised concerns about journalistic fairness and standards, and objected to the "tone" of my questions. This was one: "A firm of BakerHostetler's caliber can choose its clients. Why did Baker choose to represent Jones?"

"A hallmark of First Amendment law is that it seeks to protect speech that may be unfair, unpopular, and sometimes outrageous," Bailen wrote. (He insisted that all our exchanges be in writing.)

Neither he nor BakerHostetler had been Infowars' "counsel of record" in the Sandy Hook lawsuits, he wrote. I knew that: his involvement was strictly behind the scenes. Bailen called my reporting on his work for Infowars "inaccurate," but when I asked for specifics, he declined to provide any, writing, "My ethical obligations as a lawyer prevent me from discussing details of former client engagements, irrespective of whether any elements of them have become public."

Kyle Farrar was right. No prestigious firm wanted to be associated with Jones's battle against the Sandy Hook families. At least not publicly.

In Farrar's office in Houston, I asked about the nuts and bolts of the cases.

"You've got three elements in Texas that you'll have to prove," Bankston said. "I've got to prove that he published something, which he's done. Next, I gotta prove that it was defamatory to them, which, I mean, any reasonable-minded person sees that that's the case. And then, the big fight that I think Alex is gonna want to have, which I don't think it's a fight at all, really, is whether they're public figures or not."

Would Neil and Veronique's public statements, or Neil's testimony in Congress, make them public figures? Would the high-profile nature of the crime turn all the victims' families into "limited purpose" public figures?

It wouldn't matter, Bankston said, if the lawyers could prove that Jones acted with malice: that he knew his charges were false and made them anyway, or that he acted with reckless disregard of their truth or falsity.

"The other day I was trying to think of an act of defamation in American history which was more malicious than this," Bankston said.

"His whole schtick is to sell a product," Farrar said. "It's so malicious to tell this type of lie about somebody who's gone through what they've gone through, to sell Brain Force Plus pills."

In Austin, when Bankston and Farrar were growing up, "Jones got away with saying all this stuff before because he didn't have an audience. Who cares what some guy yelling at clouds is saying?" Farrar said. "But now his megaphone is significantly bigger. He's talking to this big audience and now he's saying this crazy stuff that has a real effect on people. It's like his rise is his downfall."

Farrar asked if we were ready for lunch, which meant piling into his white Porsche Panamera for a trip to a local taqueria.

Bankston returned to my question about what would define victory.

"For Alex Jones, it appears that the only real thing on his mind in terms of punishment is his business and his money. So I think if you put a threat to that, if you make him understand that these kinds of 'journalistic' practices have a cost and an effect, and that he won't be able to profit off of causing pain to a family, I think that's a victory too. And particularly if that message goes out to others. I think that's what a victory looks like.

"What happened to the Sandy Hook parents is really like an emblem of what all of us are fed up with," he said, and not just from Alex Jones.

"If we can win this, if we can hold him accountable, maybe we can get back to where we need to be."

THE HOUSTON LAWYERS AND I spent a lot of time talking about fake news and online lies, which in the aftermath of the 2016 election seemed overwhelming. But it was only 2018, and we had no idea how bad it could get.

When I turned up at Farrar & Ball on the morning of my final day in Houston, office manager Debbie Brooks handed me breakfast, a paper plate piled with taquitos. I parked myself in the conference room. Farrar, seated beneath one of the boar's heads, plied his laptop while Bankston dug through bankers' boxes and stacked towers of files. They were preparing for the latest phase in a big product lawsuit against 3M.

Bankston stepped out to go rummage in his office. He came back with a small stack of brightly colored cards. "These came after we filed the suits," he said, laying them in rows on the dark wood conference table.

They were notes from total strangers, intended for the families.

"Thank you for your bravery as you fight against the lies being told by Alex Jones."

"Dear Leonard Pozner, Veronique de la Rosa & Neil Heslin: Solidarity! I am so sorry for your losses. I keep your children as well as their school-mates in my memory and my heart. My youngest child was in elementary school at the time and I cannot forget that day or your children. I am firmly on your side in your fight against Alex Jones."

"I'm so sorry for all you have gone through! I'm sorry for all of this! I'm sorry you have to keep fighting this fight! I'm sorry your hearts are forever broken! I'm sending love—hugs—and positive healing energy!"

"Dear Ms. de la Rosa, My sincere condolences to you for the loss of your son. Thank you for courage in the belief in what's right & good that allows you to go forward with your lawsuit against Alex Jones. We want to give encouragement to you in your effort, just as you give encouragement to all of us who may face our own travails."

"Dear Ms. De La Rosa: Please know that many of us are grateful you are trying to hold Alex Jones accountable for the horrible painful remarks he has made for 'entertainment.' He is so disgusting and offensive that it is hard to understand."

And finally:

"Dear Leonard Pozner, Veronique de la Rosa and Neil Heslin: The remarks by Alex Jones are disgusting. Please fight him until there is nothing left."

It would take years, but that was the plan.

22

IN SPRING 2018 MY REPORTING on the Texas lawsuits brought me into contact with the Sandy Hook families for the first time. I noticed their heartbreaking tendency to ask, "How much do you want to know?" before speaking about the shooting. Lenny told me they had learned that not everybody wants to hear the painful details.

I wondered: What role does this collective turning away play in our country's embrace of conspiracy theories, or our ignorance of their impact on survivors? If we can't stand to hear about the survivors' pain, do we cede the battle for truth to the liars?

I watched a documentary released in early 2018 called *The Rise of the Crisis Actor Conspiracy Movement*, produced by Vice Media Group.[1] The film opened with a notorious hoaxer, Robert Mikell Ussery, a.k.a. "Side Thorn," and his partner Jodie Marie Mann, or "Conspiracy Granny," confronting Frank Pomeroy, the pastor of First Baptist Church in Sutherland Springs, Texas. Pomeroy's fourteen-year-old daughter, Annabelle, was among the twenty-six congregants killed when a gunman opened fire during services on November 5, 2017. Pomeroy was out of town on the Sunday the gunman arrived.[2]

"This is supposed to be a man of God, and yet he's told the whole world that twenty-six people died in his church when he knows nobody died," Ussery barked at Pomeroy when the pastor approached him in front of his church. I could not fathom where Pomeroy summoned the patience to quietly tell the couple they were trespassing, delivering the verbal warning required for police to arrest them.

"You're a filthy liar, Frank. You're a demon, Frank," Ussery shouted at him. "I'm gonna expose you motherfuckers for what you are, until the people hang you by the neck, man. You're gonna hang, traitor!"

Police arrested Ussery and Mann while the camera rolled. They later learned that Ussery, a convicted felon, had a semiautomatic pistol hidden in his pickup truck.[3]

He was jailed on weapons and other charges.[4] Mann was sent to a mental health facility for evaluation.[5]

The same film depicted Alex Jones targeting David Hogg, a Parkland survivor and March for Our Lives organizer. Jones aired a news clip of Hogg relaying his account of the shooting in a TV interview. "I know scripted PR when I hear it," Jones said. He was repeating almost verbatim the false claims he'd leveled against the Sandy Hook families a half decade earlier.

Hogg, not yet eighteen at the time, shrewdly saw Jones's shopworn claims for what they were. Hogg withdrew his offer, issued on Twitter, to debate Jones on Infowars. "He's doing that exact same stuff, and knowing that he'd done that before, I wasn't going to give him the time of day," he said in the film.

Tony Mead was filmed touring the makeshift memorials in front of Marjory Stoneman Douglas High School, practically in his backyard. "It's a horrible thing to think that seventeen children got murdered," he told the camera crew somberly. "But it's even more disturbing to think that the government would do something like this and present it as a real event, when it's not."

Mead invited the film crew into his living room, to show them his Sandy Hook Hoax Facebook page. "Lenny Pozner, one of these supposed parents of a victim at Sandy Hook, Noah Pozner, has certainly been determined to quell any information regarding the Sandy Hook incident," Mead said, preening for the camera. "And of course it's disguised by 'he's a poor victim who wants to preserve the legacy of his child.'

"His child has no fucking legacy! Nobody cares about this kid. This kid is a flash in the pan. Sandy Hook is years behind us now. The only people that care about it are the people that want the truth."

Lenny appeared in silhouette. "Facebook has not done anything to deal with this hate group," he said. "I think it'll get a lot worse. They are turning these conspiracy ideas against people, against victims. And that's a very dark evolution."

In the final moments of the film, a chyron appeared, saying Facebook had taken down the Sandy Hook Hoax page "in accordance with their Community Standards policy."

The Sandy Hook Hoax members had been targeting Lenny and the other families since 2013. It had taken Facebook five years, and the Parkland shooting, to enforce its "community standards."

I SOUGHT OUT OTHER Sandy Hook relatives to learn about their experiences with Sandy Hook deniers. I left a message for Erica Lafferty, the daughter of Sandy Hook principal Dawn Lafferty Hochsprung, at the gun safety advocacy group where she worked. Shortly afterward I got a call from a public relations agency working for Koskoff, Koskoff & Bieder, the Connecticut law firm representing the Sandy Hook families who were suing Remington Arms Company.

It turned out Koskoff was drafting a lawsuit against Jones too. Erica Lafferty was the lead plaintiff in the suit, which targeted Alex Jones, his businesses, and his collaborators, including Wolfgang Halbig.

Erica Lafferty, et al. v. Alex Jones, et al.[6] was filed on May 23, 2018, in Superior Court in Bridgeport, Connecticut. The plaintiffs included the relatives of six Sandy Hook victims and an FBI agent targeted by the conspiracists.

The Connecticut suit ratcheted up the pressure on Jones. Eight victims' families were now suing him, in two states.

Lafferty was twenty-seven and planning her wedding when she got an emergency text message on December 14, 2012. There had been a shooting at Sandy Hook, the center of her mother's professional life. "Mom was very much the type of principal who knew the name of every child in that school. She knew their siblings, and she knew their parents, and she probably could tell you how many dogs and what breed they were," Erica would say years later in a video she made for the Democratic National Committee, supporting Hillary Clinton.[7]

Erica's sister, Cristina Hassinger, a year older, was shopping with two of her children when Erica reached her on her cell. The two women met their stepfather, George Hochsprung, in Newtown and together they sprinted for the firehouse, holding hands.[8]

Pretty, with deep brown eyes and an open, expressive face, Erica has the broad smile and tough nature of her mother, the first to confront the gunman.

In 2013 Erica threw her support behind an unsuccessful bill requiring comprehensive background checks before gun purchases. She spoke at the 2016 Democratic National Convention in support of Hillary Clinton's stance on guns. She also subsequently endorsed Senator Pat Toomey, Republican of Pennsylvania and cosponsor of the background check bill, for reelection.

"I really made great efforts to ensure that this is not a partisan issue and the focus is on this being a public health and public safety issue," she told me.

Erica's role in the gun policy debate put her on Infowars' radar.

After Trump's victory Erica wrote an op-ed in *USA Today*, joining Newtown in asking the president-elect to disavow Jones and his false claims about Sandy Hook.[9]

"We cannot normalize fact-denying behavior," she wrote in the November 11, 2016, piece. "We must all realize that claiming mass shootings are elaborate, government-manufactured hoaxes is deeply hateful and hurtful to those of us living this terrible truth."

After Erica's *USA Today* piece, Infowars broadcast a five-minute rant about her, the Connecticut lawsuit said. Owen Shroyer addressed Lafferty directly, asking, "Why are you butting heads with people that want to find out the truth of what happened to your mother?"

Two more Connecticut plaintiffs, William Aldenberg, an FBI agent who responded to the shooting, and David Wheeler, whose son Ben died in the school, were targeted by Halbig, who posted photographs and video falsely claiming that they were the same person. Halbig and scores of other hoaxers seized on Wheeler's background as an actor in New York to claim that he was a paid "crisis actor" who toggled between roles as a Sandy Hook parent and an FBI agent on the day of Ben's murder.

Francine Wheeler, Ben's mother, was also a plaintiff. Dark-haired and petite, Francine is an actor and musician who met David, six years older, while performing in a variety show David coproduced in New York. The Wheelers' firstborn, Nate, and Ben, three years younger, were like "Frick and Frack," Francine told me. Nate was in fourth grade at Sandy Hook in 2012 and survived the shooting by hiding with his classmates in an equipment closet in the gym.

The Wheelers spoke frequently in public after Ben's murder and lobbied on Capitol Hill for new gun legislation. In April 2013, at the height of the gun policy debate in Congress, Francine delivered President Obama's weekly radio address[10] with David at her side.

"I've heard people say that the tidal wave of anguish our country felt on 12/14 has receded. But not for us," Francine said in the address. "In the four months since we lost our loved ones, thousands of other Americans have died at the end of a gun. Thousands of other families across the United States are also drowning in our grief.

"Please, help us do something before our tragedy becomes your tragedy."

Hoaxers sent photos they found of Francine's acting career to Jones, distorting them to make her look evil or mentally ill. When the Wheelers joined the lawsuit in 2018, Nate was fifteen and had found lies written about him and his parents on the internet.

"That required very delicate conversations with Nate that I would never want any parent to have to have with their kid," David Wheeler told me.

"I put the general population of theorists who actually believe this stuff and Jones in different buckets," he said. "I don't have a lot of anger for the general population of these people. In many cases they just don't know better, and I want them to get the help they desperately need. But these people are just being played in the worst possible way.

"Jones is the person who is profiting from this. He is the person who built a business on this."

Like Robbie Parker, David thought he could tell when Jones ranted about Sandy Hook on Infowars because he saw a spike in cruel messages on the Wheelers' social media accounts. What disturbed him most were the biblical references, warning him he would pay for his crimes when he died.

"Because of the information age in which we live, it is very important that we be careful and vigilant as a society to maintain a certain intelligence in our civic and civil discourse. Behavior of this sort must be addressed in a meaningful way," Wheeler told me.

I asked whether he was suggesting an apology, or a retraction.

"Oh hell no," Wheeler said. "Mr. Jones and his broadcast affiliates need to understand and face serious consequences for their actions, that change the way they operate in the world."

The Connecticut complaint recapped what Jones had said about Lenny, Neil, and Robbie, who had not yet joined the Connecticut suit. Infowars had mentioned only a couple of the Connecticut plaintiffs by name. But Halbig had gone after virtually all of them.

The lawsuit threw the proverbial book at Jones. It accused him and his compatriots of defamation, invasion of privacy, intentional and negligent infliction of emotional distress, and civil conspiracy. It alleged that Jones and the other defendants had violated the Connecticut Unfair Trade Practices Act, because they "unethically, oppressively, immorally, and unscrupulously developed, propagated, and disseminated outrageous and malicious lies about the plaintiffs and their family members, and they did so for profit."

"This isn't political speech; this is commercial speech, and it's painful commercial speech," Bill Bloss, a veteran product liability lawyer at Koskoff told me. "Society may have a different view than the defendants about the appropriateness of lying just to make money. We'll see."

On the Koskoff team suing Jones were Alinor Sterling, whose encyclopedic knowledge of the case made her the go-to for the rest, who included Josh Koskoff; Chris Mattei, a former assistant U.S. attorney in Connecticut; and Matthew Blumenthal, son of Senator Richard Blumenthal, Democrat of Connecticut, who had been a prominent voice in the effort to pass gun legislation after Sandy Hook. His son Matthew's work on the Sandy Hook families' lawsuit against Jones was why I asked the senator whether it mattered to him that Mark Bailen, the husband of his favored choice to chair the FCC, had represented Jones.

As he did with all of the half dozen defamation suits against him, Jones portrayed the Connecticut lawsuit as a Democratic Party effort to silence him.

Josh Koskoff, an affable and energetic litigator who was lead lawyer in the Remington case, ran KKB, a third-generation Connecticut firm. While over their history the Koskoffs had taken on big civil rights cases, for years the bulk of their caseload was made up of serious personal injury, wrongful death, and medical malpractice suits. The firm's late founder, Theodore Koskoff, defended clients ranging from Black Panther members to disgraced trial lawyer F. Lee Bailey. The late Michael Koskoff, Theodore's son and Josh's father, had teamed up with Josh's brother Jacob to write the screenplay for *Marshall*, a film about an early case in the career of U.S. Supreme Court Justice Thurgood Marshall.

Though some of the families suing Jones in Connecticut had retreated from the public eye by 2018, others remained prominent in the campaign against gun violence. Plaintiffs Nicole Hockley, whose son Dylan died in the shooting, and Mark Barden, whose son Daniel perished, led Sandy Hook Promise, a gun safety group and cause célèbre.

Lenny was a party to the lawsuit against Remington, but he had never loaned himself to the gun debate. This was on purpose.

Lenny firmly opposed conflating the battle against the hoaxers with any other cause. He believed online abuse was a discrete, increasingly pernicious problem affecting mass shooting victims across the political spectrum. It demanded focused action.

"If you're going to keep fundraising over taking guns away, that will result in a hateful response from the hoaxers," because Jones and others had been pegging their false claims to gun policy, Lenny said.

The lawyers in Texas told me they found Jones's status as a pro-Trump villain of the left a potential distraction in the Texas cases. The jury pool in Travis County, where the suits were filed, was overwhelmingly liberal. Yet Republicans dominated the state supreme court, where the case could be subject to review. Playing it safe, the Texas lawsuits made no mention of the president's ties to Jones.

My article about the Sandy Hook families' battle for truth appeared on the *Times* website immediately after the Koskoff lawyers filed the lawsuit in Connecticut.[11]

The article began with Neil, cradling Jesse's body, grateful to hold him for a final time. And then Jones, saying on air that Neil needed to "clarify" that recollection, and Shroyer's outrageous claim that it was "not possible."

To learn more about the free speech issues, I had spoken with David Snyder, executive director of the First Amendment Coalition, a San Rafael, California–based nonprofit dedicated to advancing free speech and open government. I had worked with Snyder at the *Washington Post* before he left journalism for law school at the University of California, Berkeley. In private practice, Snyder defended *Mother Jones*, *Salon*, and the *San Jose Mercury News* in defamation cases, while running a brisk pro bono campaign for greater public access to government records.

Snyder did not scoff at Jones's free speech defense.

"The law and the Constitution look with great disfavor on defamation cases, with good reason," he said, reminding me of multimillion-dollar judgments that ran solid media outlets out of business, costing good journalists, including a few we knew, their jobs. "It's not impossible to win a defamation case, but it's a lot more difficult than your typical personal injury case."

Snyder said that the families sailed over the first hurdle: they could easily prove they were harmed by Jones's false claims about the worst episode in their lives. But, Snyder told me, the Constitution tips the scales in Jones's favor. First, by requiring they prove Jones's comments were false statements of fact, not opinion. Essentially, that's the difference between Jones calling them liars (a false statement of fact, and Jones could be in trouble) or merely saying he thought they were lying while offering some basis, however weak, for that position (opinion, and a potential win for Jones).

Jones had been toeing the tightrope between statements of opinion and fact for years, and not just on Sandy Hook. After Lenny, Veronique, and Neil sued him in Texas, Jones posted a ten-minute video that was a masterpiece of double-talk. "I questioned the PR and the talking points that surrounded the Sandy Hook massacre," he said. "But very quickly I began to believe that the massacre happened, despite the fact that the public doubted it."

At the same time, Jones had an earlier video still up on his website called "Alex Jones Final Statement on Sandy Hook." "If children were lost in Sandy Hook, my heart goes out to each and every one of those parents, and the people that say they're parents that I see on the news," Jones said in that video. "The only problem is, I've watched a lot of soap operas, and I've seen actors before."

Snyder said a judge could choose which of the plaintiffs has a legitimate claim. The judge could also find that some of them are public figures, subject to a higher burden of proof, through their policy activism. Potentially, the high-profile nature of the crime could render them all public figures. "Emotionally, that doesn't seem fair, but the law can be pretty cold," he told me.

That could present another hurdle for the families. "They would have to show more than that they were harmed by the things Alex Jones said. They would have to show that he was at least negligent and did not take the steps an ordinary reporter would take to corroborate facts," Snyder said. That's the standard if the families were deemed to be private figures. If, however, they were found to be public figures, they would need to show "actual malice," as the Supreme Court defined it in *New York Times v. Sullivan*: acting with "reckless disregard for the truth," which means the reporter knew or had good reason to believe something was false but went ahead and published it anyway.

Snyder acknowledged one clear danger for Jones.

"These people are among the most sympathetic plaintiffs you'll ever encounter in a lawsuit," he told me. "If these cases were to get to a jury, I'm pretty confident a jury would try to give them what they're looking for."

I called David McCraw, the *Times*' newsroom lawyer. McCraw had come to the *Times* from the New York *Daily News* via a career in journalism starting at the *Quad-City Times* in Davenport, Iowa.

The Trump era was, as McCraw put it, "a hell of a time to be a lawyer at the *New York Times*." After a relatively sleepy couple of decades, reporters at the *Times* had been threatened repeatedly with libel lawsuits from politicians, including President Trump himself. McCraw, a talented writer and the kind of righteous legal pugilist you definitely want on your side, had made a name for himself in his response to a letter Trump's lawyers had sent to the paper, threatening to sue the paper for libel over an article Trump did not like.

The letter was extraordinarily colorful,[12] but its upshot, McCraw said, was this: "The essence of a libel claim, of course, is that a statement lowers the good reputation of another in the eyes of his community . . . Nothing in our article has had the slightest effect on the reputation that Mr. Trump, through his own words and actions, has already created for himself." Trump's threatened lawsuit never materialized.

McCraw walked me through the likely next steps in the Sandy Hook cases, quoting Texas and Connecticut defamation law. He was describing a process that, true to his prediction that day, would take years.

My eyes starting to glaze over, I asked him what his over/under was on how this ultimately would play out for the families.

"This is fake news on trial," McCraw told me, a description that would remain with me ever after.

"If this gets in front of a jury, Alex Jones is toast."

23

IN EARLY JULY 2018, Robbie Parker joined the six victims' families suing Jones in Connecticut, as did William Sherlach, husband of Mary Sherlach, the school psychologist who was slain. Now ten victims' families were suing Jones, in two states.

Robbie and Alissa had spent nearly five years edging away from the hoaxers, convinced that engaging them would invite new torrents of abuse. But the abuse continued regardless.

First there was the stranger who approached Robbie on the street in Seattle in 2016. The man had seen so many videos and social media posts labeling Robbie a fraud that he recognized him four years after Emilie's murder, on a street three thousand miles from Newtown.

Kevin Purfield, the mentally unstable man who had written them and called their friends before they left Newtown, had also resurfaced. Purfield still lived in the Portland, Oregon, area, an easy drive from the Parkers' home in Washington State. And this time, he was sending letters—so the Parkers knew Purfield had their new home address.

Purfield had been arrested in 2013 for stalking and harassing families of the Aurora, Colorado, shooting victims.[1] He was on probation when later that year he made multiple bomb threats to a courtroom complex, police precinct, and jail in Portland. He was released from prison in 2015.[2] When his probation ended, he had again begun contacting the Parkers.

"We have never sought out a restraining order or anything like that. We were advised that if we did, that might just show him that 'he is getting to us,'" Robbie told me.

That advice came from a safety expert who worked with Alissa at Safe and Sound Schools, the nonprofit she had founded with Michele Gay. The expert specialized in threat assessment, and fortunately had contacts with the Portland police department. Together, they kept tabs on Purfield, and the Parkers maintained their uneasy silence for years.

It took another grieving parent to help change Robbie's mind about joining the lawsuit.

A couple of months after the 2018 Parkland shooting, elders of the Church of Jesus Christ of Latter-day Saints asked the Parkers to talk with Ryan and Kelly Petty, fellow LDS members whose fourteen-year-old daughter, Alaina, had been murdered at Marjory Stoneman Douglas High School that February. Robbie and Alissa called the Pettys in Florida. Sitting on their bed with the grieving couple on speaker, together they offered what comfort and answers they could.

Ryan had spoken publicly about his daughter's death while running for the Broward County School Board. Alaina, slender, with brown hair and braces, had been a Junior ROTC cadet at Stoneman Douglas. She liked to target shoot on a nearby gun range with her father and brother Patrick. Her favorite weapon had been the AR-15,[3] the same style gun used to kill her.

In a wrenching meeting with the *South Florida Sun-Sentinel* editorial board in late July 2018, Ryan, a registered Republican, tried to clarify his position on guns. "We had a meeting of families and we have different views on gun control," he said, struggling with emotion as he sat facing the board in a glass-walled conference room. Ryan told me he felt judged, even ambushed, by the editorial board because of his gun policy stances. "One of my daughter's favorite things to do was go to the gun range with me," he said in the meeting with them. "You can think I'm a bad parent if you want, that's up to you. But she loved it. And I haven't been able to share that or talk about that, or remember her for that."

A photo that Ryan, a scout leader, had posted of himself shooting a rifle at a scout camp, along with questionable comments he had made on Twitter, caused an uproar during his unsuccessful campaign for the school board. "To have somebody throw something like this at me is really hard," he told the *Sun-Sentinel* editorial board. "It's personally painful—and that's okay—that I haven't been able to talk about one of the things that my daughter and I loved to do together."

The hoaxers came for Ryan Petty online. He told the Parkers of his shock when he first saw the anonymous videos on YouTube, calling Parkland a false flag and a hoax. One singled out Alaina. "The video attempted to show that Alaina was not in fact fourteen, but a thirty-five-year-old woman from another state. It was awful," Ryan told me.

For Robbie, the conversation confirmed that ignoring the abuse didn't work.

"That was a catalyst," Robbie told me. "I thought, 'These people aren't going away. It's time for us to see what we can do about it.'"

Hearing about Alaina Petty, Robbie thought also of his own girls. Samantha and Madeline were barely more than toddlers when Emilie was killed. But in 2018 "they were getting to an age where they're exposed to more things in the world, and I can't protect them as well," especially from what they find on the internet, Robbie told me. He did not want to join the lawsuit initially. But he couldn't help noticing that after the families sued Jones, the big platforms started to pay closer attention to the pain created by their failure to control their creations.

The Pettys were a good example. After Ryan complained to YouTube about the video that focused on Alaina, "they took it down immediately," Ryan told me. That's a far cry from what had happened when Robbie earlier had tried to report the scores of videos targeting him and his family.

Likewise Jones, Robbie thought. As long as Infowars was spinning money, the families' pain rolled right off him. In fact, he turned their protests into more cash, bellowing about his free speech to spur donations to Infowars, claiming he needed the money to keep the doors open.

For years, "I didn't want to admit how much this was affecting me," Robbie told me. "Now I want Jones to sit there in a court and admit how much this is affecting *him*."

THE SANDY HOOK LAWSUITS had placed a glaring media spotlight on Jones and his social media enablers at last, and Lenny worked to keep it there.

In April 2018, the same month Lenny, Veronique, and Neil filed their lawsuits against Jones, Facebook founder Mark Zuckerberg testified before Congress amid the biggest crisis in his company's history. The *New York Times*, with London's *Observer* and *The Guardian*, had written a series of stories[4] based on a trove of documents proving that Cambridge Analytica, a company controlled by right-wing megadonor Robert Mercer, improperly accessed the personal data of tens of millions of Facebook users in 2014. The harvesting amounted to the largest known leak in the company's history. Stephen K. Bannon, a top Trump campaign official and later White House adviser, sat on Cambridge Analytica's board. The company used the information it collected to construct psychological profiles of potential voters, which it then tried to sell in the run-up to the 2016 presidential election. Evidence emerged that Lukoil, a Kremlin-linked oil conglomerate, was among those interested in Cambridge Analytica's targeting of U.S. voters.

Congress summoned Zuckerberg for a couple of marathon hearings, during which he expressed regret and promised new tools to plug the data-protection gaps exploited by Russia and others.

Lenny bristled at Zuckerberg's threadbare apologies and glib non-answers. It aggravated him that even in the light of a sweeping privacy scandal, Congress seemed not to fully grasp that Facebook profited by vacuuming up and selling a vast quantity of its users' personal data, not by "connecting the world."

"All I hear are Zuckerberg's mechanical, well-rehearsed responses," Lenny told me of his main memory of that day. The scene convinced Lenny that "it's going to take another generation before they understand enough to even begin to regulate these companies."

But Zuckerberg's performance mollified others, including Facebook's investors. The company's share price rose during the hearings. "The Street is relieved," a tech industry analyst told the *Times*.[5]

That wouldn't last. The Justice Department, the FBI, the Securities and Exchange Commission, and the Federal Trade Commission all launched investigations into Facebook, while Europe imposed new regulations[6] protecting social media users' data.

Facebook's failure to protect its users and American democracy was on full, global display.

Cambridge Analytica "compounded fears that the algorithms that determine what people see on the platform were amplifying fake news and hate speech, and that Russian hackers had weaponized them to try to sway the election in Trump's favor," Karen Hao wrote in *MIT*

Technology Review.[7] A Pew Research Center survey[8] taken after the scandal broke suggested that nearly three-quarters of American Facebook users distanced themselves from the platform, including millions who deleted the app entirely.

In July 2018, during his ensuing apology tour, Zuckerberg sat for his gaffe-laden interview with Kara Swisher.[9]

Swisher told me that before her Recode podcast team began recording that day, she greeted Zuckerberg by predicting that he would wind up kicking Jones and Infowars off Facebook. He firmly disagreed. So Swisher brought it up again during the interview, telling Zuckerberg, "Make the case for keeping them, and make the case for not allowing them to be distributed by you."

Zuckerberg spoke for a few minutes, drawing a distinction between free expression and community safety.

"The principles that we have on what we remove from the service are: If it's going to result in real harm, real physical harm, or if you're attacking individuals, then that content shouldn't be on the platform. There's a lot of categories of that that we can get into, but then there's broad debate."

Pressing him, "'Sandy Hook didn't happen' is not a debate," Swisher said. "It is false. You can't just take that down?"

Zuckerberg agreed that the claim was false. "Going to someone who is a victim of Sandy Hook and telling them, 'Hey, no, you're a liar'—that is harassment, and we actually will take that down. But overall, let's take this a little closer to home . . ."

Zuckerberg was trying to pivot away from Alex Jones, toward what he thought—oh so wrongly—was a safer topic: Holocaust denial. "I'm Jewish, and there's a set of people who deny that the Holocaust happened. I find that deeply offensive. But at the end of the day, I don't believe that our platform should take that down because I think there are things that different people get wrong. I don't think that they're intentionally getting it wrong . . ."

His comments landed like a bomb. Several hours later Zuckerberg sent Swisher an email walking them back, that said in part, "I personally find Holocaust denial deeply offensive, and I absolutely didn't intend to defend the intent of people who deny that."

Within a month Alex Jones was gone from Facebook. Two years later, Facebook banned Holocaust denial content.[10]

Zuckerberg never gave Swisher another interview, nor did any other Facebook executive.

ON JULY 25, 2018, *The Guardian*, part of the reporting team that broke the Cambridge Analytica scandal, published an open letter from Lenny and Veronique to Zuckerberg.[11]

"While terms you use, like 'fake news' or 'fringe conspiracy groups,' sound relatively innocuous, let me provide you with some insight into the effects of allowing your platform to continue to be used as an instrument to disseminate hate," Lenny and Veronique's message to Zuckerberg read. They clued him in on the threatening phone calls and emails, the release of their addresses and personal information, Lucy Richards's death threats, and the fact that they live in hiding.

The Sandy Hook hoaxers use Facebook and other platforms "to 'hunt' us," they wrote.

"Our families are in danger as a direct result of the hundreds of thousands of people who see and believe the lies and hate speech, which you have decided should be protected . . . We have had to wage an almost inconceivable battle with Facebook to provide us with the most basic of protections to remove the most offensive and incendiary content."

Twisting the knife, they cited Zuckerberg's blunder with Swisher. "While you implied that Facebook would act more quickly to take down harassment directed at Sandy Hook victims than, say, the posts of Holocaust deniers, that is not our experience," the letter read. "In fact, you went on to suggest that this type of content would continue to be protected and that your idea for combating incendiary content was to provide counterpoints to push 'fake news' lower in search results. Of course, this provides no protection to us at all."

If Zuckerberg truly wanted to help them, they wrote, victims of mass tragedy should be a "protected group," making attacks on them prohibited by Facebook's own policies. That was something Lenny had been saying since 2015. Further, Facebook should provide victims and their families with dedicated staff authorized to remove hateful content immediately.

The letter to "Mr. Zuckerberg" closed with a roundhouse.

After feeling so much hope following your pledge in the Senate to make Facebook a safer and more hospitable place for social interaction, we are once again feeling let down by your recent comments supporting a safe harbor for Holocaust deniers and hate groups that attack victims of tragedy.

Our son Noah no longer has a voice, nor will he ever get to live out his life. His absence is felt every day. But we are unable to properly grieve for our baby or move on with our lives because you, arguably the most powerful man on the planet, have deemed that the attacks on us are immaterial, that providing assistance in removing threats is too cumbersome, and that our lives are less important than providing a safe haven for hate.

Sincerely,

Leonard Pozner & Veronique De La Rosa

The letter appeared on the same day as Facebook's quarterly earnings call, when the company reported that its explosive growth had flatlined amid the tsunami of terrible news.[12] Financial[13] and technology[14] reporters included Lenny and Veronique's letter in their coverage[15] of Facebook's lousy year.

Facebook's share price crashed, wiping nearly $120 billion off its market value, and $17 billion from Zuckerberg's personal fortune.[16]

Two days after Lenny and Veronique's letter to Zuckerberg, Facebook suspended Alex Jones's personal page from the platform for thirty days, citing "bullying" and "hate speech."[17]

My *Times* colleague, tech reporter Kevin Roose called it a slap on the wrist. Facebook suspended Jones's personal page but took no action against Infowars' account, which had 1.7 million followers.

Still, it appeared to be a critical moment for Alex Jones's relationship with the major social platforms. YouTube removed four videos from Infowars' channel, which had 2.4 million subscribers, and banned Jones from livestreaming for ninety days.

Spotify took down episodes of *The Alex Jones Show: Infowars*, for "hate content," after subscribers threatened to drop the service if it didn't act against Jones.[18]

"This is war!" Jones promised in a defiant video he posted to his remaining Facebook channel. But Lenny was winning.

By the end of that month, Facebook met Lenny's demands. The company established a policy against mocking crime victims and created a dedicated content team. Lenny had a list of hateful Facebook pages already compiled. Facebook removed them by the thousands.

LENNY'S FACEBOOK VICTORY COINCIDED with the first courtroom hearing in his case against Jones in Austin.

Leonard Pozner and Veronique De La Rosa v. Alex E. Jones, Infowars

LLC, and Free Speech Systems LLC was scheduled for an initial hearing on Jones's motion to have the case dismissed under the Texas Citizens Participation Act, which protects citizens' right to free speech against plaintiffs who aim to sue them into silence. On August 1, 2018, in the Travis County courtroom of Judge Scott Jenkins, Lenny and Veronique's lawsuit would be the first of the three Sandy Hook suits to reach a courtroom, a bellwether test of Jones's claim to First Amendment protection for his falsehoods.

Jones kept up his attacks on the lawsuits on Infowars, vowing to prevail. But behind the bravado he was scrambling to defend himself.

For the Connecticut case, he hired Marc Randazza and Jay Wolman of the Las Vegas–based Randazza Legal Group.[19] Randazza, a First Amendment absolutist who had amassed a list of far-right clients and ethics complaints,[20] had also defended Andrew Anglin, cofounder of the neo-Nazi website Daily Stormer.[21] In 2017, Anglin was sued for harassment by Tanya Gersh, a real estate agent in Whitefish, Montana, after he unleashed a torrent of antisemitic abuse and threats against her by posting the phone number, address, and social media profiles for her and her twelve-year-old son. Gersh, who is Jewish, was targeted after agreeing to help the mother of white nationalist Richard Spencer sell a Whitefish property she owned, after a backlash against her son in the liberal town. Gersh suggested Sherry Spencer disavow her son's views, and donate a portion of the sale proceeds to an anti-hate group. Then Richard Spencer got involved, writing an online screed in his mother's name that accused Gersh of blackmailing her. Anglin called on his followers to hit Gersh with "an old-fashioned troll storm." For two months, they sent Gersh and her son hundreds of harassing messages, including Holocaust memes, death threats, and suggestions that she kill herself.

Gersh sued Anglin with the help of the Southern Poverty Law

Center, the anti-racism nonprofit that had mobilized to counter a surge in white supremacist activity after Trump took office.[22]

Randazza argued in court that Gersh's and her son's personal information were publicly available, and Anglin's incitement was protected free speech.[23] Anglin himself was in hiding, slowing the proceedings. In 2019, Gersh won $14 million in damages. As of 2021, her lawyers were still hunting for Anglin and his assets.

In Austin, rumor had it that most law firms wanted nothing to do with defending Alex Jones's rants against parents of murdered children. Jones's divorce lawyer, Randall Wilhite, filed a placeholder response to Lenny and Veronique's lawsuit while Jones scrambled to find other counsel. He hired Mark Enoch, a Dallas lawyer who filed the motion to dismiss the case.

"This lawsuit is a strategic device used by Plaintiffs to silence Defendants' free speech and an attempt to hold Defendants liable for simply expressing their opinions regarding questioning the government. The goal of this lawsuit is to silence Defendants, as well as anyone else who refuses to accept what the mainstream media and government tell them, and prevent them from expressing any doubt or raising questions," the filing read. Enoch characterized the Pozner lawsuit as an effort "to silence those who openly oppose their very public 'herculean' efforts to ban the sale of certain weapons, ammunition and accessories, to pass new laws relating to gun registration and to limit free speech."

Jones was nothing if not consistent.

I LANDED IN AUSTIN the afternoon before the hearing and checked into an Airbnb-type place downtown. I set up my laptop on the kitchen island and pounded out a "curtain raiser" for the hearing.

The story opened with Veronique reflecting on their efforts to escape the hoaxers. By summer 2018 she and Lenny had relocated seven times.

"I would love to go see my son's grave, and I don't get to do that, but we made the right decision," Veronique said. The hoaxers posted their home addresses online "with the speed of light," she said. "They have their own community, and they have the ear of some very powerful people."

The story walked through Jones's latest court filing, in which he sought more than one hundred thousand dollars in court costs from Lenny and Veronique, should he succeed in getting their case dismissed. I tried to reach Jones and Enoch to ask about that, but they didn't return calls or emails.

The story posted online that night.[24] While the hearing seemed momentous to me, the editors viewed it as more or less an incremental step in what they rightly guessed would be a long legal slog.

I awoke the next morning to a late-night email from Elisabeth Bumiller, who had edited the story, and groaned. Such notes from editors usually meant they'd spotted a potential error or had a question, and I'd been sleeping.

But no—Bumiller was emailing a heads-up that "Alex Jones, Pursued over Infowars Falsehoods, Faces a Legal Crossroads" had gone viral. Readers had been moved by Lenny and Veronique's travails and were fuming about Jones's demand for court costs. I brought the story up on our website and read a couple of the hundreds of comments.

"This insanity has to stop," wrote a reader from Boston. "To Noah's family and the others, I've wished you peace and comfort since that terrible day; and now, I wish you justice. The people of America are on your side."

Bumiller called to assign a deep dive into Infowars, Jones's self-described "operating system of Trump." What was Infowars' business model? How much money was the hyperbolist Jones actually making, and where did it come from?

I teamed up with Emily Steel, business reporter in New York. Steel had been on a team that had won a Pulitzer for revealing the millions of dollars in secret settlements Fox News had paid to women sexually harassed and victimized by Bill O'Reilly, at the time one of Fox's star commentators.

I summoned RideAustin, the city's anti-Uber alternative, to take me to the Travis County courthouse on Guadalupe Street, a grand, gray "Depression Moderne" building with elaborate justice-themed bas-relief friezes, completed in 1931 by the Works Progress Administration.

In a dark stone entryway I met up with Mark Bankston, Bill Ogden, and Bankston's parents, Don and Susan, who had come in from Richmond for the hearing. Don, a grizzled and gruff central-casting Texas judge, was keen to talk with Mark and Ogden about the details of the case. Susan, a journalist in the mold of Texas humorist Molly Ivins, is a force of nature with wavy, wayward burnished-gold hair, big jewelry, and a big laugh. I was drawn to her profane irreverence, honed in Texas newsrooms. Years of smoking had taken a toll, and she traveled with a portable oxygen concentrator slung over her shoulder. As we filed into wooden rows in the high-ceilinged, paneled courtroom, she told me that her mother, Bonita Jean Duquesnay, had died a few weeks earlier. Bonita had doted on Mark and said that she never thought she'd see the day when his name appeared in the *New York Times*, which it had the month before her death. I settled into a row in front of the Bankstons.

Judge Jenkins warned the spectators against recording the proceedings with anything other than a notebook and pen. Bankston and Ogden viewed Jenkins, an angular man in his mid-sixties with a low-key manner and deep smile lines, as a lucky draw. Elected as a Democrat, the former plaintiff's lawyer had years before lost a child in an accident involving a drunk driver.

Jones did not appear. All his hearings would be handled by a shifting cast of lawyers with a common goal: keeping him out of the witness chair and his business information away from the Sandy Hook parents.

"The further we get, the more likely we are to get a juror who can sit there and think about the issues objectively," Randazza told me that year. "Do you want your First Amendment rights curtailed by somebody who says, 'I just can't bring myself to rule in favor of somebody who's up against the Sandy Hook parents'?"

Mark Enoch, ruddy and toothy, was a former mayor of Rowlett, Texas, a nice Dallas suburb not far from where Jones had grown up. He often represented defendants in product liability cases. In one of his biggest, he defended a helmet manufacturer against the family of a man who died of a head injury.[25]

Enoch's son helped him at the defense table, sorting poster board visual aids. Enoch seemed to like using them. In a hearing later that same month in Neil's case, he displayed a poster that depicted an evil-looking black wave against a stormy sky. Superimposed over the photo was a quote from a Reuters story about the lawsuits: "Bankston said after the hearing that he sees the case as building a wave that could topple Jones." Bankston and Ogden were delighted. Ogden took a photo of the poster with his cell phone. After the hearing, Bankston emailed Enoch, offering him a hundred bucks for the placard. He did not respond.

Enoch said in court that he had never heard of Alex Jones before Infowars hired him, but "fringe speech" like Jones's was more, not less, deserving of First Amendment protection.

The lawyers sparred over Jones's 2017 broadcast titled "Sandy Hook Vampires Exposed," which lay at the core of Veronique's defamation claim. The video was the most recent example in which Jones implied that Veronique's interview with CNN's Anderson Cooper was staged in a studio, citing as "proof" the glitch in the video that Infowars created themselves.

Enoch argued that Jones was attacking lying mainstream media like CNN, not Veronique. He said Jones was accusing CNN, the federal government, and Democrats of being the "vampires" behind the Sandy Hook plot.

Hauling up a poster board, Enoch displayed several narrow quotes drawn from Jones's several-hour "Vampires" broadcast, saying none of the statements had defamed Veronique. It was the lying mainstream media who were the vampires, he said, and it was Jones's opinion as a political commentator that Sandy Hook was a government plot.

"That's one of the pivot points for me," Judge Jenkins acknowledged. "Is it a statement of opinion, or a statement of fact?"

Enoch further maintained that Lenny and Veronique had made themselves public figures, she by testifying in favor of a ban on the assault-style rifle used to murder Noah. That meant they had to prove malice to prevail.

"It's a very interesting question of law," Judge Jenkins noted, on whether they became "involuntary" public figures by speaking publicly after their child was murdered in one of the most horrific shootings in American history.

Bankston countered that their struggle to defend themselves against online misinformation and abuse did not make Lenny and Veronique public figures. He argued that through HONR, Lenny combated damaging online falsehoods while laboring *not* to be recognized in public. Judge Jenkins, who hadn't known anything about Jones prior, looked alternately bemused and disgusted as Bankston regaled the room with the story of Jones displaying Lenny's mailing address on his broadcast and taking the call from the follower who warned, "Lenny, if you're listening, your day is coming."

Some of the spectators gasped. "Jesus," I heard Susan Bankston say, her outburst punctuated by the soft sighs of her oxygen machine.

Bankston addressed Enoch's selected quotes from Jones's "Vampires" broadcast. He reminded Enoch and the judge that Jones had used Veronique's CNN interview as the basis for a series of "monstrous" false statements casting her as a crisis actor, beginning a month after the shooting and continuing for five years.

In an affidavit, one of Bankston's expert witnesses, Fred Zipp, a former Texas newsman on the faculty at the University of Texas, had said he watched more than twenty Infowars clips that aired between 2013 and 2016, finding "a variety of factual allegations that are readily disproved by basic journalistic efforts."[26]

Zipp said Jones's statements about Veronique stood at the center of his Sandy Hook claims, falsehoods his audience clearly believed. Indeed, nearly one-quarter of Americans believed that Sandy Hook was "definitely" or "possibly" faked, according to a 2016 poll by Fairleigh Dickinson University.[27]

Near the end of the hearing, in a statement that seemed directed at me, Enoch slammed the mainstream media once more, for negative coverage of Jones's demand for $100,000 in court costs if Jenkins dismissed the case. If Jones won, Enoch said, he would accept a token dollar.[28]

The hearing adjourned in the early afternoon. The courthouse's long, dim halls discharged us blinking onto the concrete griddle of the plaza outside. Bankston and Ogden felt good. They would have to wait until fall for Jenkins's decision, but they thought he would reject Jones's motion, granting Lenny and Veronique a first victory in a long, exhausting battle.

24

MY INTERVIEW WITH ALEX JONES came, oddly enough, through his ex-wife, Kelly.

When I arrived in Austin, I got in touch with Kelly Jones, who agreed to meet me the day after the Austin court hearing. That afternoon, August 2, 2018, I traveled to meet her at what had been the couple's home and was now hers, on an arid road on the outskirts of Austin. The day was blazing hot, the solar glare on the compound's pale stone walls blinding as I got out of the car.

Kelly met me at the front gate, a heavy mechanized metal apparatus. Stepping inside the compound, I saw a life-sized baby elephant statue over her shoulder. Kelly had decorated the grounds with her collection of antique carousel creatures and ancient-looking statuary, adding to the looking-glass aura of the place.

Jones wore a chic black dress, her blond hair in an updo, her skin well cared for and luminous, but her eyes betrayed an interior agitation. "Watch for snakes," she cautioned, as we strolled a short distance down a dirt lane that led to the handsome stucco and timber ranch house at the rear of the property, farthest from the gate. Even from a distance I could see Marshy and Dobie, Kelly's overgrown "teacup" pigs, lying like mud berms in a raw dirt hole they had dug on the near side of the house, close to the coolness of the stone foundation.

We stood across from the pool, a lagoon-like sprawl of boulders and palm trees larger than most resorts', its multiple fountains hissing. A tile-roofed barbecue pavilion jutted into the water. A two-tier stone fireplace surround, all carved lions and laurel wreaths, was parked on the main pool deck as if just delivered, its poorly supported mass enough to crush multiple people if it fell.

We entered the pool house, a villa with a black carousel dragon standing guard. The decor seemed a mash-up of Tuscan and British colonial, with heavy gilt mirrors, dripping chandeliers, and a fireplace with a portrait of a tiger on its mantel. Kelly's imposing teak desk was adorned with carved elephant heads, flanked by more prancing carousel horses. The place looked unlived-in: dead gnats dusted the sink in the wet bar, and a smoke alarm with a dying battery chirped while we spoke.

Jones had a warm, engaging way, and a polish that spoke of exposure to places and experiences that did not interest her ex-husband. Kelly Nichols was born in Austin but grew up in Europe, the daughter of a diplomat. Her father "was like an agriculture counselor," she told me. "He got Prosciutto di Parma into America. He has a medal from the Italian president for it or something," she said, the fluent Italian words a departure from her slight Texas lilt.

Kelly grew up a rebel, alarming her parents, who enrolled her in a series of boarding schools in Europe and the United States. She returned to Austin to attend the University of Texas, then traveled for more than a year with PETA, staging media spectacles to draw attention to the cause. When she was twenty-nine, she returned to Austin to care for a friend dying of AIDS and earned money by waitressing and bartending in a couple of Austin dives. Then she met Alex.

It was the end of the 1990s, and a watershed moment for Alex Jones. He had his own show, a nightly orgy of conspiracy claims on Austin's community access station. He had just made his first video, *America Destroyed by Design*, about the "sacrifice of national sovereignty to global government and the U.N.," showcasing his false belief that the government plotted the 1995 Oklahoma City bombing.

The films and the cable and radio shows nurtured Jones's Infowars persona, which Kelly insists is his true self: angry, mendacious, heedless of the wreckage he creates.

Kelly was a smart, peripatetic, and lonely young woman who felt unworldly despite her travels, unsure where she fit. Alex, six years younger, seemed intensely certain of his own worth. When he told her, "I think about you all the time," she was hooked. Kelly took a job as a gofer and PR person for the cable station, then left to work for Jones.

"I was pretty irresponsibly oblivious when I was first involved with him," Kelly told me, sitting with her legs crossed in an armchair, sunlight from the windows limning her profile. "Just kind of in a 'Oh, this is cool, we're doing media stuff, whatever,' way," she said.

"I helped create the monster that he is now."

Their relationship was stormy from the start, full of signs that all was not well with her partner.

Visiting Alex's apartment for the first time, Kelly was repelled by its squalor, with clothes, trash, and used handkerchiefs strewn about. She was surprised to find Alex's father in the middle of the mess, cleaning up. That was by then a familiar role for David Jones, who had bankrolled his son's early ventures and enabled his excesses.

Jones spent about a decade at Austin community access.[1] Old-school

Austinites describe the cable outlet as the quintessence of Austin weird. The publicly funded station invited anyone, talented or simply eccentric, to learn on its equipment and make their own shows. The channels—two of them, 10 and 16—were favorite late-night stoner fodder.[2]

Jones got his own cable show in 1996, dropped out of Austin Community College, and rode his wild theories about Waco and Oklahoma City to a weekend gig on KJFK radio. In court testimony, Jones's father said he supported his son then, giving him about one hundred thousand dollars to help fund his "documentaries," capital that Jones père described as money he would have otherwise spent on college.

Passionate, paranoid, and a born performer, Alex Jones, only twenty-two then, swiftly developed a die-hard fan base, winning readers' choice awards from local publications, his father said.

But inside the cable studios, Jones was almost universally disliked, Charlie Sotelo, who was a producer at the cable outlet in those years, told journalist Harmon Leon. His colleagues saw Jones as a spoiled, self-mythologizing conversation bully who trapped them at their desks or in hallways, spewing nutty theories, needling and badgering them.

Sotelo told Leon and *BuzzFeed*'s Charlie Warzel about a fight[3] Jones got into in the station's parking lot one winter night in 1997.

Among the many fans who called into Jones's show was a young heckler who would yell into the phone and insult the host, infuriating him by calling him "Jarhead Jones." That night the man and two friends turned up at an open house Jones was hosting at the station to rib him in person. When one of them called him Jarhead, Jones invited them outside to settle things.

Jones faced off with the hecklers, insisting he could shoot them under Texas law, Sotelo told Leon. Their ringleader called Jones's bluff, then punched him in the mouth, splitting his lip. Jones windmilled, failing to connect as the man, gangly but motivated, punched Jones's face again and again. Sotelo, alarmed by the lopsided fight, yelled at Jones to go inside. But true to form, Jones wouldn't quit, even to save his own skin.

The men left before police arrived, and Sotelo watched as Jones stormed around the parking lot, concocting a wild story of a large gang and an oversized, obsessed fan with pasty green skin and eyes like a goat's pulling "a double-edged military type killing knife" on him. Sotelo called bullshit on the unspooling fable, and Jones turned on him, spitting a spray of saliva and blood that hit Sotelo's mouth and his new shirt. Jones swung at Sotelo and missed, and Sotelo delivered one more blow to Jones's face.

Police broke up the fight and ushered Sotelo inside. David Jones arrived and tried unsuccessfully to calm his son, who was shouting at the witnesses, calling them liars as they gave police their accounts of his beating. The officers threatened to arrest Jones if he didn't shut up.

In Jones's own statement, he portrayed himself as a kind of action hero, fending off circling attackers, driving one to his knees—a story substantially at odds with witness accounts.

"I am not an easy person to scare but I believe that he bears me incredible malice," Jones told the detective about the heckler. "My father Dr. David R. Jones screens calls for me and he recognizes this person's voice and mode of action. He has made death threats about me to my father. He has said he would like to sodomise [*sic*] my mother. He has [*sic*] he is coming to my house to slit my throat." Jones did not press charges.

Sotelo told Leon that the police, sounding eager, asked whether he'd swear out a complaint against Jones. Sotelo said David Jones wrote him a check for one hundred dollars. "I was sure it was for the shirt," Sotelo said. "But now I think it might have been in exchange for not pressing charges."

KELLY MET ALEX JONES in 1998, the year his radio show was syndicated on Genesis Communications Network. The conservative outfit channeled business to Midas Resources, a gold dealer catering to doomsday preppers and others hedging against their feared collapse of the banking system.[4] Genesis distributed Jones's show on about fifty small stations by 1999. News from that time supports Jones's father's proud claim, made in court, that "Alex became immensely popular very, very, very fast."

Infowars was born in this period, though the exact registration date of its website is a matter of dispute among Jones, his father, and Kelly. Alex and Kelly moved into the tiny house in Austin, with the nursery that was Jones's first studio.

Kelly ran Infowars' operations and marketing from a loft upstairs.[5] She deployed PR skills honed at PETA to get Jones onto newswires like Associated Press and Reuters. She produced his show and landed his first big guest: the actor Charlton Heston, newly elected president of the National Rifle Association.

Charlie Sotelo gives Kelly full credit for Infowars' success, saying that she "basically focused Jones's whole career."

"He wouldn't have been able to put it together without her," Sotelo told Leon.[6] "Her career was making him a thing. And she did it."

From her desk in the pool house, Kelly walked me through Infowars' early business model. She compared it to that of the televangelists of the era, who used their shows to hawk books and donations.

"In many ways, Alex invented viral marketing," said Kelly, who managed Infowars' mail-order documentary-video business. "He'd make these movies and say, 'Give these away, make copies for free, and give them to everybody you want—just go for it.' And so across the country, from the West Virginia backwoods to tiny-town East Texas, people were sharing these videos.

"It was a snail-mail YouTube kind of thing going on. And that's really how he built his core audience and created his platform."

Both Joneses are conspiracists at heart, susceptible to surreal notions of plots hatched by an ever-shifting cast of shadowy powerful figures. Kelly recalled their 2000 trip to California's Bohemian Grove with Jon Ronson, the author and filmmaker, describing it as "Alex's breakthrough moment."

"Alex snuck in there with this ridiculous giant briefcase camera that I rigged up for him." Kelly smiled. She said that she doubted Alex's bogus claims of human sacrifice at the fratty gathering at the time but kept silent.

"I turned my head a lot when I really shouldn't have. Even when he came out with this footage, which actually I edited a lot, because I had to stabilize it and make it go right side up. But, I mean, I saw the ceremony, and it was weird.

"They burn a skeleton, a child-sized skeleton, according to Alex—there was a lot of extrapolation," she said. In fact, the burning was of an effigy, equivalent to the burning of a scarecrow-style "winter witch" at a spring festival.

The edited footage became the "documentary" *Dark Secrets Inside Bohemian Grove.*

"He got a lot of legs from that, a lot of publicity, and he's getting mainstream coverage," Kelly said, reliving it. "We're publicizing events as we go along because of the PR background. And suddenly he's really breaking."

By the turn of the millennium, Alex Jones had gone from a local-media phenom to a niche national player.

Kelly told me Alex treated her poorly; she told an *Inside Edition* interviewer that he was "very cruel to me every day of our marriage," ridiculing and criticizing her, calling her fat.[7] After the 9/11 attacks, she said she was preparing to join their neighbors in a candlelight vigil when he began yelling at her that there was a conspiracy behind the attacks.[8]

Jones has said his 9/11 theory prompted many of the approximately one hundred Genesis Communications Network stations carrying his show by then to drop him, though the exact number remains unclear.[9] But Jones doubled down, and his claim to having "predicted" the attacks propelled him further into the conspiracy world's big leagues.

Kelly and Alex got married early in 2002, after Kelly became pregnant with their eldest child, Rex, born later that year. A daughter followed twenty months later, and another daughter in 2008.

"He was never involved with the kids," Kelly told me, a claim Jones disputed in court. "He gave Rex maybe one bath, but he never changed a diaper, maybe woke up once during the night."

If he wasn't sleeping, Jones was working, recording through the night without even taking a bathroom break. In the morning, Kelly would find glasses of urine on his desk and dump them.

In those early days the Joneses produced videos at a frenetic clip. Most years Jones released two films, with titles like *Police State 2000* and its three sequels; *Matrix of Evil* (2003); and the popular *Endgame* (2007), whose dark and sweeping plot predicts depopulation through mass poisoning.

The couple grew wealthy from video sales, advertising, and Jones's on-air promotion of diet supplements and other products from Youngevity, a multilevel marketing company. They bought properties and cars and a $70,000 grand piano. They stowed more than $750,000 in silver, gold, and precious metals in safe-deposit boxes, according to court records.[10]

Just before Christmas 2010, they bought the property where Kelly now lives, paying $1.1 million in a secretive deal in which they used only their middle initials and required the sellers not to release any information without their permission, citing "safety and privacy concerns," according to records.

They set up a web of companies to benefit their children and themselves. But as Kelly spent more time raising and homeschooling the kids, she had less time for the business. Alex lacked the attention span to maintain Infowars' financial records. A newly hired bookkeeper found notebooks stuffed in a drawer and worked to untangle the mess.

David Jones and Alex's mother, Carol, turned up frequently at the house and tagged along on vacations and trips to the beach, Kelly said. David advised Alex on running Infowars while maintaining his own dental empire. That changed in 2013, when father and son struck an agreement for David Jones to join Infowars full-time and "help it be a healthier and more profitable business that was being run according to the guidelines and visions that had been set up," the elder Jones said, according to court documents.

Alex and his father believed they could make a killing in the diet-supplements business. On one of the family trips to the beach, David Jones raised the topic with Kelly, who was not in favor. "My perception was that she felt that it was something that could cheapen the brand and would be the equivalent of selling snake oil or perhaps something that would take advantage of our customers," David Jones said in court.

But Alex Jones was enthusiastic. He agreed to pay his father a base salary of $300,000 to $500,000 a year, plus 20 percent of the profits from the supplements businesses they were creating.[11]

Father and son worked with an Austin lawyer, Eric Taube, to set up business entities "formed specifically to shelter the family from liability related to the nutritional supplement business," David Jones said in court.

Alex Jones and his father were involved in the new companies. Kelly was not, David Jones said in court, because she was uninterested and he did not want to work with her.

Kelly filed for divorce in 2013. Kelly, Alex, David Jones, the supplements broker Anthony Gucciardi, the Infowars bookkeeper, and several other people testified in closed proceedings about the business, which was reaping $20 million in annual revenue by then. The couple fought over who was responsible for that success and what Kelly's role had been.

Kelly's lawyers quizzed Jones under oath about the $317,000 he'd spent since they separated, buying four Rolexes in a single day,[12] jewelry for his new girlfriend (soon to be his wife), $40,000 on exotic fish and a saltwater tank, thousands on steakhouse dinners and parties in his expensive Austin penthouse rental.

Jones and his lawyers in turn accused Kelly of wasting money on the pool complex, which lawyers said in court had cost more than $500,000.

Depositions and property appraisals, psychological assessments, arrangements for their children, and the untangling of their business interests stretched on for months. The divorce was final two days after Valentine's Day 2015.

Kelly received a settlement worth several million dollars, their main residence, and most of its contents. She got three of the family's four dogs; Alex refused to relinquish Captain Fantastic, the French bulldog. Jones got the lake house and its collection of toys—including a boat, a four-wheeler, and Jet Skis—and another property. He kept his $70,000 collection of firearms and from the family's home retrieved a rug, a couple of tables, and a sterling silver cutlery service for twelve, according to court records.

During the Joneses' epic custody battle in 2017, therapists testified to Alex's narcissism[13] and paranoia, and to Kelly's "emotional dysregulation."[14] Alex's lawyers had portrayed him as a performance artist who left his bizarre and threatening persona at the office.[15] But after his volcanic courtroom eruptions undercut their arguments,[16] Kelly, who had all but lost access to their children, won joint custody. Alex contested the decision, and the battle has raged ever since.

That August afternoon, Kelly told me that legal bills had nearly exhausted her finances. The couple had spent a combined $4 million in their four-year battle over the business and custody of their children.[17]

She alleged that Alex, his parents, and his lawyers had forced her into a settlement not in her best interests. She accused Jones of psychologically harming their children, in part by putting Rex, then fifteen, on Infowars.

After the Parkland shooting survivor and gun policy activist David Hogg rescinded his invitation for Alex Jones to debate him because he realized Jones was recycling the same false claims as he did at Sandy Hook, Rex Jones taped a broadcast on Infowars demanding that Hogg debate him instead.

The five-minute segment makes for depressing viewing. Jones's son has more reasons than most adolescents to reject his old man as a deluded weirdo. Instead Rex, a kid with a sweet face and an orthodontic lisp, was striving to be just like Dad. "Mainstream media uses you as a child human shield so that you can go out and make outrageous statements," Rex said, trying to project his father's gravel menace in a voice still a child's. "The truth is, Mr. Hogg, that George Soros and the groups funding your anti-gun march are a clear and present danger to this country and draw clear parallels to Nazi Germany and the authoritarian regimes of the past."[18]

Kelly sought an emergency restraining order. "I believe that they're actually in serious physical danger as well as obvious horrible emotional danger. They need to be removed immediately," she said of the children.

"My son was a reader and loved science and was a swimmer, and he was exuberant and engaged, and he has just been forced to become this thing. And it's so heartbreaking for me. I know that he can heal from it, but I'm really worried for him. He needs somebody to intercede."

She told me that she and Alex would both appear in family court the next day for an open hearing in the same courthouse where Lenny and Veronique's hearing had been. I told her I would be there.

THE AUGUST 3, 2018, hearing supported Kelly's contention that Alex Jones bore little resemblance to his lawyers' strategic portrait of him as a "performance artist" in the studio and a stable family man at home.

On Alex Jones's legal team that day were the Wilhites, a husband-wife team from Fullenweider Wilhite, a Texas divorce firm that in a bit of Texas braggadocio calls its lead lawyers "living legends." Randall "Randy" Wilhite, the firm's graying, bookish senior partner, had been on Alex Jones's legal team during the 2017 custody trial, pressing the theory that Jones is a performance artist, harmless at home. His young wife and colleague, Alison Wilhite, appears behind Jones in a widely distributed photo from that 2017 trial, her long dark hair flowing and her face set in a severe, raised-eyebrow expression.[19] Kelly Jones had nicknamed her "Maleficent."

The courtroom was another wood-paneled affair, with a bank of windows and a mostly empty spectator gallery. I sat directly behind Jones, who sat between the Wilhites, at a table to the right of where Kelly sat. Kelly's lawyer seemed new to the case, shuffling papers and seeking help from her client in locating documents.

Kelly Jones's request for a restraining order stemmed partly from her ex-husband's "out of control" on-air behavior. The week before the hearing, Jones, ever the pro-Trump attack dog, had accused Robert S. Mueller III, the special counsel investigating Russian interference in the 2016 election, of involvement in a child trafficking ring. Jones repeatedly imitated firing a handgun, addressing Mueller as "bitch."

"It's not a joke. It's not a game. It's the real world. Politically. You're going to get it, or I'm going to die trying, bitch. Get ready. We're going to bang heads," he said.[20]

Also concerning to Kelly was one of the four videos YouTube had removed from Infowars' channel that week for violating its child endangerment and hate speech standards. Called "Prevent Liberalism," it depicted a man choking a child and throwing him to the ground.[21]

Kelly told the judge one of their children had made an anguished statement, one that I won't detail here to preserve their privacy.

Alex whispered to his lawyers, accusing Kelly of "teaching" the child self-harm.

"Pure evil," Jones told Randy Wilhite. Wilhite agreed with him: "Pure," he murmured. I stared incredulously at the back of the lawyer's head.

"This is going on the air," Jones blustered. Wilhite nodded.

The animus in the room, the disorganization, claims and counterclaims, fed the sense that the proceedings were veering off the rails. At one point Rex Jones strode into the room, wearing cowboy boots, jeans, and a purposeful expression, though his purpose was by no means clear. Alarmed, the judge, Orlinda Naranjo, halted the hearing and asked him to leave.

Jones eventually agreed to stop putting his eldest child on his show. Judge Naranjo declined to issue the temporary restraining order, and the hearing was over.

Kelly, deeply disappointed, declared the proceedings rigged against her. Accompanied by a couple of friends, she rushed into the hallway, filming her dismay with her iPhone to post on social media, her agitation bouncing off the corridors' stone surfaces.

I found Jones walking toward the exit in a heavy, rolling gait, and introduced myself. Rattled, he fumbled with the edges of his dark suit jacket, pulling it over his belly. I told him I was working on a profile of him and had been in the courtroom. He went into a stream-of-consciousness soliloquy.

"There's no context . . . They just say they removed that video for child endangerment, now I'm in court . . . She divorced me. She put me through this court process. I'm here in court for the mainstream media, because YouTube wants to virtue signal . . . The First Amendment is under attack.

"I am sorry my questioning hurt people," he said. The mainstream media "twists what I say and makes it even worse, and I am minutes away from losing my kids."

Kelly came up, still recording with her phone camera. She demanded to know what Jones was saying about her.

When sheltered on each flank by his lawyers, Alex had labeled his ex-wife "pure evil." He had accused her of attempting to harm their children and vowed to air her alleged misdeeds to millions of Infowars viewers. But now, with her standing before him, he fled. "I need to, need to go, but call me," he muttered, rolling toward a sunlit exit at the end of the corridor.

Kelly walked off to protest in front of Judge Naranjo's chambers and was eventually ejected by a bailiff.

I CALLED AUSTIN'S ANTI-UBER and gave the driver an address at an Austin industrial park. I didn't offer the business name. I'd been warned that Infowars' location was undisclosed, and to be careful.

"Oh, that's Infowars, the big secret," Ralph, the driver, said laughing. "I used to drive for UPS. Infowars didn't want the general public to know where the office was. They made us call before we made a delivery. That Jones, he's definitely got a weird gig."

I asked Ralph to let me sit in the car a block away while I texted Jones, asking to talk with him. No response. I figured I'd try the door next, and if nobody answered, I'd collect my gear downtown and catch my flight back to Washington.

After about ten minutes, Jones texted and called, giving me directions to the undisclosed location everybody knew about. We agreed that Jones would tape the conversation, and so would I.

The office park looked like most others: rows of glass storefronts hunkered behind parched, half-hearted landscaping. Warehouses and loading docks squatted across a concrete expanse out back. Jones's fellow tenants included Orkin pest control, Prometheus Security, and what looked like a mail-order fulfillment center for Kinky Friedman Cigars.

Jones's sidekick, Rob Dew, wearing a Ron Paul T-shirt and Che Guevara–type beret, let me in. Jones greeted me inside Infowars' messy reception area, a dim space hung with maps and a homely framed needlepoint that read LIBERTY OR DEATH.

Jones still wore his going-to-court clothes: dark jacket, too-snug, monogrammed white shirt, and soft-soled dress shoes. He saw me looking around. "I don't want my building described," he commanded. "We've had some real schizos come here and literally threaten us." Imagine that, I thought.

Jones told me he had guns in the building and snipers on the roof. He called me the next morning to retract that latter whopper, either in a pang of conscience or because his neighbors at Orkin might freak out.

Jones, Dew, and I walked a few paces to a conference room. It was furnished with a vintage jukebox, an "Alexapure Breeze" air filter that Jones told me was on sale that month, and a whiteboard with a scrawled list of juvenile stunts—"moving in to Starbucks," "setting up a safe space"—that the Infowars crew could perform to trigger the libs.

The room was cold enough to store meat, and I was wearing a light summer dress.

"It's freezing in here," I said lightly.

"I *liiike* it that way," Jones growled, narrowing his eyes like a community theater villain. Dew, sitting next to him, laughed, then glanced at the boss to check whether laughing was okay. I clenched my arm and stomach muscles to prevent myself from shivering so Jones couldn't congratulate himself on having scared me. He didn't. There's something kind of pathetic about him in person. A few people who know him well had told me Jones doesn't have many friends who aren't on his payroll; that his hankering for celebrity company was born of his need to find somebody on his self-perceived level to hang with. I could tell by the way Jones stole glances at Dew while he needled me that he was desperate to entertain him.

"I keep thinking there may be an honest journalist out there," Jones said, explaining why he had agreed to an interview. He held a copy of the story I had written about Lenny and Veronique's hearing.

Dew was recording before I even sat down. Jones clearly viewed our meeting as an opportunity to grill me, gather audio for his show, and recast his role in spreading Sandy Hook falsehoods.

He asked whether I'd been in the courtroom for Lenny and Veronique's hearing. "I mean, did you see the video? My Father's Day message to the families from last year, before I was sued?" he asked. His lawyer, Mark Enoch, had played the video in the courtroom that Jones released on Father's Day 2017,[22] hours before his interview with Megyn Kelly aired, in an attempt to get ahead of the blowback.

Jones was aggrieved and suspicious that the mainstream media ignored his Father's Day spin. "Pretty amazing discipline to—all in unison—do that," he said. He told me it was the "edited" Megyn Kelly interview, not his years of false claims about the Sandy Hook victims, that had prompted the families to sue.

"I just can't believe the media is so evil. Jesus, this is so organized."

He groused that I had written about his demand for $100,000 in court costs from Lenny and Veronique. Jabbing his finger at me, he said, "You have a responsibility when you wrote the blueprint article that everybody else picked up, where they said, 'Alex Jones is the scum of the earth,' 'People need to bankrupt Alex Jones,' 'People need to kill Alex Jones because he sends people to these parents' houses and he won't stop doing it.'"

He falsely accused me of failing to report that he had waived his demand for court costs. He described the gesture, made amid heavy criticism, as one of unprecedented generosity. He neglected to mention that he had also demanded $200,000 in court costs from Neil Heslin.[23]

"Can I ask you about your business?" I interrupted. "You're obviously an astute businessman. You've built quite an empire—"

"I'm not a businessman," he barked. "I'm in a total war, and my life is pledged to this and my treasure and my family. So I'm all in. This is called America. This is called the real deal. This is called Colonel Travis at the Alamo. This is commitment.

"It's in the lawsuits—the part that's totally incorrect—that I'm doing this about Sandy Hook and that I don't really believe it and I didn't believe it back then blah blah blah, and that I'm saying all these things because I got all this *money*"—he sneered at the word—"from Sandy Hook."

He talked for nearly two minutes about what he said were millions of dollars in movie, book, television, and voice-over deals he'd been offered, and refused.

"I've been taken up on the mountain. Do you understand that?" he demanded. His face looked purplish in the fluorescent light of the conference room. Blots of sweat had emerged on the front of his shirt.

"Supplements are popular, they're good, they're a fast-growing market; I use it to fund the operation. Other revolutionaries rob banks and kidnap people. Okay? I don't do that."

The families' charge that Jones profited from their pain made Jones furious. But he had strenuously resisted court orders to produce his business records. Fighting with Kelly in court two years after Sandy Hook, Jones seemed less indifferent about money.

"On paper it says I made five million, but that's not true because of the way they have the tax laws," he said in court testimony in late 2014. "If I buy a stapler, it's income, the way the business is set up. So we are trying to go off that cash basis to the accrual.

"Syndicated radio is the core, ninety percent of the driver, and then the internet interfaces to show people proof of what we are claiming on the radio," he said, describing his business model. "So the radio drives people to the web to see the proof, and then that funds the operation by them then visiting the advertisers."

For ninety minutes of our three hours together, Jones cataloged gripes over his coverage by the media like a pampered celebrity, complaining about being taken out of context and victimized by his critics.

He railed about his suspensions from YouTube and Facebook, companies he said were in league with "Chi-coms" that had already taken over Hollywood, major universities, big banks, and the *Washington Post*. He called his threat against Mueller "satire," a cowboy spoof backed by spaghetti western music. Jones is not a stupid person, and his lawyers have imbued him with a reptilian sense of the legal lines. He knows satire and wild Hillary-is-a-pedophile charges are protected by the First Amendment. When making a handgun gesture, telling Mueller "you're gonna get it," he prefaced that phrase with the word "politically," in case anyone accused him of threatening real violence. What flummoxed him now, and endangered his business, was that the social platforms didn't care about his hair-splitting. He'd become a reputational liability for them, and they wanted him gone.

Stomping around the conference room that early August afternoon, waving his arms, Jones repeated his bigoted, phony claims about 9/11 and Oklahoma City, the Muslim takeover of America. He's big on spy and special ops lingo like "intel," going "operational," and "CIA misinfo." He moved on to his shopworn theories about Bill Gates's plan to depopulate the world, and the government planting feminizing chemicals in the water supply. "You say you have sons? Are they as masculine as their grandfather?"

I asked him if he's always like this—whether dinnertime in the Alex Jones household sounds like this.

"I am here giving you the unfiltered truth of my soul. Who I am. What I stand for. Not the distorted thing. Not the twisted thing. The truth.

"I did say years ago I believed Sandy Hook happened. I did say I was sorry for how things got twisted and could be hurtful. But—I mean, I barely covered Sandy Hook. I mean, my god, we had a big law firm, spent a ton of money, to go back and actually get every reference to it, and I was shocked by how much the media has been deceptive."

I asked what Jones liked to do in his off-hours, whether he had any hobbies. "I go home, I swim in the pool with my kids, I play with them, I hang out with my friends. I went out last night with my wife, one of our friends, and his girlfriend.

"I have a great life, let me tell you."

That comment reminded me of a crack Jones had made during his divorce deposition in 2014, when Kelly's lawyers questioned him about blowing more than three hundred thousand dollars on parties, jewelry, and boozy dinners out.

"It's a free country," he told them then. "I had to restart my life."[24]

Neil Heslin, pursued by the people Jones emboldened, was struggling to restart his life then too. Hoaxers had publicly savaged his character and someone had shot a gun into his house. It would be four years before Neil could bring himself to take down the bare Christmas tree[25] that he and Jesse had planned to decorate before his boy was killed, and longer than that before he could alter Jesse's bedroom, where his toy Hot Wheels tracks still snaked across the floor.

"I HATE COVERING SANDY HOOK!" Jones blurted. "I've hated covering it for four and a half years, okay? I was covering a giant phenomenon of people not believing media anymore because they've been caught in governments' lying so much!"

How tiresome Jones was, and how dangerous. A charismatic, irresponsible man-child with an entourage of paid enablers and an audience of millions.

"You're not really the *New York Times*. You're here to create a narrative," he said, narrowing his eyes, one of them so bloodshot it looked infected. "You're here to get a legal strategy just like Megyn Kelly did, so that you can try to get a false verdict to set the precedent to overturn the First Amendment."

Jones so wanted to unnerve me, to impress the lickspittle Rob Dew by making me flinch. But I noticed he was moving away, not toward me, parading around the room, seeking the corners. If he wanted to intimidate me, why not lean over the desk and get in my face?

It struck me later that Jones may have needed that distance. People often ask: How can Jones or anyone attack the victims of mass murder? The answer: from afar. Jones doesn't talk to actual people. He preaches to an invisible audience, an abstraction that makes it easier to absolve himself when its members act on the hatred he sows.

The remove of the internet nurtures groupthink bubbles whose members reward one another for attacking victims in the name of "truth." It endows Kelley Watt, mother of two, with the temerity to demand Lenny dig up Noah's body and "prove to the world you lost your son."

"When anybody's behind a machine, whether it's a gun or a computer or a car, a dehumanization takes place that makes it easier to commit an act of violence," Veronique had told me that year.[26]

Wielding swift, world-spanning communications technology, Alex Jones had become a menace. But in the conference room that day, he was the man behind the curtain, a moneyed, coddled con man, and a coward.

The following morning I was back in Washington and Jones was on the phone. "I think I need to clarify a few things with you, now that I look back," he said. He talked for nearly two more hours, telling me he viewed our interview as "the hill I die on," a last-ditch effort to see himself fairly portrayed in the media.

Jones likened himself to Tom Robinson, the falsely accused man in *To Kill a Mockingbird*, convicted by people too lacking in courage to assert his obvious innocence.

"The only folks who are gonna give Tom the benefit of the doubt are those who have less to lose," Jones told me. "There's no way you can side with a Black man accused of this act, when you can show how much a part of the system you really are by saying, 'Hang him.'"

25

A WEEK AFTER MY INTERVIEW with Alex Jones, Big Tech pulled the plug on Infowars.

Beginning the first week in August 2018, the biggest social media companies dropped Jones and Infowars from their platforms.[1] Apple deleted all but one of Infowars' six podcasts from iTunes. Google nuked Jones's YouTube channel, which had amassed more than 2.4 million subscribers; its 36,000 videos had racked up more than 1.6 billion views since 2008. Audio streaming sites[2] Spotify and Stitcher, the professional networking site LinkedIn, and the porn-streaming site YouPorn dumped him.[3] "Hate has no place on YouPorn," the company said. Pinterest, the innocuous virtual pinboard for recipes and pet videos, took down Infowars' page.

Twitter was the last big platform to act, kicking Jones off Twitter and Periscope, its live video-streaming app, in early September.[4]

It's not clear what motivated the companies to act en masse. But as more platforms booted Jones, the rest clearly feared being the last. The platforms said Jones had violated their policies against hate speech[5] but said nothing about his spreading conspiracy theories.[6] In its statement, Facebook specifically ruled out "false news"[7] as the reason for taking down four of Jones's pages, with millions of followers. Facebook said the pages instead violated its policies by "glorifying violence" and "using dehumanizing language to describe people who are transgender, Muslims and immigrants."[8]

Facebook's content experts had recommended booting Jones and Infowars in keeping with its policy banning "dangerous individuals and organizations." But years later, *BuzzFeed News* learned that Zuckerberg disagreed that Jones was a "hate figure." So he created a new policy[9] that amounted to a reprieve: Jones and Infowars earned a permanent ban, but Facebook would leave up posts from users who spread Infowars' and Jones's lies. Insiders told *BuzzFeed* that Zuckerberg's capricious rewriting of his own team's rules was another example of his reluctance to engage Trump-world conservatives, timidity that slowed Facebook's efforts to curb the messaging of extremist groups implicated in the January 6, 2021, attack on the Capitol.

Lenny told *BuzzFeed* that eventually "Zuckerberg has to be held responsible for his role in allowing his platform to be weaponized and for ensuring that the ludicrous and the dangerous are given equal importance as the factual."

Still, in summer 2018, the social behemoths danced on a red-hot public griddle. After years of inaction they turned on Jones and his cohorts with breathtaking speed. Lenny made the most of it. He called out WordPress.com, one of the world's largest blogging platforms, which had made it easy for people like Jim Fetzer and Maria Chang to attack the Sandy Hook families. For years Lenny had asked Automattic, which operates WordPress.com, to remove copyrighted content about Noah and other children from hoaxer blogs. Automattic had steadfastly refused, sending Lenny form letters saying it believed the hoaxers' unauthorized posting of Lenny's family photos of Noah was "fair use of the material," defining "fair use" as "criticism, comment, news reporting, teaching, scholarship, and research." WordPress.com further warned him that it could collect damages from people who filed phony copyright claims.

Lenny gave an interview to the *New York Times* criticizing the company.[10]

"The responses from their support people are very automated, very generic, very cold and there's just no getting through to them," he said. "They have taken this incorrect interpretation of freedom of speech to an extreme."

Lenny's timing again proved spot-on. Matt Mullenweg, Automattic's chief executive, had just boasted in an interview with Kara Swisher that his WordPress super-platform had "always had a robust terms-of-service team."[11]

"We do avoid hate speech," he said. "Egregiously fake or harmful things we're pretty good at getting off the system."

The *Times* reporters went to Automattic with Lenny's complaints. The company responded in a statement: "While our policies have many benefits to free expression for those who use our platform, our system, like many others that operate at large scale, is not ideal for getting to the deeper context of a given request.

"The pain that the family has suffered is very real and if tied to the contents of sites we host, we want to have policies to address that."

The resulting *Times* article ran under the headline THIS COMPANY KEEPS LIES ABOUT SANDY HOOK ON THE WEB. (The New York Times Company is a small investor in Automattic.)

Afterward "I got an email from someone informally, that said, 'Let's talk on the phone,'" Lenny told me. "It was Matt, the founder."

Lenny supplied WordPress.com with a list of repeat offenders.

"A huge amount of content, thousands of blogs disappeared," Lenny told me.

After Mullenweg intervened, Lenny said, "WordPress took down blogs that were ten years old. I still work with them."

HUNDREDS OF NEWS STORIES about Infowars' deplatforming tarred Jones with spreading the Sandy Hook hoax. He reacted as he had in our August 3 interview: complaining that he didn't invent the hoax, protesting that now he believed kids died, saying he hadn't mentioned Sandy Hook in years. The HONR volunteers found this amusing. Alex Jones was learning, as they had, that debunking is hard.

Jones had boasted that once upon a time, Fox News, TV, and Hollywood had all come calling, offering rich deals that he passed up to build Infowars. Now his role in tormenting Sandy Hook families clung to him like a stench. If those opportunities had ever existed, they were probably gone now.

But President Trump and his acolytes still had Jones's back.

Jones slapped a red BANNED label over the Infowars logo. Every day on his show he slammed "evil" big media and tech, and implored the president to help him. Trump tweeted, "Social Media is totally discriminating against Republican/Conservative voices. Speaking loudly and clearly for the Trump Administration, we won't let that happen."

Congress summoned Twitter co-founder Jack Dorsey and Sheryl Sandberg, Facebook's chief operating officer, to Capitol Hill. Tech companies' power to censor online political speech concerned Americans across the political spectrum. But many conservatives saw blasting the companies for alleged bias as a useful distraction from the fact that 2016 misinformation from Russians and others had mostly benefited Trump, not the left.

"I will be delivering information to the president TODAY on the Sherman antitrust actions," Jones said, red-faced and breathless, at the top of his August 23, 2018, broadcast. "These are top law firms, at the highest levels of the Republican Party. So Facebook, Google, Apple, all you guys, you're getting hauled up before Congress right now," he growled.

"We've got your ass." Jones jabbed his finger and slapped the desk.

"You think you're going to shut us down? You think you're going to kick me off the air? I'll just hand-deliver it to everybody. I'll just become a lobbyist for freedom, the worst you've ever seen!"

Jones went to Washington for the hearings on September 5. True to his word, he proved a terrible lobbyist for internet regulation.

Jones capered in the hallways, saying "beep-beep-beep" in an imitation of a Russian bot.[12] He stalked in and out of the hearing room with his cell phone held aloft. He performed for his film crew outside the Capitol, lobbing verbal broadsides at Big Tech and Democrats, and got into a minor dustup with police. He was going for a repeat of his 2016 Republican convention tour de force, but this time he wasn't a rising, far-right provocateur. He was a hate-mongering tormenter of the families of murdered children, cut off from his social media supply.

In a hallway outside the Senate Intelligence Committee hearing where Dorsey and Sandberg testified, Jones confronted Senator Marco Rubio, Republican of Florida, who was talking with a gaggle of reporters.

Jones planted himself to Rubio's immediate right and uncorked a stream of high-decibel adolescent insults. Sometimes Jones's confrontations oozed theater, everyone in on the joke. This was different. Wild-eyed and quivering, Jones was a better-dressed version of his younger self in the Austin cable station parking lot decades before, taking a beating but unable to stop flailing.

Rubio was half Jones's size, but this was his turf. He glanced at the fuming talk show host, rolled his eyes, and shared a laugh with the reporters.

"Marco Rubio the snake, little frat boy here." Jones mocked Rubio's laugh, swiveling between the senator and the cameras.

Rubio turned to him. "Who are you, man? Who is this guy? I swear to God I don't know who you are, man," Rubio said, laughing. A few reporters, keen to dispatch Jones and get back to their questions, offered "Infowars."

Upstaged, Jones yammered, "Bigger than Rush Limbaugh, tens of millions of views. He knows who Infowars is, playing this joke over here."

He touched Rubio's arm, daring him. "The deplatforming didn't work," he told him.

A man from Rubio's security detail pushed Jones's hand roughly away. "Don't touch me again, man," Rubio told him, as the body man spoke into a microphone in his fist.

"I just patted you nicely—you want me to get arrested! It's not enough just to take my First Amendment?" Jones challenged, growing excited.

"You're not gonna get arrested. I'll take care of it myself," Rubio told him.

"Oh! You're gonna beat me up!" Jones exclaimed, as Rubio dismissed him and turned back to the reporters. "Oh, he's so mad! You're not gonna silence me. You're not gonna silence America. You are literally like a little gangster thug."

Turning to the cameras, he said, "Rubio just threatened to physically take care of me." Nobody bothered with him.

Rubio wrapped it up and headed back to the hearing room. "You guys can talk to this clown," he said.

In fact, deplatforming Alex Jones did work. By the time our profile of Alex Jones and his business, "Conspiracy Theories Made Alex Jones Very Rich. They May Bring Him Down," ran in the *Times* a few days later, on September 7, 2018, Jones's world had changed dramatically.[13] In Texas, Judge Jenkins had denied Jones's motion to dismiss Lenny and Veronique's lawsuit, clearing the way for a slow march to a jury trial.[14] Jones's removal from most social platforms caused daily traffic to Infowars' website to plunge by half, from 1.4 million visits a day to 715,000.[15] Casual fans abandoned Jones, leaving an audience of his most zealous, paranoid supporters.

Jones began making regular trips to Washington to grandstand, bully, and rally President Trump's conspiracy-minded supporters. Every day on his show, Jones told them that Democrats and their Deep State enablers were plotting to seize power from Trump—and that they must be stopped, by all means necessary.

SPRING OF 2019 BROUGHT new sorrow to the Sandy Hook families. Early in the morning of March 25, Jeremy Richman, whose daughter Avielle had died in classroom 10 with Jesse Lewis, was found dead, a suicide. Richman was forty-nine, a neuroscientist married to a fellow scientist, Jennifer Hensel. After their daughter's death they founded the Avielle Foundation, which sought answers for violence through the study of brain chemistry and health.

Jeremy had testified before the Sandy Hook Advisory Commission, and a plea he made to them appears in its final report. "We would often hear people say, 'I can't imagine what you're going through. I can't imagine losing your child,'" he said. "The fact was that they were imagining it. They were putting themselves into our shoes, for at least a second. And as hard and as horrible as it sounds, we need people to imagine what it is like . . . Without that imagination, we'll never change."[16]

David Wheeler, whose son Ben was killed at Sandy Hook, counted Jeremy and Jennifer as his closest friends among the parents. "He was by so many measures a remarkable and rare individual. You forgave Jer for wanting to be the smartest guy in the room, because he was the smartest guy in the room ninety-nine percent of the time," he told me. "And when he wasn't, he was humble about it."

The Avielle Foundation kept an office in the Edmond Town Hall. At seven in the morning on March 25, electricians working in the hall's grand theater found Jeremy's body.[17] Earlier that month Jeremy had hosted a discussion there with University of Houston researcher Brené Brown about embracing vulnerability and drawing strength from struggle. Scarlett Lewis was there.

"When I heard Jeremy died, I doubled over," Scarlett told me, sitting in her kitchen, the warm place cluttered with Jesse's artwork where we'd had many conversations over the years.

On the wall in one corner is a chalkboard covered in protective clear plexiglass, where Jesse had scrawled the phonetically spelled words "nurturing healing love." Scarlett found the message when she returned to her house after Jesse's funeral. Like Jesse's note saying "Have a lot of fun" that J.T. found in his bedroom after Jesse's death, the words seemed an exhortation. For Scarlett they defined the mission of the Jesse Lewis Choose Love Movement, which distributes a free life skills program to schools for teaching character social and emotional development.[18] Scarlett had poured herself into it, and by that time it was reaching more than a million children in schools around the world.

Like the Avielle Foundation, Choose Love aspired to end violence through a clearer understanding of its origins. But with Jeremy gone, Scarlett was afraid.

"He had the knowledge, skills, and tools to seemingly overcome this. And then, when you saw what happened to him, of course it comes up in the back of your mind: 'Oh my God, what does that mean for me?'" Jeremy's wife, Jennifer, had written something similar. Jeremy "is the person who had every tool in the toolbox at his disposal," she wrote, and yet "he succumbed to the grief that he could not escape."[19]

The Richmans were among the families targeted by Wolfgang Halbig. It was Avielle whom Halbig insisted was a living Newtown child, a girl who had sung at the Super Bowl in 2013. The couple had joined seven other victims' families, including the Parkers, in suing Alex Jones and Halbig in Connecticut.

Scarlett and I talked about the fear in all of us, of suffering unsurvivable loss.

"After Jesse's murder, I never thought about killing myself. I had J.T., who needed me," Scarlett said. "But I did lose the fear of dying."

She showed me to the door. She had taped a scrap of paper to its pale yellow frame, printed with a daily affirmation:

I handle every situation with patience and understanding.
I am a good person.
I love myself and everyone loves and adores me.

Beneath it gleamed a new dead-bolt lock.

On April 10, 2019, less than a month after Jeremy's death, Robbie Parker flew from Portland to Washington, D.C. He had been invited to testify before a Senate panel about his family's torment by conspiracy theorists.

Robbie had never testified before Congress as several other Sandy Hook relatives had. His statement that day would be his first time speaking in an open forum about Emilie since the impromptu news conference in his Newtown church parking lot, the night after her murder, and the first time he addressed the abuse that followed. He knew his appearance would draw more attacks, but he was done ignoring the hoaxers.

Washington had changed since some Sandy Hook families had testified in the Capitol in 2013, weeks after the massacre. Though their effort to pass new gun legislation had failed, the bill and the atmosphere had been at least nominally bipartisan, lawmakers of both parties respectful of the families and their loss.

But the capital since then had grown into a viciously polarized place, where for the sake of political advantage the party in power had made common cause with the conspiracists.

Robbie's invitation to testify came through Koskoff, Koskoff & Bieder, the law firm representing him and seven other victims' families who were suing Alex Jones in Connecticut. Senator Richard Blumenthal, whose son Matthew worked on the Jones lawsuit at Koskoff, sat on the subcommittee holding the hearing. But Senator Ted Cruz, Republican of Texas, chaired the panel and ran the show that day. With the 2020 election looming, Cruz and fellow Republicans staged the hearing to bolster Trump's—and Alex Jones's—false claim that social media had colluded to silence conservatives.

Even by Washington standards, Cruz stood out for his situational principles. As a corporate lawyer running for Senate in 2010, Cruz championed Texas tort reform while crossing the border to New Mexico to defend record-shattering personal injury awards.[20] Decrying Wall Street's influence in Congress that year, Cruz told a touching story of how he and his wife, Heidi, drained their personal piggy banks to pay for his campaign—while failing to disclose a low-interest loan for the same purpose from Goldman Sachs, his wife's employer.[21]

In 2016, with Cruz and Trump neck and neck for the Republican presidential nomination, Trump launched a series of withering personal attacks against the Texas senator and his family. Among them was a conspiracy theory pushed by Trump and Alex Jones that Cruz's father had been with Lee Harvey Oswald before Oswald assassinated John F. Kennedy.[22]

Cruz and his campaign denounced Trump as "utterly amoral," a "pathological liar" whose nomination would be "a sure disaster for Republicans." But that was then. Six months later, with Trumpism sweeping the GOP, Cruz endorsed his nemesis[23] and emerged as one of the new president's chief Senate sycophants.

In 2018, Cruz supported Alex Jones's claims too.

"Am no fan of Jones—among other things he has a habit of repeatedly slandering my Dad by falsely and absurdly accusing him of killing JFK—but who the hell made Facebook the arbiter of political speech?" Cruz wrote[24] on Twitter after Facebook sanctioned Jones. "Free speech includes views you disagree with. #1A."[25]

So it was that at 2:30 p.m. on a mild spring afternoon, Cruz gaveled to order his subcommittee hearing, "Stifling Free Speech: Technological Censorship and the Public Discourse."

The hearing exemplified how dangerously inured many Republicans had become to the disinformation flowing from the White House. Trump broke social media's rules daily then, sowing lies and hate that the companies refused to sanction. Yet here was Cruz, his family smeared by Trump on Twitter, aping the president's false claims that social media censored only conservatives. No one yet knew where such servility would lead. On that day my unease focused on Robbie.

Cruz sat at the center of a semicircular dais, backed by two staffers whose stubble beards matched his. Most of the subcommittee members' chairs were empty. They showed up only when it was their turn to talk.

The somber, wood-paneled chamber was similar to where Neil had testified six years earlier, with the gilt-framed portrait of him and baby Jesse resting on his knee. This time only a sprinkling of spectators attended the hearing. The panel was slated to hear from Neil Potts, a public policy director at Facebook; Carlos Monje Jr., a public policy director at Twitter; a conservative filmmaker angry that his anti-abortion movie was given an R rating; an anti-abortion lobbyist; two professors; and "Mr. Robbie Parker, Father of Sandy Hook Victim," who appeared next to last on the witness list, behind the lobbyist. Robbie sat in the third row of the gallery, elbows on his knees, holding his printed testimony and patiently waiting his turn.

"Free speech is foundational to our Constitution," Cruz began in his stentorian courtroom tenor. "Since the beginning of journalism, there has been bias in the media. From the first journalist carving on stone tablets, media bias and bias of human beings has been a fact of life. But today we face an altogether different threat. The power of big tech is something that William Randolph Hearst at the height of yellow journalism could not have imagined.

"Polling shows roughly 70 percent of Americans receive their political news from social media," he said. "Over and over again I've heard from Americans concerned about a consistent pattern of political bias and censorship on the part of big tech."

After several flowery minutes Cruz got around to admitting that he had zero evidence conservatives were being systematically censored. "Much of the argument on this topic is anecdotal. It's based on one example or another example," he said. "There is a reason for that: because we have no data."

The hearing seemed destined to fill that data vacuum with hot air.

Cruz ceded the floor to Senator Mazie K. Hirono, Democrat of Hawaii, the subcommittee's ranking member. Hirono had come to Congress as a representative in 2007. Her kindly expression and mild voice belied her reputation for deft verbal barbs.

"There are many areas where the Senate should be conducting

oversight of the tech industry. Baseless allegations of anti-conservative bias is *not* one of them," Hirono said. "If anti-conservative bias by the tech industry has been proven false, why in the world are we holding a hearing on it in the United States Senate? The answer is simple, people. It's politics."

Potts and Monje then spoke in turn. They told the senators that one billion posts appear on Facebook, and five hundred million tweets go up on Twitter each day, most from users outside America. They admitted that they could not consistently enforce their own bans on lies, hate, and inciting violence, much less censor for Republican ideology.

Cruz and his allies declined to explore that central problem. Instead they spent more than forty minutes grilling the tech executives about anti-abortion messages, including a blocked tweet from a political action committee that funds their campaigns. It quoted Mother Teresa, saying that abortion was "anti-woman."

"Is Mother Teresa hate speech?" Cruz repeated the question five times.

Hirono complained about Cruz's filibustering. "I'll tell you what. When you chair the committee, you can decide on the time," he told her.

"I hope that day comes soon," Hirono replied.

Senator Marsha Blackburn, Republican of Tennessee and one of Trump world's more rabid online provocateurs, had questions about tweets denouncing Planned Parenthood's trafficking in "baby body parts."

Cruz called on Josh Hawley, Missouri's ambitious new Republican senator.

Hawley addressed Twitter's Monje: "Let me just say that I enjoyed your testimony about Mother Teresa and why she was or was not engaging in hate speech. I thought your inability to answer that question was absolutely hilarious and really indicative of the lack of transparency that your company is engaging in."

Monje noted that the Mother Teresa tweet was still posted to Twitter.

"But you wouldn't say whether or not it was hate speech," Hawley persisted ridiculously. "The record will reflect that you would not say whether or not it was hate speech."

Less than two years later Hawley would join Cruz in a craven effort to cast doubt on the 2020 presidential election results. On January 6, 2021, the day of the Capitol insurrection, a photographer captured Hawley pumping his fist and giving a thumbs-up to angry "Stop the Steal" demonstrators as they amassed at the Capitol.[26]

After nearly two hours, Blumenthal entered the room for his turn at the microphone. Five years prior, Pat Llodra had asked his office to help Lenny fight a wave of online hate. Blumenthal quizzed the social media executives about the political leanings of Silicon Valley.

"I have not done a survey of Silicon Valley. I see a lot of Democrats. I think there are a lot of libertarians as well," Facebook's Potts offered.

"A good answer to an unfair question," Blumenthal said. "I should disclose that my son, Matthew Blumenthal, is a trial lawyer in Bridgeport with a firm named Koskoff, Koskoff & Bieder that represents Robbie Parker, who also happens to be one of my constituents," Blumenthal said, not realizing that Robbie no longer lived in Connecticut.

"Robbie Parker is going to be testifying, assuming we are done with this panel in time," Blumenthal said.

I watched Robbie take a pen and begin to edit his written statement, cutting it down to fit the dwindling time remaining.

After more than two hours, Cruz introduced Robbie Parker. Robbie wore a black suit and a tie with rose-colored stripes. He offered the panel a tight smile, then shook his head, pained, when Cruz mangled the pronunciation of Emilie's name.

Choking up at times, Robbie told the panel about Emilie and her sisters, how hard it had been for him and Alissa to support them while grieving themselves and while being attacked on social media. He told them about the posts on his Facebook account telling him to watch his back and that he would burn in hell, and the companies' "programmed responses or cold silence" when he reported it. He recalled the visit from the FBI, after Kevin Purfield wrote to the Parkers at home.

"It takes a certain type of person to walk into a school and carry out what he carried out on those teachers and children. However, it takes a very different kind of person to witness that event, then regurgitate demonstrably and undeniably false information about that event while simultaneously attacking victims' families for profit. I understand that motivation. And I cannot accept it."

Robbie noted that the companies had begun to respond and remove some content targeting the families.

They did it, he thought, "because it became more generally known that such abuse was happening, and people were shocked and horrified.

"In other words, only when they realized it would tarnish their brand."

Robbie lost his composure at the end as he wondered aloud about how much the anger provoked by the online attacks had distracted him from attending to Alissa and the girls.

"We live in a beautiful country at a beautiful time when we can share our opinions, not just with the people at our dinner table that we know agree with us, but to the whole world," he said, emphasizing the word *beautiful*, growing tearful when he said it. "Allowing hateful, false information to proliferate on the internet doesn't just corrupt our national dialogue. It has real consequences for real people. I believe that social media companies have a duty to moderate content. Not to limit, but rather to protect the First Amendment in the way in which it was intended."

Hirono let Blumenthal ask the first questions of Robbie. But Robbie had already said his piece.

The preening Cruz would spend the coming election year working social media's refs in the guise of defending free speech. Democrats, too, played that game, mindful of the platforms' persuasive power. This worked well for the companies, hiding behind the First Amendment when called out on their egregious corporate misconduct.

Of all those on the dais that day, only Robbie spoke with clarity. He demanded these rich, private companies do their jobs and stop miscreants like Jones, who profited from his pain.

It was nearly 6:00 p.m. when the hearing finally ended. Most of the senators had left before Robbie testified, and the room was empty.

I greeted Robbie, and we gathered a couple of plastic water bottles and paper coasters emblazoned with the Senate seal as souvenirs for his girls.

We left to grab dinner. The earthy scent of spring thaw rose as we walked through platted gardens between the Senate office building and Union Station.

Taking our seats in the soaring terminal's noisy grill café, Robbie told me that in the midst of the posturing on the panel, the two social media executives had offered him their condolences. Neil Potts, the man from Facebook, had told him, "'It has taken us way too long to do what we needed to do and what was right, and I'm sorry.' I felt like he was very sincere."

We talked for a long time about the terrible days after the shooting. Robbie told me about the interconnected fraternity of violence survivors, and how they pay it forward by connecting with the people left reeling after each new tragedy, as Robbie and Alissa had done with the Pettys after Parkland.

Grief is a journey, Robbie told me, made harder by people who expect you to reach a destination that, if you've ever lost a child, remains forever out of reach.

"Our culture, media stories, the way they're written about tragedies, everything always has to end on a positive, you know?" he asked.

I knew.

"I get it," he said. "But what that does is shelter people from our pain."

I thought about the hearing and how everyone had gone, leaving Robbie, whose world was still the same. The partisan demagoguing did not surprise me. But the careless disregard for Robbie's time, and the political weaponization of his life's worst moments, felt awful.

After dinner Robbie said he planned to walk to the Library of Congress, aggregator of the world's written truth, one of the most beautiful buildings in Washington. "I would love to take a picture of Alissa's book there," he told me. But it was nearly 8:00 p.m., and the library was long closed.

We said goodbye. I watched Robbie stride across the broad stone plaza, past the fountain in its center, broken and stained.

26

DANE COUNTY COURTHOUSE
MADISON, WISCONSIN
OCTOBER 15, 2019

LENNY POZNER RAISED his right hand, swore to tell the truth, then told the jury about his last normal day as a parent, seven years earlier.

In his spare, factual way, Lenny recalled for the jury that sunlit morning in Newtown, December 14, 2012. "It was a regular morning. We got ready for school. I drove my three kids to school. It was a regular day. We played music in the car and joked around, and I dropped my three kids off at the car line."

We had spoken more than once about that morning, and as Lenny testified, I imagined again the pale winter light filtering through the trees on either side of the road, and the dappled images of Sophia, Arielle, and Noah in the clean winter air.

They had motored along Dickinson Drive toward the school, passing dark firs and leafless trees under a clear sky. The December holidays glittered before them like an untouched morning drift, ready for diving in. Reaching the drop-off line, Sophia and Arielle sprang from the back seat of the car. Noah tumbled out last, wearing his Batman hoodie and cartoon-printed superhero sneakers, a tangle of first-grade boy and belongings.

"I don't remember if I said 'I love you' that day, but I said, you know, 'Have a great day,'" Lenny testified, his voice mild and low. "It was cold, but he jumped out not wearing his jacket," Noah's beloved brown bomber. "He had one arm in one sleeve and his backpack on his other arm, and he was kind of juggling both and walking into the school that way.

"And that's—that's the last visual that I have of Noah."

In the jury box, a young woman in an ivory sweater wept.

On the blustery morning of October 15, 2019, Lenny became the first Sandy Hook parent to confront an online tormentor in court.[1] He was suing James Fetzer for defamation in Fetzer's home state of Wisconsin. Fetzer, the disgraced former University of Minnesota Duluth professor and coeditor of *Nobody Died at Sandy Hook: It Was a FEMA Drill to Promote Gun Control*, had spent the past six years of his retirement savaging Lenny's and Veronique's reputations. Fetzer, who had never met the couple, derided them as profiteers and liars while promoting a book filled with lies about Sandy Hook, including on Infowars and in an open letter to President Trump. By the time Lenny filed his lawsuit, the PDF version of *Nobody Died at Sandy Hook* had been downloaded more than ten million times.

Lenny had gone to enormous lengths to convince Fetzer, James Tracy, Wolfgang Halbig, Tony Mead, Kelley Watt, and the rest of Fetzer's ragtag "research community" that the massacre had happened and that his son was indeed dead. He had offered himself up to the Sandy Hook Hoax Facebook group, answering their questions and enduring their abuse. He had given them Noah's birth and death certificates and his school reports. He had met with the Connecticut medical examiner and shared with them the stark details of Noah's autopsy examination. Despite it all, they clung to their denial, accusing him of lying and forgery. The more he fought back, removing their websites and videos and their book, the more vicious they grew.

Alex Jones was this crowd's megaphone. Squirming under legal pressure, Jones had been insisting on his show that he didn't know Fetzer. That was false. Jones and Fetzer first met in the mid-2000s and had shared the marquee for at least two big 9/11 conspiracy conferences, one of them captured on C-SPAN.[2]

"These people are the staples in the hoax world; they are the content creators," Lenny told me about Fetzer and his ilk. "That's what made them truly dangerous."

Lenny had appealed to Fetzer directly, emailing him in February 2016 to demand that Fetzer take down a blog post about Noah, according to court documents.

Fetzer replied within hours. "Well if this isn't my lucky day! Hearing from one of the world's great liars and frauds makes my day. Since there is nothing 'harassing' about my post other than that it exposes you as a hypocrite and con-artist who is doing what he can to avoid being exposed, where you no doubt fear that, if the public becomes aware that Sandy Hook was an elaborate hoax, you may have to give back the money you have defrauded from the public and might even be prosecuted.

"I would like nothing better than to have the opportunity to engage you in legal action, which ought to be of enormous interest to the people of the United States."

Lenny granted Fetzer's wish and sued him.

"I needed to defend my son," Lenny told the jury that day. "He couldn't do that for himself, so I needed to be his voice."

Back in the courtroom, James Henry Fetzer, defendant, slouched in a wooden chair near his attorneys, listening with his chin on his chest and his jowls splayed out, his fingers laced over his belly, thumbs twiddling. White-haired and tousled, in his late seventies, Fetzer wore baggy khaki pants, a brown tweed sport coat with elbow patches, and white leather New Balance sneakers.

Fetzer was born in 1940 in Pasadena, California, and grew up in Altadena. His father worked as an accountant for Los Angeles County, and his mother was a homemaker. His parents divorced when he was young, and he moved in with his father and stepmother after his mother died by suicide when he was eleven.

In a profile of Fetzer in 2006, reporter Mike Mosedale summarized what Fetzer told him about his mother's suicide, writing that it "had rid him of illusions about mortality and made it easier for him to confront unpleasant truths."[3] Fetzer received his degree in philosophy from Princeton University in 1962, graduating magna cum laude, according to a University of Minnesota curriculum vitae that he wrote himself, listing awards he has received, including a high school prize for leadership in 1958. His senior thesis at Princeton, interestingly, was on "the logical structure of explanations of human behavior." He served for four years in the Marine Corps, stationed in Asia for a time, and left with the rank of captain.

Fetzer attended Indiana University, receiving a master's degree and, in 1970, a PhD in history and philosophy of science. While at Indiana, Fetzer divorced his wife of four years, with whom he had a son. His first teaching gig was as an assistant professor at the University of Kentucky, where he met and married his current wife, Jan. He remained at Kentucky for seven years but to his fury was denied tenure. He bounced around for a decade seeking a tenured position, including at the University of Virginia and the University of South Florida, without success. By his own admission, Fetzer was volatile and confrontational, bore-sighted on the merits of his own positions in academic debate. But colleagues from that time say he was smart, even brilliant, and a prolific writer, cranking out books and articles on artificial intelligence, computer and cognitive science, and philosophical reasoning.

In 1987, Fetzer wrote himself into a tenure-track job. His machine-learning scholarship helped him win a full professorship in philosophy at the University of Minnesota Duluth, where he ascended to chairman of its small philosophy department a year later.

He cultivated a hippie intellectual persona, wearing plaid flannel shirts and hiking boots, his bushy muttonchops extending to the corners of his mouth. In 1996 he was appointed a Distinguished McKnight University Professor and awarded a $100,000 McKnight Foundation research grant.

Fetzer's fellow awardees studied autoimmune disease, childhood emotional development, and the effects of poor vision on the brain. Fetzer purported to show that the U.S. government carried out and covered up the assassination of President John F. Kennedy.

Oliver Stone's 1991 film *JFK* spurred Fetzer and thousands of other Americans into a new vocation: assassination researcher. In a scorching critique of Stone's mangling of the historical record, conservative columnist Richard Grenier coined a lasting definition of conspiracy theories as "the sophistication of the ignorant."[4] Yet the popular reaction to *JFK* helped prompt legislation establishing the Assassination Records Review Board, which cleared more than four million pages of previously sealed documents for release to the public.

In those years more than 80 percent of Americans believed that Lee Harvey Oswald had coconspirators, compared with about 60 percent today. But Fetzer advanced a plot so improbable that more seasoned researchers worried he'd discredit the entire inquiry.

Traveling to conferences, posting to online chat boards, and working the media, Fetzer postulated, variously, that shots came from as many as four buildings; that Kennedy's assailants included his limousine driver; that Kennedy's brain was swapped for another before his autopsy; and that Abraham Zapruder's eyewitness film was "a complete fabrication." Former presidents Gerald Ford and Lyndon Johnson were both in on the plan, Fetzer maintained, saying Johnson told a mistress that "the oil boys and the C.I.A. had decided that Jack had to be taken out." Fetzer pushed his views with a bullheaded imperviousness to countervailing documents and facts that presaged his pursuit of the Sandy Hook conspiracy fifteen years later.

In 1998, Fetzer went on the road to promote a book he'd edited, titled *Assassination Science: Experts Speak Out on the Death of JFK*, similar in format to the Sandy Hook opus he would coedit more than a decade later. Pushing his weird theories at a conference in Dallas that year, Fetzer refused to yield the floor to questions, so the organizers cut power to his microphone.

"The Kennedy assassination conspiracy theories, which became an overwhelming obsession, were probably his turning point," said Eve Browning, who worked with Fetzer in the philosophy department in Duluth and later at the University of Texas at San Antonio. Fetzer's views devolved from "strange, to weird, to wacko, to ghoulishly delusional."

Thirty years older than Alex Jones, Fetzer followed a similar track, broadcasting his dark prophecies on Duluth public access TV and in the alternative press, publicly denouncing as frauds, liars, or spies those who disagreed with him. Fetzer held a prominent perch in the state university system and was a guest on shows hosted by Fox's Bill O'Reilly and MSNBC's Jesse Ventura. "Americans are entitled to know the truth about their history," Fetzer told a credulous reporter for the university newspaper, the *UMD Statesman*, which sponsored one of his lectures. He regurgitated that argument after Sandy Hook.

Fetzer's national notoriety crested in the mid-2000s, when he and Steven E. Jones, a physics professor at Brigham Young University, co-founded Scholars for 9/11 Truth. Jones was a dean of the 9/11 truth movement, having written a scientific paper arguing that it was not the airplane strikes but a "controlled demolition" that destroyed the World Trade Center towers. In its first six months Scholars for 9/11 Truth drew about two hundred members, including some fifty college professors and a few former government officials. A half million people visited its website.[5]

Jim Fetzer and Steven Jones met for the first time at "9/11: Revealing the Truth—Reclaiming Our Future," a truther conference at an Embassy Suites in Chicago in June 2006.[6]

For Jones, a quiet-natured, fifty-seven-year-old scientist and a former George W. Bush supporter, the Chicago conference was his first trip outside Utah since advancing his theories. The trip plunged him into the company of five hundred die-hard conspiracy theorists, whose avatar was Alex Jones, Austin radio host and filmmaker.

Steven Jones pleaded for more government transparency, including raw data from government studies, to test his claims. Fetzer and Alex Jones were all about fist-pounding assertion and bellicose calls to action.

Journalist John Gravois asked Fetzer in Chicago whether it bothered him that most scientists had declined to entertain the scholars' theories. "I don't think it's a problem, because we have so much competence and expertise among ourselves," Fetzer replied.

Alex Jones—no relation to the physicist—delivered a keynote address, hailing the whooping crowd as patriots "defending the very soul of humanity against the parasitic controllers of this world government, who are orchestrating terror attacks as a pretext to sell us into even greater slavery!"

In an eerie foreshadowing of the internet-fueled calls to insurrection that followed the 2020 presidential election, Fetzer urged Americans to arm themselves. "Let me tell you, for years I've been waiting for there to be a military coup to depose these traitors!"

Days later Steven Jones and Fetzer sat on the dais as Jones welcomed another sold-out truther crowd, this time for an "American Scholars Symposium" in Los Angeles.

C-SPAN covered the event's keynote panel presentation. A youthful Alex Jones addressed a rapt crowd from a lectern emblazoned with INFOWARS.COM. C-SPAN's chyron identified Jones as a "Documentarian & Radio Talk Show Host."

"Long before 9/11 took place, I was fully aware of the historical fact that governments—Western governments, Eastern governments, the U.S. government, elements of it, normally small criminal elements, have staged terror attacks or have manufactured different pretexts to invade countries, attack nations, or enslave their own domestic populations with police state measures," he said. "However you look at 9/11, we know the official story is a fraud end to end."

Fetzer, the least-known of the panelists, spoke last. His reading glasses low on his nose, barking so close to the microphone that his hard consonants sounded like drumbeats, Fetzer read his top ten reasons why "the hijackers are fake." He demanded to see bodies and autopsy reports, insisting most of the hijackers were "alive and well" in the Middle East. Watching C-SPAN's coverage of the broadcast, I realized Fetzer was advancing the same bogus theory he did on Sandy Hook, if you substituted "Newtown" for "Middle East."

Fetzer raised his voice, working to upstage his fellow panelists. The Bush administration, he warned, was planning another "fake" terrorist attack as reason to "suspend the Constitution and turn the American military on the American people . . . Do not look away!" The crowd cheered.

"Those preparations have been made," Alex Jones chimed in. "A dictatorship is being set up."

As Fetzer gained national notoriety, his academic career was circling the drain.

In early 2004, Fetzer met with the university administration to discuss allegations from a female colleague who said Fetzer had made repeated, unwanted sexual advances.[7] Fetzer acknowledged a conflict with a female coworker but fought the sexual misconduct allegations. He lost. The university suspended Fetzer without pay for the remainder of the spring 2004 semester. He agreed to retire upon returning to campus and left in late May 2006, just before the 9/11 conference in Chicago.

Fetzer and his wife moved from Duluth to a sleepy subdivision of cookie-cutter houses in Oregon, Wisconsin, about twenty minutes from Madison. Fetzer told me their daughter and her husband had persuaded him to move closer to their home near Madison, Wisconsin, because they planned to start a family, calling it "the most touching moment of my life."

On his way out the door, the university gave Fetzer a onetime payment of his $102,000 annual base salary plus a retirement-related sum of $13,000. He was given the lifetime title of Distinguished McKnight University Professor Emeritus and the permanent use of his university email address.

Fetzer routinely invokes the university's name and his McKnight emeritus title to lend credence to his lies about Sandy Hook and the families. He attached a copy of his university CV to an amicus brief he filed in Texas, supporting Alex Jones after Lenny sued him for defamation.

Just before the trial began in October 2019, a disclaimer appeared on Fetzer's university CV page, titled "The University of Minnesota

Duluth's position on James Fetzer's conspiracy theories." Calling Fetzer "a UMD Philosophy Professor Emeritus and conspiracy theorist," the statement stressed that Fetzer's "theories are his own and are not endorsed by the University of Minnesota Duluth or the University of Minnesota System. As faculty emeriti, Fetzer's work is protected by the University of Minnesota Regents Policy of Academic Freedom, which protects creative expression and the ability to speak or write on matters of public interest without institutional discipline or restraint."

A University of Minnesota Duluth spokeswoman declined my requests to talk with the administration about Fetzer.

"We were disappointed with our administration's lack of courage," Eve Browning said. "Universities are like a lot of organizations: not very good at dealing with bad actors, and tend to shelter them rather than face bad publicity. And he won a lot of arguments by yelling and threatening lawsuits."

A couple of months after Fetzer retired, Katherine Kersten, columnist at the Minneapolis *Star Tribune*, reviewed his weirdest claims with prescient alarm. Marveling that the man had taught, in his words, "critical thinking, logic and scientific reasoning" to students for decades, she warned that Scholars for 9/11 Truth includes dozens of professors like him, teaching at brand-name institutions.

"The climate in which this paranoia was born persists," she wrote. "In this poisoned atmosphere, the number of those who share Prof. Fetzer's nightmares will likely continue to grow."

And so it did, thanks to the magic of the internet.

THE AMERICAN JUSTICE SYSTEM provides an imperfect remedy for people seeking relief from online terror. But for now, the courts are all we have.

In 2015, Fetzer and about a dozen other hoaxers had compiled their false theories into *Nobody Died at Sandy Hook*. Its entries include "Is Lenny Pozner Noah's Father?," "Lenny Pozner's Vicious Attack," and "Noah Pozner's Death Certificate Is a Fake."

The book's 2016 edition ended with an editor's note. "Lenny Pozner (or whatever his real name may be) has been abusing copyright claims to take down research on Sandy Hook, which involved an elaborate scam, fleecing the American public of millions of dollars in donations.

"We are in the process of solving a crime, where exposing that Noah Pozner did not die at Sandy Hook has become a crucial element of its resolution."

Lenny's pro bono legal team was led by a husband and wife, Jake and Genevieve Zimmerman. The Zimmermans are friends and colleagues of Kyle Farrar of Farrar & Ball, where Bankston was representing Lenny against Alex Jones. Genevieve is a partner and personal injury lawyer in the Minneapolis firm of Meshbesher & Spence, and worked with Farrar, Bankston, and Ogden in their medical devices lawsuit against 3M. Jake has his own firm, focusing usually on intellectual property litigation. The Zimmermans' daughter and son are a couple years younger than the children killed at Sandy Hook. When the Zimmermans heard that Farrar & Ball was representing Lenny, Veronique, Neil, and Scarlett, they asked what they could do to help. In Wisconsin the Zimmermans had teamed up with "the two Emilys," Emily Feinstein and Emily Stedman, litigators with the Madison-based firm of Quarles & Brady. Emily Feinstein had reached out to Genevieve when she saw Lenny's complaint against Fetzer, filed in Madison. Most lawyers on the team had young children and represented Lenny with a ferocity born of imagining their own families in his position.

Fetzer insisted on representing himself in pretrial proceedings, using his pro se status to probe for material to pad his theories. Fetzer deposed H. Wayne Carver, the retired Connecticut medical examiner, via video. He presented more than twenty exhibits, in a Perry Mason performance that had Carver rolling his eyes and suppressing laughter. Fetzer cagily asked for the victims' crime scene photographs, "under suitable conditions of confidentiality." Carver, wise to him, told him to ask the state. Fetzer asked Carver whether he had been in contact with the state police about Sandy Hook prior to the crime. "That's absurd. Of course not," Carver told him. Fetzer, asking for Halbig, wanted to know whether Carver had used a Porta Potty at the scene. "I must have peed someplace," he replied.

Observing this, Team Zimmerman sought to shut down any avenue for Fetzer to parade his nonsense in a courtroom by "finding a fixed position from which we could establish reality," Jake Zimmerman said.[8]

The Zimmermans narrowed Lenny's case, tethering the defamation claim to four separate statements in Fetzer's book insisting that Noah's death certificate was a forged document fabricated by Lenny.

They sought a summary judgment from Dane County Circuit Court judge Frank Remington. They would submit to the judge evidence that Noah had lived and died, Fetzer was required to submit evidence to the contrary, and the judge would rule.

Lenny and Veronique authorized the lawyers to compile the short, precious paper trail Noah left behind. Noah's birth certificate, his inky footprints with toes like peas in a pod, noting "Baby Pozner A," seven pounds and two ounces, was born at 8:34 a.m. on November 20, 2006, a minute before Arielle, Baby Pozner B. Noah's immunization records. Records from trips to the Danbury Hospital emergency room when he was three, once for a barky cough and once for three stitches when he whacked his forehead on the corner of the kitchen counter while running through the house.

Lenny took blood tests to prove he was Noah's father. Of course his DNA matched Noah's, gathered from a swab Carver had created on the Saturday morning he performed Noah's postmortem. The lawyers submitted Carver's postmortem report, noting the deformed bullet caught in the folds of Noah's Batman hoodie, and his grievous wounds. They ordered a copy of Noah's social security record, which arrived bearing a gold foil seal and a red acetate bow. They produced an affidavit from Sam Greene, the funeral director, saying he had completed Noah's death certificate himself and that Lenny never touched it.

Fetzer produced no admissible evidence. Judge Remington had to rule that Fetzer's four statements were false and had defamed Lenny. A jury would decide how much, if anything, to award Lenny in monetary damages.

On October 14, the trial began.

The first time I saw Lenny in person was in the Madison courtroom. He was in his early fifties then, thinner than in his photos with Noah, with short, graying hair, close-cut beard, and rectangular glasses with hints of red in their metal frames. His charcoal, lightweight suit jacket and pale blue tie suggested he had traveled from warmer climes, and contrasted with the jurors' flannel shirts, sweaters, and fleece vests.

The jurors included a grade school teacher, an actuary, a computer tech with shaggy dark hair, and a mom of two from the nearby farm town of Mazomanie. There was a University of Wisconsin student with doubts about the moon landing, and a gray-haired woman who in 1988 was working in Madison's nearby City-County Building when a troubled young man burst in and opened fire, killing the Dane County coroner and a coworker. Several in the jury pool had been dismissed for saying they thought Fetzer was, in the words of one, "nuts."

Veronique did not travel to Wisconsin for the trial, but she gave Lenny a handwritten list of memories to share with the jury. "He was a joker who would tell his sisters that while they slept at night, he would go to work as a manager in a taco factory . . . He would ask me, 'If God created the universe, who created God?' . . . He believed in superheroes."

The courthouse seemed too nondescript a setting for this landmark on Lenny's long journey. No ornate wood dock or grand statue of blindfolded Lady Justice, just a utilitarian, blond-paneled room, in a buff-color modern building of energy-saving design. Lenny betrayed no outward emotion. He had told Noah's story so many times over the years that he could recite it without reliving it, most of the time. It upset him more to appear in person before Fetzer, sitting ten feet in front of him, and the half-dozen other conspiracy theorists sitting in the gallery, including Kelley Watt and Tony Mead.

Judge Remington greeted the jury both mornings of the trial with folksy commentary on the traffic. Calm and deliberate, with the stooped, lean look of a distance runner, he came from a locally prominent legal family.

Fetzer had hired real lawyers for the trial. Richard Bolton and Eric Baker came from a firm that had defended the Freedom from Religion Foundation. Fetzer of course viewed his case as a watershed moment in First Amendment jurisprudence.

Judge Remington had advised Fetzer to invest in professional counsel, having tired of the antics of Fetzer's DIY legal team, led by Alison "Sunny" Maynard, a disbarred attorney and a hoaxer, a.k.a. Sonja Mullerin.

Fetzer had insisted on deposing Lenny on videotape. The deposition was sealed by the court, but Fetzer helped Maynard and at least three other hoaxers access it illegally.[9] Wolfgang Halbig circulated photoshopped images from the deposition online, comparing Lenny's image to earlier photographs, claiming the man in court was an impostor.

The judge charged Fetzer with contempt and fined him $7,000. He ordered him to either bring the cash to court at the start of the trial or "bring your toothbrush, because you'll be going to the Dane County jail." He also ordered him to "try to put the genie back in the bottle," and get Lenny's deposition back from whomever he had given it to.

Before the trial began, Judge Remington offered the jurors the option to remain anonymous. All of them took it.

"PART OF THIS DISPUTE, as you know already, has already been resolved by the court here, so the only real remaining question is what kind of damages this caused to Mr. Pozner, and by 'damage,' we mean money, because that's all we can do," Genevieve Zimmerman told the jury in her opening remarks.

"I expect that Mr. Fetzer's lawyers are going to attempt to convince you that our—that, first of all, that maybe the lies weren't that big of a deal. Or maybe, maybe the argument is going to be that Mr. Pozner was so damaged by the death of his son that the damage that Mr. Fetzer caused was minimal and perhaps ought to be excused. Keep those ideas in your mind as you hear the testimony that's presented to you."

Fetzer's attorney didn't have a lot to work with. Bolton tried to use Lenny's work to clean garbage off the web as evidence against his claim of emotional distress. Lenny "says that this stuff is really distressing to him and he seeks it out, and I think that that—you'll find that that's inconsistent with his—the injury that he's claiming," Bolton said. Bolton argued that Fetzer had nothing to do with Alex Jones, not least because Fetzer is "a distinguished professor and researcher.

"The book itself is a book with I think thirteen-some authors, at least six of them PhD scholars," Bolton said. "*Nobody Died at Sandy Hook*, while it may be provocative in many respects, I think you'll find that it is, in fact, a serious book of academic research."

The trial, from jury selection to verdict, took only two days.

The jury heard from Roy Lubit, the New York forensic psychiatrist who testified that the hoaxers' campaign to invalidate Lenny had caused a second traumatic injury, the first being Noah's murder.

Jake Zimmerman played unedited recordings of the death threats left on Lenny's voicemail by Lucy Richards, who said she'd heard about Lenny from Fetzer's website.

Fetzer took the stand. He drew an admonishment from the judge for attempting to relitigate the case, referring to the four defamatory statements as "allegedly" defamatory, and criticizing the judge's ruling as "a mistake," because his defamatory statements are true. He also acknowledged sharing Lenny's sealed deposition with the hoaxers.

On the wall of the courtroom, Lenny's lawyers had projected a photo of Noah that Lenny had taken on a family vacation to the beach. Noah stood against a glossy green hedge, engrossed by something outside the frame. His hair looked damp from swimming or running, his cheeks pink and downy in the final summer of his life.

"Can you tell us what you remember about your son's funeral?" Jake Zimmerman, Lenny's lawyer, asked him.

Lenny spoke in a deliberate, low tone. Even in casual conversation he pauses often, weighing words.

"The funeral home was pretty much standing room only. There were that many people there.

"And—and before the funeral, we had—we had a private viewing where we opened the coffin, and—and I got a chance to say, you know, one last goodbye to Noah."

"How did you say one last goodbye to Noah?"

"I wanted to hold his hand, but I couldn't."

"Why? Why couldn't you hold his hand?"

"Noah was shot in his hand and his face, so part of his body was covered. I remember, well, saying goodbye to him and kissing him on his forehead in a familiar way that I've always done, and that was the only part of him that was not covered. And that was the last time I saw him."

"What did you do after the funeral service?"

"Oh, we—we—we went to the funeral—or to bury Noah, to the cemetery."

In the jury box, a young woman with curly hair looked at the ceiling, working on her composure. A man in the uppermost row had placed his hand on his face, sliding his fingers until they covered his mouth. In the front row, the woman who had survived a shooting remained stoic until the end of Lenny's recollection, when she, too, wiped her eyes.

OUTSIDE, A BITTER WIND sprang up, hurling ruby leaves and dust against the windows of the hallway outside the courtroom, while the jury deliberated. Lenny, exhausted, retreated to his hotel to wait.

Kelley Watt had arrived for the full second day of the trial. She had brought her iPad along and was showing photographs of Noah and his half brother to reporters, insisting they were the same person. "We were here to testify," Tony Mead told me, but nobody had asked them. Fetzer said his attorneys refused to put them on the stand. Jake Zimmerman told me later he would have loved it if they had. He would have shown the jury the Vice Media video of Mead, saying of Lenny, "His child has no fucking legacy! Nobody cares about this kid. This kid is a flash in the pan." He'd share the taunt Kelley Watt had posted to Noah's Google+ page: "I want to hear the 'slaughter' and I won't be satisfied until the caskets are opened."

Fetzer took his friends out to dinner, to thank them for having his back. Watt couldn't eat. "I never trust the meat at an Asian buffet," she told me.

The jury deliberated for five hours. They awarded Lenny $450,000, more than Fetzer's net worth. Several months later Fetzer leaked Lenny's sealed deposition transcript, and the judge awarded Lenny another $650,000 in sanctions, boosting the total to $1.1 million.

Fetzer was late returning to the courtroom. After dinner with the hoaxers, he had stopped to visit his daughter and grandchildren. He burst through the courtroom door just after the announcement.

"You're kidding," he huffed to the room. "Unbelievable! I'm staggered."

FETZER HAD AGREED TO meet me the next day. But that morning he called and tried to cancel, all swagger gone. "My family liked my work on JFK. They were happy with my work on 9/11," he told me. "But Sandy Hook for them has been too much.

"They want to live their private lives, and they're entitled to it."

Lenny too, I thought. I told Fetzer I was already en route to his house. There was silence on the line. Fetzer was warring with himself over whether to obey his irate wife and daughter or to stoke his ego with another interview.

"Call my cell when you get here," he said in a low tone. "I'll come out and we'll talk in the car."

Fetzer eased arthritically into the passenger seat of my rented Chevy Malibu. We sat in front of his house while he talked. And talked, and talked, Monday-morning quarterbacking his defense, unspooling his plans for appeal and First Amendment martyrdom.

"The person you've seen in court is not the same person whose photograph has appeared worldwide," he told me. "Why in the world would he be concerned about keeping it confidential? Unless it's not the same person.

"Wolfgang was all over it. He actually wrote to four different offices of the FBI—what are the consequences of impersonating Leonard Pozner at a sworn deposition? He was so incensed."

As Fetzer started to reel off years' worth of Sandy Hook theories, I looked out the windshield and down the street. The leaves on a young tree were turning fiery red; the neighbors had stacked festive gourds around their mailbox.

I asked Fetzer about the beautiful sapling. "Maple," he said, and without missing a beat continued his monologue.

"One of the things, by the way, was the very peculiar claim from this doctor that Leonard Pozner could recover from the death of his son, but that these few words in the book are going to cause permanent damage? I mean, that's absurd on its face."

"Absurd" is one of his favorite words. But Fetzer's most florid prose evaded him that day. He didn't look well. His hands shook, and he had a slight paralysis of one side of his face. Through the windshield I saw a mom and two cute little kids emerge from their house down the street. A boy and girl, they wore cheerfully colored jackets that flapped open as they scrambled into a minivan. Empathy for Fetzer deserted me. I wanted to escape this creep, this deluded bore, filling my space with his lies and bad breath.

Fetzer's flip phone rang. I overheard his daughter telling him to end the interview, that "we're not healing." I was grateful for the interruption.

I asked to use the Fetzers' bathroom. We walked to the front door, and Fetzer made an obsequious plea on my behalf to his stone-faced spouse. Jan Fetzer stepped aside without looking at me. When I returned, she was sitting with her back to me at the near end of the dining room table, household bills spread in small piles around her, resolutely punching numbers into a plastic calculator. Fetzer sat at the other end of the table, hunched once more over his laptop.

Before leaving Madison I met Lenny for lunch at a farm-to-table place near his hotel, walking distance from the courthouse.

Lenny thought the verdict fair. He had heard Wisconsin juries don't tend to award enormous judgments. Jake Zimmerman knew collecting on the award would amount to drawing blood from a stone, but Lenny was determined. Over the coming months he would try to force the sale of the Fetzers' pop-up camper. The sheriff seized it and put it up for auction, but no one bid.

We talked for a while about Halbig. The Connecticut parents were suing him too, but Lenny worried he would be dropped from the suit, as the lawyers focused on the deep-pocketed Jones.

More than a year after we spoke, Halbig's homeowners insurance company offered a settlement. The terms were not publicly disclosed, but the money wasn't much.

In early 2020 Halbig was arrested for distributing Lenny's personal information online, a misdemeanor. Police arrived in the middle of the night and hauled Halbig in for some truly grim mugshots that circulated for months. Halbig was not prosecuted, though. He had emailed Lenny's information to law enforcement, suggesting he believed he was reporting a crime.

Lenny's thoughts returned to the Fetzer trial. It had gone smoothly, but it was exhausting, and of course, Fetzer would appeal.

Lenny told me he'd omitted part of his testimony about Noah's funeral, when he described kissing Noah that one last time. He had intended to say more. But he could not share the moment when his lips touched his son's brow, expecting Noah's softness and stirring, his rising sweet scent, but finding only the cold and unyielding truth of his loss.

We walked out into the square near the courthouse and said

goodbye in front of Lenny's hotel. The sun cast irregular stripes across the pavement, and a crisp breeze carried the overripe scent of fallen leaves.

Walking to my car, I read a text message from Norm Pattis, a trial lawyer in New Haven who represented, off and on, Alex Jones.

"You saw that $450 judgment v truthers?" he wrote. I texted back that I was in the room.

Pattis typed, the ellipsis bubble pulsing on the screen.

"The cynic in me thinks: Fifty Shades of Victimhood. Posner [*sic*] claims PTSD over a whacky book?

"I want that appeal."

27

IN MID-MARCH 2019, a year after Lenny and Neil filed their lawsuits, Mark Bankston deposed Jones in Texas. Farrar & Ball posted the videotaped exchange on—where else?—YouTube. Dressed in an olive-drab polo shirt and sporting a close-cropped beard, Jones planted his elbows on a table set against a marbled-gray backdrop. From the time he took his seat, Jones alternated between bemused cooperation and adolescent defiance, widening his eyes theatrically, sighing heavily, and audibly cracking his neck during breaks between questions.

Lenny had sued Jim Fetzer and won in less than a year, a remarkable achievement. But suing Alex Jones had presented a far more difficult challenge. By mid-2021 the Sandy Hook families' lawsuits against Jones in Texas and Connecticut had consumed more than three years, with a trial date for any of them likely another year away. Jones had deep pockets and employed a revolving cast of nationally known free-speech absolutists, culture warriors, and publicity seekers, supplemented by a shifting cast of second-string lawyers.

More than five hundred thousand people viewed Jones's three-hour deposition video, a purgative accounting of his workaday dishonesty, and his high-wire struggle to defend it under oath without perjuring himself. Claiming not to know even the date of the shooting, Jones questioned the facts in front of him and his questioner's motives, seeking to lure Bankston into his gravity-free realm where theater supplants truth. But this was Bankston's turf, and for the most part, the lawyer tethered Jones to it.

Space prevents me from presenting Jones's depositions in their entirety. But Farrar & Ball posted the full versions of all of Jones's depositions and those of his associates, including Rob Dew and Paul Joseph Watson, to its YouTube channel.[1] Together they provide an aerial tour of America's scorched, post-truth landscape.

"One of the things that you've tried to make clear is that you're not the one who started the theory that Sandy Hook was a false flag, correct?" Bankston asked Jones, who answered, "Yes."[2]

Bankston played a ninety-second montage of Jones's remarks during his December 14, 2012, broadcast, which Infowars later posted under the header "Connecticut School Massacre Looks like False Flag Says [*sic*] Witnesses."

"You've heard me say look for a big mass shooting at schools," Jones said in the clip. "You've heard me. We've gotta find the clips. The last two months I've probably said it twenty times."

In another clip Infowars flashed an early, incorrect account of the shooting from the U.K.'s *Daily Mail* on the screen, while Jones said, "And that is an inside job right there, either way you cut it."

In the final clip Jones says: "Now that Obama's coming in with gun control, magically these shootings are popping up. People gotta find the clips the last two months! I said they are launching attacks, they are getting ready. I can see them warming up with Obama. They've got a bigger majority in the Congress now, in the Senate. They are going to come after our guns. Look for mass shootings. And then magically it happens. They are coming, they are coming, they are coming!"

Bankston turned to Jones. "The truth is, Mr. Jones, you were the first person in the world to make the false flag theory about Sandy Hook, and you did it before the bodies were even cold," Bankston said, his even tone turned lecturing. "That's the truth."

Jones, belligerently: "No. That's not true."

Jones told Bankston that he doubted the official narrative of all major events "because there's such a long history of governments and corporations and legal groups engaging in fraud. And I said that before you played the clip."

TOWARD THE END OF the deposition, Bankston drilled in on Jones's accountability for the families' hurt.

Jones presented his theory of the case.

"It really is the fact that we've allowed the government and institutions to become so corrupt that people lost any compass of what's real. And I myself have almost had, like, a form of psychosis back in the past, where I basically thought everything was staged, even though I've now learned a lotta times things aren't staged. So I think as—as a pundit and someone giving opinion that my opinions have been wrong, but they were never wrong consciously to hurt people."

So Jones telling millions of Americans that Sandy Hook was as "phony as a three-dollar bill," his mocking portrayals of Robbie and Veronique as actors faking their grief, his insinuations that Neil had lied about seeing a bullet hole in his son's head, his efforts to direct hate toward Lenny were simply his opinion, warped by exposure to government lies. Jones didn't do it for money, fame, or entertainment. He did it, he claimed, because he was a victim of "the trauma of the media and the corporations lying so much."

Alex Jones, self-described God-and-country patriot, tough-talking, gun-toting hater of snowflake libs, was claiming he maligned the families of murdered children because he was "like a child, whose parents lied to him over and over again. Well pretty soon they don't know what reality is."[3] It sounded risible. In fact, it was an early hint at a chilling new legal strategem deployed by conspiracy theorists sued for defamation, including those sued for spreading lies about the 2020 election. In a twist on an insanity defense, it posits: If people are too deluded, too immersed in falsehoods to distinguish between truth and reality, can they be held accountable for the harm they cause?

I WATCHED JONES'S DEPOSITION again in early fall 2021. At that point, the nation was experiencing another surge in COVID-19 infections, with 2,000 new deaths being reported each day. Since the start of the pandemic in early 2020, more than 700,000 people in the United States had died from the virus.[4]

Most of the people who died that summer were unvaccinated, among them those who believed the falsehoods Jones and many others had spread: that the vaccines were poison, evil, government tyranny.

On August 21, even Trump was booed at an Alabama rally for telling people to take the vaccine. "Shame on you, Trump," Jones said on Infowars, calling the former president a "dumbass."[5]

By September, scores of Americans were calling poison control centers after overdosing on ivermectin, a common livestock dewormer that they wrongly believed was a treatment for COVID.[6] On his show, Jones ripped open the packaging of what he said was ivermectin for humans.

"See this, Fauci? You see this, Bill Gates? I'm gonna kill those prions, you bastard murderers. You gonna hit me with a bioweapon, you monster? You wanna suppress me? You want to kill me? You son of a bitch! You goddamned demon. You think I'm easy to kill?"

Jones pushed a pill into his maw, washing it down with what looked like a beer.

"You leftists taking all the shots and dying are the dumbasses!"[7]

Taking a break from watching Jones's deposition, I checked email and saw one from a colleague, forwarding a note from a friend unnerved by the virus's resurgence.

"I despair that the country is losing its ability to self-correct," my colleague's friend had written. "It all began after Sandy Hook."

INTERNET-BORNE CONSPIRACISM had metastasized, spreading over the years from Sandy Hook to encompass virtually every major trauma. On January 6, 2021, conspiracists loyal to President Trump acted on his most audacious claim: that Democrats had colluded in a grand scheme to steal the presidency from him.

The lawyers in Texas, Mark Bankston and Bill Ogden, told me that the attack marked a sickening turning point for them.

Upon filing the families' lawsuits against Jones in 2018, "I was so filled with enthusiasm, and thinking I could make a big difference," Bankston said. Now he's dejected: "What I saw between 2018 and the present was how so many people decided that the leading issues of the day were fair game for this kind of manipulation.

"A large portion of our political culture has perhaps correctly deduced that there are things that are way more useful, more potent, and more powerful than truth."

I was writing this final chapter while the United States marked twenty years after the September 11, 2001, terrorist attacks. The media that weekend was awash in coverage, some pondering the ways in which the 9/11 "truther" movement provided a tool kit for the wave of political conspiracists who followed. Kevin Roose, the *Times'* technology columnist, did a deep dive on the 2005 homemade video project *Loose Change*, which eventually reached 100 million people.[8]

The video's "DNA is all over the internet—from TikTok videos about child sex trafficking to Facebook threads about Covid-19 miracle cures," Roose wrote, all of it urging skeptics, as the *Loose Change* filmmakers did, to dig in and research the event themselves.

That call to action was echoed by Alex Jones, who helped produce a subsequent, slicker version of *Loose Change*.

Then came the internet, with its power to stir mass movements and link the rare conspiracists prone to violence into a mob.

"A malign force seems at work in our common life that turns every disagreement into an argument, and every argument into a clash of cultures," former president George W. Bush said on September 11, 2021, in Shanksville, Pennsylvania, where Flight 93 crashed to earth.

"So much of our politics has become a naked appeal to anger, fear and resentment. That leaves us worried about our nation and our future together," he said.

Bush drew a link between the foreign extremism responsible for September 11 and the dangerous beliefs underpinning the Capitol insurrection.

"There is little cultural overlap between violent extremists abroad and violent extremists at home," Bush said. "But in their disdain for pluralism, in their disregard for human life, in their determination to defile national symbols, they are children of the same foul spirit, and it is our continuing duty to confront them."

When Joe Uscinski, the University of Miami professor and *American Conspiracy Theories* coauthor, began studying conspiracy theories in 2011, he set a Google Alert for the term. "Then it was like five articles a day," appearing in the media about the phenomenon, he told me. "Then you get to 2016, and it's between fifty and one hundred a day."

Even as president, Trump inspired conspiracy-minded distrust of the government he led. His handling of the coronavirus pandemic provided just one example. Every day for months, he minimized the coronavirus pandemic, promoted quack cures, and politicized the government's response, until its every aspect drew skepticism and partisan resistance. As deaths mounted, he used his online platform to blame the machinations of an imagined "Deep State" for the leadership failures that led voters to reject him for a second term. His refusal to accept his loss in 2020 raced from unthinkable to a threat to our democracy in less than a year.

Though the Stop the Steal mob nearly killed some of them, Republican Party leaders still tend the conspiracy theory's embers, ensuring it flares anew with every vote.

As I wrote this, Republican-led legislatures in several states were embarking on "reviews" of 2020 election results, following false allegations of voting irregularities raised in Arizona, where a faulty, partisan review of Maricopa County's results touted by Trump was condemned as a sham by Republicans and Democrats alike.

"For those who are pushing the fraud narrative, the actual truth is beside the point," Nate Persily, a Stanford University law professor, elections expert, and scholar of democracy, said in the *Times*.[9]

"The idea that the election was stolen is becoming a tribe-defining belief. It's not about proving something at this point. It's about showing fealty to a particular description of reality."

Never before has a major American political party so assiduously nurtured this destructive force. Yet many of Trump's followers, including the Sandy Hook conspiracists in this book, held a range of political views before Trump appealed to something darker that united them.

The anti-vaxxers booing Trump in Alabama underscored that conspiracists' first loyalty is to their beliefs, a situation easily exploited by a demagogue of any political persuasion.

"A lot of times what you find is that people are concentrating on the specific theories, like, 'Oh my god, this theory is going viral. How can we refute that theory?'" Uscinski said. "When really what's going on is that it's just a visible manifestation of the underlying worldview."

Even QAnon, typically described as a mass delusion afflicting the far right, transcends politics, Uscinski told me. "In our models, partisanship and ideology drop out as predictive of QAnon beliefs," he said. "So what we're left with are dark personality traits." These traits are led by what psychologists call the "Dark Triad": narcissism, psychopathy, and Machiavellianism, meaning the willingness to manipulate others to gain a certain result. Those three, along with others like dogmatism and poor conflict management skills—screaming, throwing things, even slapping or punching in an argument—distinguish people likely to spread antisocial conspiracy theories.

John Kelly of Graphika, who maps the viral spread of misinformation across the web, said few theories have united disparate conspiracy-driven groups like coronavirus vaccine skepticism. Anti-technology types, antisemites, far-right politicos, and hippie alternative-medicine fans all found something to hate in the government's vaccination campaign.

The Center for Countering Digital Hate, a nonprofit working to disrupt the spread of online hate and misinformation, studied online anti-COVID vaccine content over six weeks in early 2021.[10] The center found that 65 percent of anti-vaxx messages were attributable to just twelve people, all with enormous social media followings. On Facebook the "Disinformation Dozen" were responsible for nearly three-quarters of vaccine misinformation. The group spans the political spectrum. In second place, behind diet supplements and snake oil peddler Joseph Mercola, is Robert F. Kennedy Jr., son of the assassinated Democratic presidential candidate, whose Children's Health Defense group targeted African Americans and Latinos with anti-vaxx messages. Behind Kennedy, in third place, are Ty and Charlene Bollinger, pro-Trump conspiracists who hawk books and DVDs touting their loopy claims, including that vaccines fulfill Microsoft billionaire Bill Gates's plan to inject people with microchips.

Deplatforming these repeat offenders is "the most effective and efficient way to stop the dissemination of harmful information," the center said. Yet Facebook, Google's YouTube, and Twitter had to that point failed to enforce their own policies prohibiting COVID misinformation, and most of their accounts remained active. By the end of 2021, that had belatedly begun to change.[11]

It's no accident that Alex Jones failed to make it into the Disinformation Dozen. Deplatformed in 2018 and 2019, his social media presence remains a faint shadow of what it was, despite constant efforts to sneak back on. By late 2021 less than 1 percent of all traffic to Infowars' website came from social media, according to an analysis for the *Times* by Similarweb,[12] an internet tracking company.

Deplatforming blunts misinformation superspreaders' influence and access to funding. White nationalist Richard Spencer, who rode Trump-era bigotry to stardom, had his social media accounts yanked in the aftermath of the 2017 neo-Nazi violence in Charlottesville. Sued for his role in that violence, Spencer told a judge in 2020 that he was having so much trouble raising money online he couldn't afford a lawyer.[13] Spencer's National Policy Institute has closed. He lives in Montana, shunned even by his neighbors.[14]

To be sure, deplatforming is a pretty blunt weapon. Platforms take ages to do it, it creates "free speech" martyrs, and it can drive extremists onto anything-goes platforms like Gab and Parler, where they can become further radicalized.

Experts argue for a finer-grained approach, like imposing time delays between the submission and appearance of a post; disabling the comments function to avoid its becoming a misinformation forum; and, perhaps most important, disabling recommendation algorithms for those accounts so that false content isn't pumped into the feeds of susceptible people.

J. M. Berger, a writer and researcher on extremism, terrorism, and propaganda, told me he recommends deplatforming as a part of a "drug cocktail" of measures, including civil lawsuits and prosecutions, which "are good at stopping groups or movements with strong organizational elements, but less effective at diffuse movements or those that straddle the line of the law."

NEARLY TEN YEARS AFTER Lenny saw the first conspiracy posts about Sandy Hook, his combination of content removal and lawsuits had slowed their spread to a crawl. His use of the bully pulpit and his content-removal demands got the platforms' attention. Facebook designated him a "trusted partner"; and YouTube, a "trusted flagger," whose reports of offensive content lead to swift takedowns. After a virtual meeting with Lenny in mid-2019, Twitter also made changes. The company told me that feedback from Lenny and others "provided an important perspective to inform changes to our rules and enforcement, including expanded guidance in our hateful conduct policy that specifically prohibits the denial of violent events, including abusive references to specific events like Sandy Hook."

Lenny said that's still not enough. But every Sandy Hook conspiracist I spoke with said that it's nearly impossible to post lies about Sandy Hook on social media, and have them stick around for more than a day or two.

America's immersion in harmful online delusion has the world's sociologists, psychologists, and political scientists scrambling to understand and counter it.

Hany Farid, at the University of California, Berkeley, has been advising Congress on reforming Section 230 of the Communications Decency Act of 1996, which protects social media companies from being sued for any defamatory content they distribute.

Farid acknowledges that the big platforms have grown beyond any reasonable hope of comprehensive policing: in 2019 on Facebook, 510,000 comments were posted and 293,000 statuses were updated *every minute.*[15] Farid's favorite congressional proposal targets only the misinformation identified by the companies' secret algorithms, then fed to users to boost engagement and maximize profits. If that recommended content proves defamatory, the same libel laws that govern publishers like the *New York Times* would apply, and the social media companies could be sued.

"It's so elegant," Farid said of the potential solution. "The bill literally can be one sentence."

Graphika's Kelly offers a different solution. He told me he wants to go after those who "fly on the jet stream of misinformation into high political office," using fraudulent accounts and interactions.

Following the Russians' lead, a growing number of think tanks, political action committees, and consultants seek to create the illusion of support or outrage by manipulating conversation and trends on social media. Such tactics are an amped-up, cyber version of Astroturf campaigns, in which lobbyists orchestrate letter-writing campaigns to Congress or op-ed pages that simulate a grassroots upwelling. In 2008, John McCain's presidential campaign drew criticism for offering people incentives to post seemingly spontaneous comments and letters online that were actually written by the campaign.[16] Politicos have upped the ante, distributing blogs and content to like-minded people; creating or buying fake social media accounts and users, a.k.a. bots; and instigating and amplifying viral misinformation campaigns. Tens of millions of bots populate Facebook and Twitter, some of them using stolen identities, my *Times* colleagues wrote in 2018.[17]

"Right now, coordinating and controlling a phalanx of online assets to deliver an effect by pretending to be something they're not is against Facebook policy," Kelly told me. "Why can't doing something like that simply be against the law?"

The First Amendment does not protect "fighting words," which the Supreme Court defined in 1942 in *Chaplinsky v. New Hampshire* as words which, "by their very utterance, inflict injury or tend to incite an immediate breach of the peace. It has been well observed that such utterances are no essential part of any exposition of ideas, and are of such slight social value as a step to truth that any benefit that may be derived from them is clearly outweighed by the social interest in order and morality."

Kelly views artificially manipulating online truth as akin to commercial fraud or criminal conspiracy. "I'm not a legal expert," he said. "But if Congress could pass something like that, it would help."

Sander van der Linden, a social psychologist who directs the Cambridge Social Decision-Making Lab at the University of Cambridge in the U.K., has focused on conspiracism's front end: potential believers.

Van der Linden has developed a new tool to build up people's resistance against false news before they encounter it online. He and his colleagues worked with the U.S. Department of State's Global Engagement Center and the Department of Homeland Security's Cybersecurity & Infrastructure Security Agency to build Harmony Square, a free, ten-minute online game that invites players to wreak disinformation havoc on a fictional, peaceful small town.

The game draws on "inoculation theory," which holds that people exposed to a weakened version of misinformation-spreading techniques are better able to spot it and avoid sharing it in real life. As players travel through four levels of the game, they create a misinformation campaign, using techniques for spreading political misinformation around elections: "trolling, using emotional language, polarizing audiences, spreading conspiracy theories, and artificially amplifying the reach of their content through bots and fake likes."

The researchers found that people who played the game found the manipulative social media posts they encountered less reliable and had more confidence in their ability to spot them. Most important, they were far less likely to share bogus information they found online. The positive effect was the same whether players identified as liberal or conservative.

Van der Linden and his colleagues developed Bad News, a disinformation simulator in which players use bogus tweets to gain followers, that has been translated into fifteen languages. They created Go Viral, a game to strengthen people's resistance to pandemic misinformation, which is part of the British government and World Health Organization's Stop the Spread campaign. The researchers are working with Facebook's WhatsApp to distribute the games, and with YouTube on a pilot program to turn them into animated videos that would appear in the ad space when users click on a video flagged as harmful.

"People don't like to be duped, so if you warn them in advance so they can defend themselves instead of approaching people and saying, 'This is what you need to believe,' they feel special and in the know," van der Linden said of the games' appeal to would-be conspiracists.

"These are purely prophylactic," van der Linden told me. Unfortunately, "There's no known intervention I know of that can essentially de-radicalize people who have become extremists."

In hours of conversations with conspiracy theorists, including someone in his own family, van der Linden said he's tried "pre-suasion," a gateway to persuasion in which "you affirm and validate before you engage."

With this family member, who no longer believes implausible conspiracy theories, van der Linden acknowledged that some conspiracy theories are real, then, after asking permission to express his point of view, offered some that seem obviously false.

"Autonomy and a sense of agency and power are important for a conspiracy theorist. So it's a slow-burn approach—you need to go slow and take one step at a time."

Even then, it often doesn't work.

"Conspiracism is a monological belief system. If you convincingly say this part isn't true, they come up with a bigger conspiracy that contains the smaller. They're going to do everything they can to protect their worldview, even when they acknowledge your facts."

Lenny calls this "the conspiracy blob": people so armored against the intrusion of truth that they repurpose undeniable facts to embroider the original theory. We were talking about Robert David Steele, a former CIA officer and frequent Infowars guest who denied the existence of COVID-19 even as he died from it in August 2021. "We will never be the same because now we know that we've all been lied to about everything" were among Steele's last words. "But, now we also know that we can trust each other."[18]

The conspiracists quickly adjusted. They claimed Steele had been murdered in the hospital as payback for his truth-telling. Lenny told me he died with Sandy Hook hoax material still on his website.

IN PERHAPS THE MOST HONEST moment of his Texas deposition, Jones summarized it well: a significant slice of the population distrusts not only some but all official narratives.

In hundreds of hours spent in and around the people affected directly by Sandy Hook, and hundreds more talking with and listening to believers in conspiracy theories and their loved ones, I tried to learn what drives them. For some, it is mental illness, but dismissing all conspiracists as mentally ill would itself be a form of denial. Conspiracists act on an impulse common in all of us, historians of mass movements warn. They are Hannah Arendt's "atomized, isolated individuals," ripe for joining a movement that affords them fellowship with other souls "obsessed by a desire to escape from reality because in their essential homelessness they can no longer bear its accidental, incomprehensible aspects."[19] That describes Lucy Richards, housebound, poor, and angry, conspiracy internet her main source of stimulation and companionship, who defended her online friends by threatening a grieving father's life. And it describes Infowars provocateur Joe Biggs, jobless after years of military service, arrested for leading the Proud Boys, his new brothers in arms, inside the Capitol on January 6.

Conspiracists' dives into unreality often come after a shock or prolonged setback, the agitation of uncertainty prompting retreat from a confounding reality. Whether roiled by personal calamity or by large-scale events like 9/11, the pandemic or a mass shooting, they seek order amid frightening instability. This describes Pizzagate gunman Edgar Maddison Welch, a young father beset by guilt over a car accident in which he nearly killed a child, duped by Alex Jones's call to free enslaved children.

For many, Trump hastened that escape, by serving up a tweeted litany of lies more accessible and palatable to his followers than objective truth.

"When no clear, authoritative source of truth exists, when uncertainty rages, human nature will lead many people to seek a more stable reality by wrapping themselves in an ever-tighter cloak of political, religious, or racial identities," J. M. Berger wrote.[20] "The more uncertainty rises, the more alluring that siren call becomes."

The more time people spend in this alternative world, the harder it is for them to leave. Turning back would prompt attacks by the group whose approval the conspiracist cherishes. Acknowledging error could bring shame at their gullibility and a reckoning with the pain they've inflicted. As Wolfgang Halbig put it: "Dear God, if I am wrong, I've hurt many, many innocent people."[21] He doesn't say that in contrition, though. It's part of his pitch.

To better understand what drives a longtime conspiracist, I called Kelley Watt's daughter, Madison. She was in grade school when her mother focused on liberal plots to indoctrinate public schoolchildren while her family unraveled.

Madison didn't see much hope for changing her mother's mind. "The only thing that could make her question it would be if that inner circle of hers started to doubt or chip away, but even then it would be hard," she told me.

"There's a great deal of narcissism in this idea that 'everyone's got it wrong and we're in this select group of people that knows.' It would explode her own persona to allow any doubt to come in. Her whole identity has been built on this for so many years. She's invested so much."

Madison recalled years inhabiting the dreams her mother had for her, a shy, bookish little girl dressed in fur and entered in beauty contests. Once, while her mother shopped at Dillard's, Madison crawled into a hulking four-poster bed on display and fell asleep. Finding her, Kelley was so taken by the tableau that she bought the bed. After the Watts lost their house, mother and daughter slept in the bed together, in an apartment so tiny they put it in the dining room. Kelley Watt still has it. She crawled into it with her iPad and wrote a chapter for Fetzer's book

Nobody Died at Sandy Hook. Watt analyzed the gunman's bedroom, believing it too empty, too lacking in decor, for a kid that age to have lived there. She was keen for her family to read her contribution, proud to be published in a book alongside six PhDs.

"I think she feels bad that she in some way hasn't accomplished something," Madison told me. "It's really important for her to be seen as someone really intelligent and good at research."

Her mother believes she'll eventually be proven right, Madison said. "And if that happens, even in the hereafter, then that's her claim to fame."

As the Sandy Hook lawsuits crawled forward, as another anniversary of a horrible event neared, Jones used his broadcasts and his lawyers to transmogrify the legal proceedings into another conspiracy plot. At least three of the lawyers leading Jones's defense—Robert Barnes, Marc Randazza, and Norm Pattis—appeared on his show, sitting beside Jones while he spouted toxic nonsense.

As the Yale historian Timothy Snyder wrote after the January 6 riot, "When we give up on truth, we concede power to those with the wealth and charisma to create spectacle in its place."

Team Jones first fought to have the Sandy Hook lawsuits dismissed on free speech grounds, maintaining they were, in legal parlance, strategic lawsuits against public participation, or SLAPP suits. Anti-SLAPP statutes in more than thirty states[22] aim to prevent powerful interests from using the legal system to silence critics, from reporters to environmental protestors. They lost those motions, in part because of Jones's misconduct.

As the three cases in Texas and the one in Connecticut moved toward trial, Jones, loath to shed any light on his business operations, refused to produce financial records and other documents ordered by the courts as part of the discovery process. In Texas in 2019, Jones was fined

$8,100 for failing to comply with discovery in Scarlett Lewis's suit. Later that year he was found in contempt in Neil's case for the same thing and fined another $26,000. Two months later he was fined another $100,000. Jones and his staff repeatedly failed to show up for depositions. For a deposition on Infowars' corporate operations he twice sent Rob Dew, who answered, "I don't know," to most of the questions.

In Connecticut in May 2019, Jones provided the Sandy Hook families' lawyers with nearly sixty thousand emails, a dozen of which included attachments containing child sexual abuse. The lawyers turned the material over to the FBI, which determined that the material wasn't Jones's. It was sent from outside Infowars and not opened by anyone.

On his show Jones accused the Connecticut lawyers of planting pornography in a plot to destroy him.

"Let's zoom in on Chris Mattei," Jones said, referring to a Koskoff lawyer on the case. "I'm done. Total war. You want it. You got it."

Jones, with lawyer Norm Pattis sitting beside him, pounded on a paper copy of Mattei's Wikipedia page. He ranted for some twenty minutes, referring to the lawyer, a former assistant U.S. attorney in Connecticut, as "a nice Obama boy," and a "jerk-off son of a bitch." He offered $1 million for information on Mattei's role in the plot, saying he wanted his "head on a pike."

Connecticut Superior Court judge Barbara Bellis called Jones's behavior on the broadcast "indefensible, unconscionable, despicable and possibly criminal."[23] She denied Jones's motion to dismiss the case under Connecticut's anti-SLAPP statute. Jones appealed to the Connecticut Supreme Court, which ruled against him.

Jones's lawyers motioned to depose Hillary Clinton in an apparent effort to confirm Jones's suspicions that she was behind the Connecticut lawsuit. The motion, a public document, revealed details from a Sandy Hook relative's sealed deposition, drawing another rebuke from the judge. Jones's lawyers sought Judge Bellis's recusal, citing a "cloud of apparent bias and antagonism." That failed, and they appealed.

Jones had pledged a fight to the death, to defend before a jury his constitutional right to say whatever he wanted. But in the end he lost that opportunity, by denying the authority of the justice system. His money, lawyers, even his reliable dad couldn't save Alex Jones from himself. By the end of 2021, Jones had lost all four Sandy Hook lawsuits. It was a clean sweep, a resounding victory for the families of Noah Pozner, Jesse Lewis, Dawn Hochsprung, Ben Wheeler, Daniel Barden, Dylan Hockley, Avielle Richman, Vicki Soto, Mary Sherlach, and Emilie Parker.

Judges in Texas and Connecticut ruled Jones liable by default, for refusing to provide critical evidence ordered by the courts. In Texas, lawyers called a default ruling "death penalty sanctions": a rare punishment reserved for only the most intractable defendants.

Texas came first. In a scalding ruling on September 27, 2021, Texas District Court judge Maya Guerra Gamble wrote that for more than three years Jones and his legal team exercised "flagrant bad faith and callous disregard for the responsibilities of discovery under the rules," abuses that characterized every case Jones was involved in. In rejecting lesser sanctions, the judge considered "Mr. Jones's public threats, and Mr. Jones's professed belief that these proceedings are 'show trials.'"

In Connecticut the Jones defendants weren't just careless, Judge Bellis wrote in her November 15, 2021, ruling. "Their failure to produce critical documents, their disregard for the discovery process and procedure and for Court orders is a pattern of obstructive conduct that interferes with the ability of the plaintiffs to conduct meaningful discovery and prevents the plaintiffs from properly prosecuting their claims."

Pattis, Jones's lawyer, said he would appeal.

On his Infowars show a couple of hours after he lost in Connecticut, Jones seemed not to get it.

"We need to defend all of our speech rights to say whatever we wish. That's the First Amendment."

Then he asked his fans to send money.

MARK BANKSTON CALLED THE RULINGS "the biggest exhale of my life."

"For years, I had been consumed with the idea that I can't let these families down. That I can't let the country down," he told me. "One of the purposes of a lawsuit like this is to send a message to anyone following in Alex Jones's footsteps, to make it clear what our society will not allow. Thanks to the bravery of these parents, that message has been sent."

In trials set for 2022, ten years after the Sandy Hook shooting, a jury will decide how much Jones should pay the families for the torment he caused. But the answers for why so many Americans chose to believe him and join in that torment still defy reckoning.

LENNY WAS CHILLINGLY PRESCIENT when he predicted nearly a decade ago that what the Sandy Hook families were enduring would become a feature of life in the digital age. The struggle to defend objective truth against people who consciously choose to deny or distort it has become a fight to defend our society, and democracy itself.

The nature of conspiracy theorizing has changed, from relatively harmless speculation about the Bermuda Triangle or Bohemian Grove to sinister theories that place democratic governments at the center of dark plots to control, sicken, and murder their own citizens. These online accusations are now often invented and spread by demagogic leaders around the globe, who use social media platforms to undermine trust in the very institutions that keep the powerful in check—elections and the courts, competing branches of government, and objective journalism. These political opportunists play to constituencies willing to relinquish objective truth for attractive, fantastical Deep State schemes in which political opponents are pedophiles and satanists, and society's most vulnerable morph into villains deserving vigilante justice. As the world saw during the Capitol insurrection on January 6, 2021, some believers in these myths have proved themselves increasingly willing to use violence.

Societal chasms between adherents to truth and consumers of fantasy are widening, aided by those who deliberately manipulate social media channels and discourse. By the end of the second decade of this new century, foreign adversaries and domestic extremists, the traditional culprits, had been joined by a significant swath of the Republican Party.

Alex Jones has said he helped secure a donation from Publix supermarket heiress Julie Jenkins Fancelli to pay for Trump's rally in Washington[24] on the morning of the insurrection, when Trump sicced his followers on Congress, and said he had been asked by the White House to lead the march to the Capitol. Jones was center stage at a rally the night before the Capitol attack, and he and Roger Stone participated in meetings with Trump allies and aides inside and near the Willard Intercontinental Hotel, near the White House. Stone hyped his plans to appear in the January 5 and 6 rallies online, raising money for "security." He swanned around Washington protected by the Oath Keepers, one of whom was later indicted for taking part in the insurrection. Stone says he left the city before the attack. Jones covered the riot all day, traffic to his website exploding. A number of Jones's Infowars associates and extremist friends from the Proud Boys and Oath Keepers invaded the Capitol. Jones himself stayed safely on the perimeter, even at one point ineffectually trying to call off the breach of the building, Dan Friesen, who watched Infowars that day, told me. "His grift doesn't really work if the Capitol is on fire. I don't think he wanted that, because he'd get kicked off his payment processors."

Infowars cameraman Sam Montoya stormed the Capitol with a shouting mob on January 6, plugging Infowars merch on the way:

"Support the war, Infowars-Store-dot-com, we got sales to go! We got sales for days! Please, it's because of your support that we're able to get into things like this and show you the real news that the media won't, that they will spin tomorrow. They're gonna spin it. But guess what? Guess what? We are in the Capitol, baaaaby! Yeah!"

Montoya was filming the mob bashing through a door leading to the House of Representatives chamber when Ashli Babbitt, thirty-five, was fatally shot by a U.S. Capitol Police officer. Infowars aired the graphic footage under the headline PATRIOTS STORM CONGRESS RAW FOOTAGE INCLUDES EXECUTION OF ASHLI BABBITT, ending the video with an ad for Infowars "Alpha Power" supplements.

Montoya was arrested in April 2021, his video cited by the FBI as evidence against him.[25] In late November, the House committee investigating the January 6 attack subpoenaed Jones and Roger Stone.

Fortunately for the country, the conspiracist-in-chief lacked the intellectual or organizational heft to forge a sustained conspiracy of his own. Once inside the Capitol, the mob Trump mobilized smeared feces, vandalized, pilfered, and posed. But they failed to seize the proverbial radio tower and take over. Vice President Mike Pence dithered, then refused to act on a highly dubious legal blueprint crafted by conservative lawyer and Trump enabler John Eastman, claiming Pence could delay certification of the vote in Congress, keeping Trump in power.[26]

State election officials resisted Trump's threats and refused to throw the race. Courts rejected frivolous lawsuits alleging voting fraud filed by "Kraken" conspiracists Rudy Giuliani and Sidney Powell. Trump lost his social media megaphone, his most vital link to the masses. His dreams of a coup dissolved in a narcissistic rage.

But what Trump started, he or others may yet finish. Hours after their fellow lawmakers ran for their lives, Josh Hawley and Ted Cruz went onto the Senate floor to challenge an election they knew Trump lost. These two ambitious cynics had postured and preened during the hearing Robbie attended in 2019, ignoring his warnings about the dangers of viral lies.

The insurrection was "the blossoming of a rotten seed that took root in the Republican Party some time ago and has been nourished by treachery, poor political judgment, and cowardice," Senator Ben Sasse, Republican of Nebraska, wrote in *The Atlantic*.

"My party faces a choice: We can dedicate ourselves to defending the Constitution and perpetuating our best American institutions and traditions, or we can be a party of conspiracy theories, cable-news fantasies, and the ruin that comes with them. We can be the party of Eisenhower, or the party of the conspiracist Alex Jones."[27]

Sadly, the Infowars' theme line remains apt: "There's a war on for your mind."

That war is not over.

Indeed, from the makers of voting machines smeared by charges of election fraud to the targets of cable TV calumny, the Sandy Hook families have been joined by many others in their battle for truth.

EPILOGUE

NEIL HAD NEWS.

"I'm getting married!"

I had never heard him so ebullient, the fatigue in his voice replaced by buoyant optimism. Neil had known Terry Chavez since they were students at Shelton High School in Connecticut. She and Neil had kept in sporadic touch over the years on Facebook, of all things. In the first years after Jesse's death, Terry had come to a couple of the fundraisers Neil and Scarlett had organized for the Choose Love Movement. Terry was engaged to someone else then, and Neil was still overwhelmed.

In late 2019, Neil had wished Terry a happy birthday on Facebook, then called her for a quick hello that lasted five hours. Other marathon conversations followed. On November 5, 2019—Neil remembers that it was a local election Tuesday—they met in Newtown.

They visited Jesse's and some of the other children's graves, and the new Sandy Hook Elementary School. They went out to dinner. Neil admitted it was an unusual first date. But he told me he needed to show

Terry, or maybe to warn her, "that this is my life, and this is what you're dealing with." Terry, an occupational therapist in an elementary school, who works with disabled and troubled children, got it.

Neil told her that evening that he was ready for a relationship. They began going out on weekends, taking day trips. But for months, Neil didn't talk about Terry or introduce her to anyone he knew. He didn't even tell Scarlett about her. "My heart was in the right place, but I didn't know the right thing to do," he said.

"I was afraid that these people, the followers of Jones and Halbig, would latch on to her and contact her."

Terry had no idea. She thought Neil was ashamed of her. They broke up for a time, the widening pandemic making it difficult to date anyway. After months apart, Neil finally sat down with Terry and told her what was up. "I asked her, 'If these fucking nuts try to say something to you, we'll deal with it together, okay?' I didn't want to scare her away."

He didn't. Terry comes from a big family, and they swept Neil in, feeding him on holidays he usually spent alone, accepting him as he was. "I haven't been this happy in twenty years," he told me. We both choked up.

"I'm blessed to have her by my side. She has my back, and I love her."

Neil was banging around the house in Shelton as we spoke by phone, chasing his dogs Sparky, Murphy, and Max out of his workspace. He was trying to make his home ready for Terry on an income hammered by the quarantine shutdown. Installing new lighting, replacing flooring, removing black marks from walls—it occurred to me that he could have been describing a psychic refurbishment.

Running down his to-do list, he paused, then plunged ahead.

"I'm cleaning out Jesse's room," he told me. "It was exactly the way it was when he left—I mean, when he got killed."

Jesse's bedroom was a couple of feet from Neil's, in the front corner of the house. Neil rarely went in there, except to pile up donations of teddy bears and blankets, candles and toys, obscuring Jesse's bed. He had boxed a few of Jesse's toys once, then given up.

But as another anniversary of the shooting passed, "I got it in my head that it was time," Neil said. He told Terry, who insisted on helping.

On a sunny Sunday morning in mid-January, they began. Neil showed Terry a lock of Jesse's hair from his first trip to the barber, on December 10, 2009, and another tuft Neil snipped before they closed Jesse's casket three years later. He gave Terry one of Jesse's rubber ducks and one of his coloring books filled with exuberant drawings.

Neil took up Jesse's plastic backpack, with its bright Disney colors, and opened it for the first time since the shooting.

"His homework wasn't there. He had it with him, and he must have already turned it in. There were some books Vicki Soto gave to him that day, a plastic bag with books to read. That's what they did that morning from nine to nine-thirty—they swapped their books and turned their paperwork in," Neil recalled.

"And there was some projects. A little thing he made with paper and a picture of him standing next to a plastic Santa at Scarlett's, with his little leather jacket on. It was in a construction-paper frame."

At the bottom was an empty juice box and a plastic bag with the husks of the oranges Neil had packed for Jesse's snack that morning.

It was hard. But with Terry it was bearable because, for the first time since Jesse had died, "it's not just me, it's us," Neil said. "It's all about her and I now. A team."

When Jesse's room was empty, Neil would repaint and furnish it as a small sitting room for them.

With President Biden in office, "I'm going to push for stronger laws for things like Jones and Halbig, where there's penalties for this," he told me. "The way Trump carried on with that election, he has done the same thing Alex Jones has done: he created a conspiracy saying, 'This election is a fake, it's a fraud.' He's a hoaxster."

Neil would soon be sixty years old. He was keen for a fresh start.

"Now that I'm thinking about it, would you like a little teddy bear that has Jesse's name on it? I'll make a little box for you with some of his soldiers. Not a big box. Just some stuff I'd like you to have."

In the final months of writing this book, I lost touch with Robbie for a time. He resurfaced briefly to say he was working through some things. In therapy, he had been struggling to understand a volcanic anger that before Emilie's death dwelled so deeply within him that Alissa and the girls had never seen it. When Robbie was six, he had been sexually abused by a person of authority in his church—a close friend of his family's. Emilie's death slashed open those wounds, making Robbie lash out some days and feel emotionless—"matte gray"—on other days. He considered writing a memoir as a way to better understand himself. He wrote the first draft as a conversation between two fictional women whose children had died. They communicated by letter and in verse, voicing Robbie's pain from a vast literary distance. Alissa asked him: What if a publisher tells you to get rid of the fictional characters and step to the center of your story? What would you do then? Robbie inched off the sidelines into the first person and recast his story.

Robbie sent me a draft of his work. He had written about what he called the "huge chasm," the incomprehension that isolates people suffering catastrophic loss. He described an evening in the neonatal ICU, Robbie and his team struggling to resuscitate an infant born too soon. With medication they kept the tiny boy's heart pumping long enough for his mother to leave the operating room. Robbie gently removed the tubes and lines. Entering her room, he placed the dying baby in his mother's arms, then left the parents to grieve.

Hours later the child's mother pleaded, "Why did this happen?" Robbie walked her through the clinical explanation.

"I presumed that my clinical experience and comprehension allowed me to be thoughtful and even compassionate," he wrote. "The reality was I didn't know the first thing about what they experienced that day or every day afterward. I was oblivious of the huge chasm that existed between my interpretation of the events and theirs."

Five years later Emilie was murdered, and Robbie stood on the other side of the chasm, reflecting on his former self. He realized that the grieving mother's plea for answers deserved, in his words, much more than a logical response. "She needed someone that could speak heart-to-heart and not head-to-heart," he wrote. "Unfortunately, she got me, and I was unaware of how to do that for her."

Robbie and Alissa feel that chasm widen the moment they tell a new acquaintance that they have three daughters, the eldest of whom would be eight, or twelve, or fifteen, had she lived. Robbie calls it "hitting them with the information," because people often react as if struck. "I don't know what to say," they often protest. Robbie wants his book to provide some of the words.

Robbie is a lifelong Texas Rangers' fan, and not long after the family moved to Newtown, they went to Fenway Park to watch the Rangers play the Red Sox. Robbie and Emilie staked out the Rangers' batting practice before the game, and Robbie nabbed a home run ball for her. He posted a photo on Facebook of Emilie grinning, holding it aloft. The first season after her murder, the Rangers invited the Parkers to opening day. Robbie threw out the first pitch, into the glove of catcher Iván "Pudge" Rodríguez, a childhood idol. For four hours in the stands on a sparkling day, life was normal again.

Reporters who were there wrote a narrative of the Parkers "moving forward." This tendency toward happy endings concerns Robbie. "As a culture, all of us need to garner more genuine empathy, the kind of empathy where we push past our comfort zones, so we can be present with someone else in their pain," he told me.

"When someone distances themselves by telling me 'You are so much stronger than I am,' or puts a silver lining or a nice bow around my experience, they are denying themselves a chance at real growth within themselves and missing out on making a sincere connection with another person.

"We all need more genuine connections, trying to understand each other better. Right?"

Robbie's comment put me in mind of Lenny's long-ago evening with the Sandy Hook Hoax Facebook group, when one member told Lenny that the shooting deniers did so because they can't face the fact that "this could happen to them"—to their own children.

None of us wants to believe it could have been our loved ones there, as the gunman arrived at the school in Sandy Hook, or at the nightclub in Orlando, the outdoor concert in Las Vegas, or the Walmart in El Paso. It could have been our family to whom we said goodbye on September 11, 2001. In that sense, we are all deniers.

I told Robbie that I had gotten another letter from Kevin Purfield, the man who had stalked the Aurora victims' families and written to the Parkers for years. It went on for twenty-six pages, delusion filling every sheet and scrawled into the margins. Purfield was writing this time from jail: he had been arrested for stalking the editor of the *Oregonian* newspaper.[1]

I read Robbie a couple of excerpts. He sighed in exasperation.

"I'm at the point where if you showed me what a conspiracy theorist wrote, but without attaching the name, I could guess who it is. I'll bet I could bat .400."

VERONIQUE IS IN HER early fifties now, with a no-nonsense manner and a flat, almost midwestern accent, though she spent her childhood speaking French. Sophia and Arielle resemble her. Over three years our conversations had grown more informal. But Veronique still keeps many rooms locked, protecting the vulnerable edifice that is her existence as the mother of a murdered child.

We met at her home in South Florida—she has since moved—in a gated community of curving streets and modern bungalows on the banks of a lagoon. It was a brilliant late-December day. Veronique had escaped the granite stillness of that worst Connecticut winter. But all Decembers weigh heavy, the more recent ones bringing awareness that Noah has been dead longer than he was alive.

We sat together in a small, airy living room. Veronique rested her hands on her knees, and I could see the tattoo on the inside of her right wrist: Noah's name flanked by wings, the dates of his birth and death beneath it. Outside the window birds swept over the water and landed on a lawn glistening with droplets of dew. Sophia and Arielle darted in to say hello but did not linger.

Veronique cannot watch videos of Noah, seeing him move and hearing his voice. But she displays photographs of her little man all over the house. Her favorite was taken on vacation in Seattle a couple of months before his death. Noah grins in the foreground, stretching away from Arielle, who is wrinkling her nose over his shoulder as she encircles his neck with both arms. Veronique showed President Obama the black-and-white image when he visited Newtown for the prayer vigil on December 16, 2012. He wrote, "We will always remember Noah!" and signed his name.

For a very long time, "there was this level of disbelief or denial: 'No, this didn't really happen,'" she had told me once. "You feel like you're living in a photographic negative of what your life used to be. You're trying to make out the old familiar outlines, but really it's just completely alien and you just keep wanting to wake up, and you never do."

She recalled a doctor in the crisis unit that came to Newtown to counsel the families. She was from sub-Saharan Africa and told Veronique that in her culture people believe children choose their parents. When a child dies, the parent spends an indefinite time trapped: "Not quite dead, but you're not quite alive," Veronique said. "Until you go either way. And you *can* go either way."

But Sophia and Arielle had also chosen Veronique, so she cooked the food and did the laundry and drove the car, sunlight glinting through the windshield with a warmth she did not feel as she ferried the girls shopping or to playdates. Noah's palpable absence in the back seat prompted Sophia to instruct, "Mommy, please don't play any music that makes you cry," meaning every one.

"You wonder, what would they have been like if this had not happened?" Veronique mused. "I guess that's been studied as a discipline—intergenerational trauma. It's passed on almost like the color of your eyes or the color of your hair, burned into your DNA."

The girls talk with their counselors about Noah's death, but not with Veronique. "I'm thinking at some point, when they're middle-aged women and I'm an old lady, that they will want to have that dialogue," she told me. But for now, "we're not the right people to talk about it with each other."

Veronique disappeared into a sunny back bedroom off the living room. She returned with Noah's brown corduroy bomber jacket. It hung buttoned to its neckline, its sleeves stuck stiffly out on a hanger too large for it. On the flap of its right pocket a nickel-sized white button bore a scarlet cartoon heart and the message I AM LOVED.

Noah wasn't wearing his jacket when the gunman came, and it had gone missing. Veronique begged the state trooper assigned to her to track it down. They found it on the coat hook where Noah had hung it.

"I know it seems agonizing, but it's also comforting. It's a tactile, very concrete way of sort of keeping him here," Veronique said. She rested it on the arm of the love seat where I sat.

"That's the connection that we need, because we live in a world of five senses. We're trapped here for now."

Nearly a decade after Sandy Hook, "we have a thirty-thousand-foot view of the damage that lies and hoaxes can cause in a society," Veronique said. As a nurse, she was particularly horrified by the human toll exacted by mythmaking around the coronavirus, calling it a "mass hallucination, spreading like the virus itself."

She had been talking with Sophia and Arielle about it. "They're a lot more cognizant because they know what they've lived through. They see it as pathological, almost like those people are caught in a video game or a movie script of their own design. You're not truly living if you're trapped in a reality of your own making."

When Veronique was in high school, kids were summoned to lectures about the dangers of cults.

"That was before the internet. Now kids can be sitting in their bedrooms and be indoctrinated. We have people in leadership right now trying to construct an alternate reality and force us to live in it.

"The warping of the facts is having horrible consequences. History is not going to look kindly on us," Veronique said. "Not at all."

IN LATE 2021 LENNY bought a house, the first house he has owned since the one in Sandy Hook.

Lenny had lived in about a dozen apartments in the last decade, moving mostly because trolls kept posting his home address. Now the hoaxers' voices had quieted enough for him to sink deeper roots. Still, he asked me not to name his new town.

Lenny still keeps Noah's pajamas in his dresser drawer. He also has Noah's camo-patterned Crocs, his Batman costume, and the green-striped hoodie he wore in the last photo taken of him, lighting a Hanukkah candle the night before his death.

A few years ago, Lenny met up with Vanessa, the early HONR volunteer who had seen her own son's face reflected in Noah's. They met at her in-laws' house, on the beach in Florida. Lenny brought along a couple of big bags of Noah's things. He had not been able to part with them earlier; he kept digging through them, seeking Noah's scent in the fabric. But now he wanted someone to have them. Would she like anything, maybe for her own son? They went through the bags together, trying not to cry. Vanessa chose some pants and a pair of Noah's pajamas with Spider-Man on them, and carried them home.

"So many people all over the world were distraught over the horrific killings of these beautiful little children," Vanessa told me. "I feel blessed to have gotten to know Lenny and his family, and to have helped in whatever way I could."

Lenny's original mission, tracking Noah's scent through cyberspace, wrestling his image away from the deniers and profiteers, was drawing to a close. As HONR's executive chair, Alexandrea Merrell was guiding its evolution into a resource for all victims of online abuse. HONR now helps the targets of anonymous stalkers, people falsely accused of crimes, and women whose former partners posted intimate imagery without their consent.

Lenny, deeply committed to HONR's mission, found the work difficult at times. Some survivors wanted HONR to restore their privacy by removing every reference to a murdered loved one, even news coverage. Some suffering people turned into trolls themselves, intent on shaming a murderer's family. Lenny understood the urge to turn online artillery in the enemy's direction. The temptation was enormous, and he, too, at times had given in. But he wanted something more now. Vengefulness would violate the entire enterprise, crossing delicate lines between policing abuse and outright censorship, between free speech and incitement. These were the boundaries the Sandy Hook lawsuits sought to make clear, as a legacy.

One evening I called when Lenny had the girls over. Sophia and Arielle had declared themselves vegetarians. Lenny wasn't sure what to feed them and had bought a ton of potatoes at Costco. He was in a good mood, and we laughed a lot.

Lenny held real power. The social platforms scurried to remove what little malicious Sandy Hook content appeared, to avoid another public drubbing. In a segment of *This American Life*, radio producer Miki Meek equated Lenny to Walter White in *Breaking Bad*, calling him "the one who knocks." Lenny loved that.

Merrell had been nudging Lenny toward his next chapter. Should he unmask himself, become a TED Talk sensation? Share his expertise at the ubiquitous misinformation conferences? Work for one of the social media companies, shaming them for pay?

We meandered together through what lay ahead in the lawsuits, and Jones's bizarre past maneuvers.

Lenny didn't care. Like the big platforms, he had kicked Alex Jones to the curb.

"Jones has admitted that he was completely wrong about how he reported on the Sandy Hook tragedy and how he accused families of faking their grief. He claimed he was insane when he made those statements."

We did the math. By late 2021 it had been nearly three years since Alex Jones uttered the words "Sandy Hook" on Infowars, and longer than that since he had mentioned Noah. Jones's name rarely appeared in the news without a mention of the torment he had inflicted on the Sandy Hook families. Lenny had turned the tables, binding Jones forever to his repugnant claim about an American tragedy, on an internet that never forgets.

"I'm ready to move on," Lenny told me.

"I've won."

ACKNOWLEDGMENTS

I am forever grateful and indebted to the many people who made this book possible by sharing their experiences, wisdom, and advice.

My profound gratitude goes first to the courageous people at the center of this story. Lenny and Veronique, Neil and Scarlett, Robbie and Alissa—I can never adequately thank you for trusting me with your history and your memories of Noah, Jesse, and Emilie. While I have not stood on your side of what Robbie calls the huge chasm, I have tried over these years to treat this precious gift with utmost respect and care. My heart is full and my admiration for you total.

I am deeply grateful to other Sandy Hook family members, victims and survivors of other mass tragedies, first responders, and many others for their wisdom. Francine and David Wheeler, Bill Cario, Mary Ann Jacob, Andy and Barbara Parker, Lori Haas, Caren Teves, Anita Busch, and many others shared experiences, recollections, and facts that vastly improved my understanding. Rob and Debra Accomando, Monte Frank, Pat Llodra, Kenneth Feinberg, Scott Jackson, Monsignor Robert Weiss, Rabbi Shaul Praver, Colleen Murphy, and Dan Malloy, thank you for helping me to more accurately describe the community and state's response to the tragedy and its aftermath. Experts on trauma recovery and reporting helped illuminate the terrible toll on the families and the secondary trauma inflicted by those who denied their loss. They include Alice Forrester, chief executive of the Clifford Beers Clinic in New Haven, Connecticut; Roy Lubit, forensic psychiatrist; Joanne Cacciatore of the Arizona State University School of Social Work; and Bruce Shapiro of the Dart Center for Journalism and Trauma at Columbia University. I am beholden to Joe Uscinski of the University of Miami; Hany Farid of the University of California, Berkeley; Joanne Miller of the University of Delaware; John Kelly of Graphika; Sander van der Linden of the University of Cambridge Social Decision-Making Lab; Nicholas DiFonzo of Roberts Wesleyan College; Kathryn Olmsted of the University of California, Davis; and others for sharing their knowledge of American conspiracism's history, causes, and potential responses. Dan Friesen, who with Jordan Holmes created the *Knowledge Fight* podcast, generously shared his insights into Jones's career and historical forebears, and his library of past Infowars broadcasts.

I am grateful for conversations with people, most of whom requested anonymity, who have struggled to retrieve loved ones from a netherworld of online unreality, and shared their stories with me in an effort to help others. Targets and debunkers of conspiracy theorists informed and enriched this narrative, including Deborah Lipstadt, James Alefantis, Jessikka Aro, Raymond Gutjahr, Charles Frye, Keith Johnson, Doug Maguire, Tiffany Moser, and many unnamed others.

Lawyers involved in the Sandy Hook and other lawsuits and experts in media and First Amendment law provided invaluable guidance: Mark Bankston, Bill Ogden, Kyle Farrar, Genevieve and Jake Zimmerman, Josh Koskoff, Chris Mattei, Alinor Sterling, Emily Feinstein, James Libson,

Staci Zaretsky, Floyd Abrams, the *Times*' indefatigable David McCraw, and David Snyder, executive director of the First Amendment Coalition and my friend.

To the many people who out of concern for their personal safety or livelihoods, or both, contributed to this book under condition of anonymity, I am so thankful.

I am deeply grateful to *New York Times* Washington bureau chief Elisabeth Bumiller, who in 2018 recognized the through line from Sandy Hook conspiracism and its superspreader Alex Jones to a White House inclined to "alternative facts," and supported years of coverage. Washington investigations editor Dick Stevenson guided the marvelous reporter Emily Steel and me during our initial dive into the world of Jones and Infowars and deftly edited multiple stories thereafter. Jonathan Weisman, keen editor, writer, author, and friend, brainstormed multiple stories tracing conspiracism's impact on the 2020 election and its ominous aftermath, while providing inspiration and room for this project. Thank you all.

Given the current struggle in our culture to agree on even established truths, the time and care taken by colleagues and friends who helped research, edit, and refine this book was vital. *Times* editor Justine Makieli fact-checked, end-noted, and shared insights that elevated the book. Chris Cameron, *Times* colleague and researcher, sifted through documents and dug into Jones's business model; *Times* preeminent researcher Kitty Bennett provided research for our initial investigation of Jones, and pretty much every story thereafter. Lynette Adams transcribed long interviews (from Newfoundland) on tight deadlines, and Ron Skarzenski talked me off the ledge while resolving tech issues that always seemed to arise at the worst moments.

The deep reporting of my incredible colleagues at the *Times* and at many other outlets informs this book, as the endnotes attest. Kara

Swisher, Cecilia Kang, Jim Rutenberg, Jack Nicas, Dai Wakabayashi, Kashmir Hill, Mark Walker, Charlie Warzel, Jon Ronson, Will Sommer, Liz Wahl, and many other reporters and friends, thank you for your contributions, brilliant insights, and navigational help. A special thanks to Kristin Hussey, reporter on the *Times*' coverage team for the Sandy Hook shooting and its aftermath, for reams of research, hours of interviews, sharp and heartbreaking observations, camaraderie, and razor wit. To the accomplished writers and careful readers who fielded ideas, offered advice, and plowed through my drafts: Peter Bodo, Christopher Buckley, Nora FitzGerald, Christine Guinness, Shawn McCreesh, Elizabeth Nevin, Maura Reynolds, Scott Shane, Edward Stringer, Mike Tackett, Melanie Trottman, and Glyn Vincent, I owe you each a gigantic favor (and some of you, return postage). David E. Hoffman, I am beyond grateful for your edits, advice, and constant encouragement. A shout-out to dear ones in Blue Hill, Maine, where a large part of this book was born under Ethelbert's portrait at Arcady: Thank you for listening and for your keen interest in the families' welfare.

How to thank the gifted people who brought this book into the world?

My agent, Gail Ross of Ross Yoon, had pushed me for years before this project to write a first book. Gail seized on the importance of this topic from my *Times* reporting, and threw her full formidable self into championing it as a book, from proposal (honed with Dara Kaye) to publication, offering herself as sisterly sounding board and staunch advocate.

John Parsley, editor in chief at Dutton/Penguin Random House, publisher Christine Ball, and their incredible team: senior production editor LeeAnn Pemberton; Amanda Walker, executive director for publicity; Stephanie Cooper, marketing director; meticulous copyeditor Maureen Klier; and associate editor Cassidy Sachs conveyed at every step their deep understanding of what the Sandy Hook relatives' battle for truth means for all of us. John Parsley was my editor on the book. His empathy for the families and determination that Lenny's warning for society ring clear in the book was a constant, from long phone calls during the pandemic to his brilliant observations in the text. Over more than two years, as political conspiracism's dark implications played out in real time and the project's scope expanded, John's commitment to the fullest possible accounting never diminished. Thank you to Leita Walker of Ballard Spahr, who combined detailed legal reviews with astute suggestions. Cassidy Sachs was my most vital link with LeeAnn and her production team, a calming, empathetic guide to a new world for me. When I asked if you had ever been an animal trainer, I meant it.

I am filled with gratitude and love for my family. They weathered my angst and serial interruptions to daily life with enthusiasm and a righteous belief that this story needed to be told. Charlie and David, Mom, Paul, and Kath, I love you.

To my beloved Paul, first and frequent editor of every chapter, wellspring of ideas, sturdy optimism and steadfast love: Dear, you were right. It got done.

NOTES

PROLOGUE

1. Barack Obama, "Remarks by the President at Sandy Hook Interfaith Prayer Vigil," transcript of speech delivered at Newtown High School, Newtown, Connecticut, December 16, 2012, https://obamawhitehouse.archives.gov/the-press-office/2012/12/16/remarks-president-sandy-hook-interfaith-prayer-vigil.
2. State of Connecticut, Division of Criminal Justice, *Report of the State's Attorney for the Judicial District of Danbury on the Shootings at Sandy Hook Elementary School and 36 Yogananda Street, Newtown, Connecticut*, November 25, 2013, https://portal.ct.gov/-/media/DCJ/SandyHookFinalReportpdf.pdf.
3. Dave Altimari and Jon Lender, "Sandy Hook Shooter Adam Lanza Wore Earplugs," *Courant Community*, January 6, 2013, https://www.courant.com/community/newtown/hc-xpm-2013-01-06-hc-sandyhook-lanza-earplugs-20130106-story.html.

CHAPTER ONE

1. John Pirro, "Stars Rise on Firehouse Roof," *Connecticut Post*, January 2, 2013, https://www.ctpost.com/local/article/Stars-rise-on-firehouse-roof-4160022.php.
2. Scarlett Lewis with Natasha Stoynoff, *Nurturing Healing Love: A Mother's Journey of Hope and Forgiveness* (Carlsbad, CA: Hay House, 2013), 11–12.
3. National Fire Protection Association, "US School Fires, Grades K–12, with 10 or More Deaths," https://www.nfpa.org/News-and-Research/Data-research-and-tools/Building-and-Life-Safety/Structure-fires-in-schools/US-school-fires-with-ten-or-more-deaths.
4. Joseph Popiolkowski, "Exhibit at Cleve Hill Teaches About 1954 School Fire That Claimed Lives of 15 Students," *Buffalo News*, March 7, 2015, https://buffalonews.com/news/local/education/exhibit-at-cleve-hill-teaches-about-1954-school-fire-that-claimed-lives-of-15-students/article_298dfd65-ae2e-510a-ba9e-3ca1b07b6464.html.

5. Kendra Bobowick, "Yale Laboratory Comes Down at Fairfield Hills," *Newtown Bee,* August 13, 2010, https://www.newtownbee.com/08132010/yale-laboratory-comes-down-at-fairfield-hills/.
6. *Newtown Bee,* "The Demolition of Yale Laboratory at Fairfield Hills," YouTube, accessed October 17, 2021, https://www.youtube.com/watch?v=OWJccPhjYpw.
7. Mike Lupica, "Morbid Find Suggests Murder-Obsessed Gunman Adam Lanza Plotted Newtown, Conn.'s Sandy Hook Massacre for Years," New York *Daily News,* March 25, 2013, https://www.nydailynews.com/news/national/lupica-lanza-plotted-massacre-years-article-1.1291408.
8. Scarlett Lewis's account of the day of the shooting is drawn from interviews with the author, and from her book, *Nurturing Healing Love: A Mother's Journey of Hope and Forgiveness* (Carlsbad, CA: Hay House, 2013).
9. Associated Press, "SHES Students to Neighbor: 'Our Teacher Is Dead,'" YouTube, 1:28–1:30, https://www.youtube.com/watch?v=Uoh6HurUf4M&list=PL-NBW0UIz_mYoeu1_RpvsUVQ2LI6bYhQ4&index=83.
10. Erin Burnett, "Out Front," CNN, video 0:56–1:25, https://www.youtube.com/watch?v=z6gJqul1gIw&list=PL-NBW0UIz_mYoeu1_RpvsUVQ2LI6bYhQ4&index=81.
11. Michelle Parks, "High Profile: Scarlett Lewis," *Arkansas Democrat-Gazette,* December 11, 2016, https://www.arkansasonline.com/news/2016/dec/11/scarlett-maureen-lewis.
12. Dr. Janet Robinson, "The Story of Sandy Hook," TEDxSHS, May 31, 2014, YouTube, accessed July 6, 2021, https://www.youtube.com/watch?v=resEfJYS-BQ.
13. "Slain Teacher's Family Speaks: She Died 'Protecting Those Babies,'" CBS News, December 17, 2012, https://www.cbsnews.com/news/slain-teachers-family-speaks-she-died-protecting-those-babies/.
14. State of Connecticut, *Shooting at Sandy Hook Elementary School: Report of the Office of the Child Advocate,* 2014, pp. 48–80, https://portal.ct.gov/-/media/OCA/SandyHook11212014pdf.pdf.
15. Alison Leigh Cowan, "Adam Lanza's Mental Problems 'Completely Untreated' Before Newtown Shootings, Report Says," *New York Times,* A17, November 21, 2014, https://www.nytimes.com/2014/11/22/nyregion/before-newtown-shootings-adam-lanzas-mental-problems-completely-untreated-report-says.html.
16. Dave Altimari, "Supreme Court Orders Documents Seized from Sandy Hook Shooter Lanza's Home Released; Key Win in Courant's Yearslong Fight for Records," Courant.com, October 23, 2018, https://www.courant.com/news/connecticut/hc-news-sandy-hook-supreme-court-decision-20181023-story.html. After a five-year court battle with the State of Connecticut, the *Hartford Courant* gained access to Adam Lanza's journals and other documents that state police removed from the Lanza house. The *Courant* argued that public access to the information was vital for understanding a mass killer's state of mind and could help prevent future tragedies.
17. Andrew Solomon, "The Reckoning," *New Yorker,* March 10, 2014, https://www.newyorker.com/magazine/2014/03/17/the-reckoning.
18. Kristin Hussey and Lisa W. Foderaro, "New Sandy Hook School Is Ready Nearly 4 Years After Massacre," *New York Times,* July 29, 2016, https://www.nytimes.com/2016/07/30/nyregion/new-sandy-hook-school-is-ready-nearly-4-years-after-massacre.html.

CHAPTER TWO

1. Alissa Parker, *An Unseen Angel: A Mother's Story of Faith, Hope, and Healing After Sandy Hook* (Salt Lake City, UT: Ensign Peak, 2017), 36–38. Alissa and Robbie Parker's account of the

day of the shooting is drawn from Alissa's book and the author interviews with Robbie Parker.

2. Joanne Cacciatore and Sarah F. Kurker, "Primary Victims of the Sandy Hook Murders: 'I Usually Cry When I Say 26,'" *Children and Youth Services Review*, September 2020, https://doi.org/10.1016/j.childyouth.2020.105165.
3. "*60 Minutes* Reports: Tragedy in Newtown," CBS News, December 16, 2012, https://www.cbsnews.com/news/60-minutes-reports-tragedy-in-newtown.
4. Elizabeth Williamson, "A Lesson of Sandy Hook: 'Err on the Side of the Victims,'" *New York Times*, May 25, 2019, https://www.nytimes.com/2019/05/25/us/politics/sandy-hook-money.html.
5. Parker, *An Unseen Angel*, p. viii.

CHAPTER THREE

1. Mary Ann Jacob, "I Think of People Who Died at Sandy Hook Every Day," *Vogue*, May 24, 2016, https://www.vogue.com/projects/13439280/gun-control-sandy-hook.
2. The account of Arielle's classroom activities on the morning of the shooting and her escape are drawn from interviews by the author and: Kaitlin Roig-DeBellis with Robin Gaby Fisher, *Choosing Hope: How I Moved Forward from Life's Darkest Hour* (New York: G. P. Putnam's Sons, 2015), pp. 86–94.
3. Cara Buckley, "A Painful Duty: Consoling a Town Preparing to Bury Its Children," *New York Times*, December 17, 2012, https://www.nytimes.com/2012/12/17/nyregion/in-newtown-finding-words-for-a-mother-burying-her-boy.html.
4. Anderson Cooper, *Anderson Cooper 360*, transcript, CNN, December 21, 2012, https://transcripts.cnn.com/show/acd/date/2012-12-21/segment/02.

CHAPTER FOUR

1. Kaster Lynch Farrar & Ball LLP, "Alex Jones/Sandy Hook Video Deposition, Part I," videotaped deposition, March 2019, https://www.youtube.com/watch?v=I7siWJ86g40, 8:24-9:06.
2. Callum Borchers, "Alex Jones Is a Narcissist, a Witness Testifies. And He's Undermining His Own Attorneys," *Washington Post*, April 20, 2017, https://www.washingtonpost.com/news/the-fix/wp/2017/04/20/alex-jones-is-a-narcissist-a-witness-says-and-possibly-the-worst-client-ever.
3. Penny Owen, "From the Ashes: Volunteers Rebuilding Davidian's Church," *The Oklahoman*, December 20, 1999, https://www.oklahoman.com/article/2679266/from-the-ashes-volunteers-rebuilding-davidians-church.
4. Ira Glass and Jon Ronson, "Beware the Jabberwock," *This American Life*, podcast audio, March 15, 2019, https://www.thisamericanlife.org/670/beware-the-jabberwock.
5. Sam Howe Verhovek, "Branch Davidians Shed No Tears for McVeigh," *New York Times*, June 13, 2001, U.S., https://www.nytimes.com/2001/06/13/us/branch-davidians-shed-no-tears-for-mcveigh.html.
6. Peter Fimrite, "Masked Man Enters, Attacks Bohemian Grove; 'Phantom' Expected Armed Resistance," SFGATE, January 24, 2002, https://www.sfgate.com/bayarea/article/Masked-man-enters-attacks-Bohemian-Grove-2881742.php.
7. Barack Obama, "Statement by the President on the School Shooting in Newtown, CT," December 14, 2012, https://obamawhitehouse.archives.gov/the-press-office/2012/12/14/statement-president-school-shooting-newtown-ct.

CHAPTER FIVE

1. Wilson Andrews, Bonnie Berkowitz, Alberto Cuadra, Emily Chow, Laris Karklis, Dan Keating, and Katie Park, "U.S. Mass Shootings in 2012," *Washington Post,* December 14, 2012, https://www.washingtonpost.com/wp-srv/special/nation/us-mass-shootings-2012/.
2. Joshua L. Powell, *Inside the NRA: A Tell-All Account of Corruption, Greed, and Paranoia Within the Most Powerful Political Group in America* (New York: Twelve, 2020) p. 5.
3. Eric Lichtblau and Motoko Rich, "N.R.A. Envisions 'a Good Guy with a Gun' in Every School," *New York Times*, December 21, 2012, https://www.nytimes.com/2012/12/22/us/nra-calls-for-armed-guards-at-schools.html.
4. Wayne LaPierre, NRA press conference, December 21, 2012, transcript, *New York Times,* https://archive.nytimes.com/www.nytimes.com/interactive/2012/12/21/us/nra-news-conference-transcript.html.
5. Sebastian Murdock, "Exclusive: NRA Official Sought Sandy Hook Hoaxer to Question Parkland Shooting, Emails Show," *HuffPost*, March 27, 2019, https://www.huffpost.com/entry/exclusive-nra-sandy-hook-hoaxer-parkland-shooting_n_5c8aa54de4b03e83bdbe59eb.
6. Anti-Defamation League, "The Oath Keepers Anti-Government Extremists Recruiting Military and Police," 2015, https://www.adl.org/sites/default/files/documents/assets/pdf/combating-hate/The-Oath-Keepers-ADL-Report.pdf.
7. Dmitriy Khavin, Haley Willis, Evan Hill, Natalie Reneau, Drew Jordan, Cora Engelbrecht, Christiaan Triebert, Stella Cooper, Malachy Browne, and David Botti, "Video: Day of Rage: An In-Depth Look at How a Mob Stormed the Capitol," *New York Times*, June 30, 2021, https://www.nytimes.com/video/us/politics/100000007606996/capitol-riot-trump-supporters.html.
8. Alex Seitz-Wald, "Newtown Truthers: Where Conspiracy Theories Come From," *Salon*, January 16, 2013, https://www.salon.com/2013/01/16/newtown_truthers_where_conspiracy_theories_come_from/.
9. Fairleigh Dickinson University's Public Mind Poll of 863 registered voters was conducted nationally by telephone with both landline and cell phones from April 22 through April 28, 2013, and has a margin of error of +/–3.4 percentage points; https://portal.fdu.edu/newspubs/publicmind/2013/guncontrol/final.pdf.

CHAPTER SIX

1. Skinner conditioned pigeons to peck at a target, which he used to try to develop a pigeon-guided missile. According to his preliminary tests, the pigeons were nearly perfect.
2. Max Read, "Behind the 'Sandy Hook Truther' Conspiracy Video That Five Eight [*sic*] Million People Have Watched in One Week," *Gawker*, January 15, 2013, https://www.gawker.com/5976204/behind-the-sandy-hook-truther-conspiracy-video-that-five-million-people-have-watched-in-one-week.
3. Joanne Cacciatore and Sarah F. Kurker, "Primary Victims of the Sandy Hook Murders: 'I Usually Cry When I Say 26,'" *Children and Youth Services Review*, September 2020, https://doi.org/10.1016/j.childyouth.2020.105165.
4. Alex Seitz-Wald, "Sandy Hook Truthers Are Not Giving Up," *Salon*, March 20, 2013, https://www.salon.com/2013/03/20/sandy_hook_truthers_are_not_giving_up/.
5. Anderson Cooper, *Anderson Cooper 360*, transcript, CNN, January 11, 2013, https://transcripts.cnn.com/show/acd/date/2013-01-11/segment/01.

6. Ben Zimmer, "Plots, Politics and the Meaning of 'Crisis Actors,'" *Wall Street Journal*, March 2, 2018, Life & Work, https://www.wsj.com/articles/plots-politics-and-the-meaning-of-crisis-actors-1520008999.
7. Jeffrey S. Morton, Patricia Kollander, and Thomas Wilson, "Letters: Why James Tracy, FAU's Conspiracy Theorist, Should Resign," *Palm Beach Post*, April 29, 2013, https://www.sun-sentinel.com/opinion/fl-xpm-2013-04-28-fl-online-letter1-20130428-story.html.

CHAPTER EIGHT

1. Dawn Johnson, Twitter post, December 14, 2018, 9:03 a.m., https://twitter.com/sundene/status/1073579179361349632.
2. Veee Whooo, Twitter post, December 14, 2018, 8:10 a.m., https://twitter.com/missveewhoo/status/1073565880645226504.
3. Olivia Searcy, Twitter post, December 14, 2018, 7:03 p.m., https://twitter.com/LivSeeds/status/1073730119737446400.
4. Kathryn Orr, Twitter post, December 14, 2018, 8:48 a.m., https://twitter.com/kcorr54/status/1073575483877388288.
5. Jim Miani, Twitter post, December 14, 2018, 7:36 a.m., https://twitter.com/JimMiani/status/1073557421098967040.
6. Caroline J. Kistin, Twitter post, December 14, 2018, 9:31 a.m., https://twitter.com/CJKistinMD/status/1073586171933048832.

CHAPTER NINE

1. Kris Cambra, "H. Wayne Carver II," n.d., accessed August 1, 2021, https://medicine.at.brown.edu/article/h-wayne-carver-ii/.
2. Joseph E. Uscinski, ed., *Conspiracy Theories and the People Who Believe Them* (New York: Oxford University Press, 2019), 257–265.
3. Richard Hofstadter, "The Paranoid Style in American Politics," *Harper's*, November 1964, https://harpers.org/archive/1964/11/the-paranoid-style-in-american-politics/.
4. Kathryn S. Olmsted, "Conspiracy Theories in U.S. History," in Uscinski, *Conspiracy Theories and the People Who Believe Them*, 285.

CHAPTER TEN

1. Alexandra Stevenson, "Facebook Admits It Was Used to Incite Violence in Myanmar," *New York Times*, November 6, 2018, https://www.nytimes.com/2018/11/06/technology/myanmar-facebook.html.
2. "The Father Taking On the Sandy Hook Trolls," interview of Lenny Pozner by Matthew Bannister, BBC *Outlook*, 2016, https://www.bbc.co.uk/programmes/p04kfy5f.
3. David Owens, "Sandy Hook Denier Back in State to Face Harassment Charge," *Hartford Courant,* February 24, 2017, https://www.courant.com/breaking-news/hc-sandy-hook-denier-harrassment-0225-20170224-story.html.
4. Ira Glass and Jon Ronson, "Beware the Jabberwock," *This American Life*, podcast audio, March 15, 2019, https://www.thisamericanlife.org/670/beware-the-jabberwock.
5. Alexis Krell, "He Shot His Ex-Wife During an Argument at Their Tacoma Home, Then Shot Himself, Charges Say," *News Tribune*, June 14, 2019, https://www.thenewstribune.com/news/local/crime/article231520323.html.

CHAPTER ELEVEN

1. Southern Poverty Law Center, "Willis Carto," n.d., accessed August 2, 2021, https://www.splcenter.org/fighting-hate/extremist-files/individual/willis-carto.
2. Keith Johnson, "Interview: Father of Little Boy Murdered at Sandy Hook Speaks," *American Free Press*, June 14, 2014, https://americanfreepress.net/afp-interview-father-of-little-boy-murdered-at-sandy-hook-speaks/.
3. Evelyn Schlatter, "Buyer Beware: Veterans Today and Its Anti-Israel Agenda," Southern Poverty Law Center, January 6, 2011, https://www.splcenter.org/hatewatch/2011/01/06/buyer-beware-veterans-today-and-its-anti-israel-agenda.
4. Lenny Pozner, "Our Grief Denied: The Twisted Cruelty of Sandy Hook Hoaxers," Courant.com, July 25, 2014, https://www.courant.com/opinion/op-ed/hc-op-commentary-pozner-sandy-hook-newtown-hoax-07-20140725-story.html.
5. Jill Duff-Hoppes and Robert Perez, "Before the Danger, There Are Signs," *Orlando Sentinel*, April 25, 1999, https://www.orlandosentinel.com/news/os-xpm-1999-04-25-9904230670-story.html.
6. Reeves Wiedeman, "The Sandy Hook Hoax," *New York*, Intelligencer, September 5, 2016, https://nymag.com/intelligencer/2016/09/the-sandy-hook-hoax.html.
7. Lauren Ritchie, "Sandy Hook Conspiracy Theorist Needs a New, Less Harmful, Hobby Commentary," *Orlando Sentinel*, April 5, 2019, https://www.orlandosentinel.com/opinion/os-ne-lauren-ritchie-sandy-hook-wolfgang-halbig-20190405-story.html.
8. Kaster Lynch Farrar & Ball LLP, "Alex Jones/Sandy Hook Video Deposition, Part I," YouTube, 37:13, https://www.youtube.com/watch?v=I7siWJ86g40&t=479s.
9. Newtown Connecticut Board of Education, minutes of the Board of Education Meeting, May 6, 2014, https://www.newtown.k12.ct.us/_theme/files/Board%20of%20Education/Board%20Minutes/2013-2014%20Minutes/5-6-14%20minutes_appr_att.pdf.
10. Nanci G. Hutson, "Newtown School Board Greets Sandy Hook Skeptics with Silence," *Connecticut Post*, May 7, 2014, https://www.ctpost.com/local/article/Newtown-school-board-greets-Sandy-Hook-skeptics-5458643.php.

CHAPTER TWELVE

1. Bill Loomis, "1900–1930: The Years of Driving Dangerously," *Detroit News*, April 26, 2015, https://www.detroitnews.com/story/news/local/michigan-history/2015/04/26/auto-traffic-history-detroit/26312107/.
2. Digital Millennium Copyright Act of 1998: U.S. Copyright Office Summary, December 1998, https://www.copyright.gov/legislation/dmca.pdf.
3. Tom Dreisbach, "Alex Jones Still Sells Supplements on Amazon Despite Bans from Other Platforms," March 24, 2021, NPR, https://www.npr.org/2021/03/24/979362593/alex-jones-still-sells-supplements-on-amazon-despite-bans-from-other-platforms.
4. Lenny and Veronique Pozner, "Sandy Hook Massacre 3rd Anniversary: Two Parents Target FAU Conspiracy Theorist," Sun-Sentinel.com, December 10, 2015, https://www.sun-sentinel.com/opinion/commentary/sfl-on-sandy-hook-anniversary-two-parents-target-fau-professor-who-taunts-family-victims-20151210-story.html.
5. Paula McMahon, "Fired FAU Prof James Tracy Testifies About Feud with Sandy Hook Victim's Parents," *SunSentinel*, December 1, 2017, https://www.sun-sentinel.com/local/broward/fl-reg-james-tracy-fau-testifying-20171130-story.html.

6. James Tracy, "FAU Professor Questions Whether Sandy Hook Massacre Was Staged," Sun-Sentinel.com, December 14, 2015, https://www.sun-sentinel.com/opinion/commentary/sfl-former-fau-professor-questions-whether-sandy-hook-massacre-was-staged-20151214-story.html.
7. Kevin Carey, "Academic Freedom Has Limits. Where They Are Isn't Always Clear," *Chronicle of Higher Education*, January 15, 2016, https://www.chronicle.com/article/academic-freedom-has-limits-where-they-are-isnt-always-clear/.
8. Christine Hauser, "ABC's 'Pink Slime' Report Tied to $177 Million in Settlement Costs," *New York Times*, August 10, 2017, Business, https://www.nytimes.com/2017/08/10/business/pink-slime-disney-abc.html.
9. Masson v. New Yorker Magazine Inc., n.d., Oyez, accessed March 11, 2021, https://www.oyez.org/cases/1990/89-1799.
10. Hustler Magazine, Inc. v. Falwell, n.d., Oyez, accessed August 2, 2021, https://www.oyez.org/cases/1987/86-1278.
11. Legal Information Institute, "47 U.S. Code § 230—Protection for Private Blocking and Screening of Offensive Material," LII / Legal Information Institute, 2018, https://www.law.cornell.edu/uscode/text/47/230.
12. Electronic Frontier Foundation, "Section 230 of the Communications Decency Act," 2019, https://www.eff.org/issues/cda230.
13. Jordan Hoffner, Peabody Awards acceptance speech, 2008, accessed August 2, 2021, https://peabodyawards.com/award-profile/youtube.com/.
14. Nitasha Tiku, "Twitter CEO: 'We Suck at Dealing with Abuse,'" *The Verge*, February 4, 2015, https://www.theverge.com/2015/2/4/7982099/twitter-ceo-sent-memo-taking-personal-responsibility-for-the.
15. Charlie Warzel, "'A Honeypot for Assholes': Inside Twitter's 10-Year Failure to Stop Harassment," *BuzzFeed News*, August 11, 2016, https://www.buzzfeednews.com/article/charliewarzel/a-honeypot-for-assholes-inside-twitters-10-year-failure-to-s.

CHAPTER THIRTEEN

1. David Barstow, Susanne Craig, and Russ Buettner, "Trump Engaged in Suspect Tax Schemes as He Reaped Riches from His Father," *New York Times*, October 2, 2018, https://www.nytimes.com/interactive/2018/10/02/us/politics/donald-trump-tax-schemes-fred-trump.html.

CHAPTER FOURTEEN

1. Seaborn Larson, "Downtown Stabbing Suspect Takes Plea Deal; Prosecutors Offer 10-Year Prison Term," *Great Falls Tribune*, February 14, 2018, https://www.greatfallstribune.com/story/news/crime/2018/02/14/downtown-stabbing-suspect-takes-plea-deal-prosecutors-offer-10-year-prison-term/338253002.
2. Paula McMahon, "Woman Accused of Threatening Sandy Hook Parent Jailed After She Was a No-Show for Court," Sun-Sentinel.com, April 3, 2017, https://www.sun-sentinel.com/local/broward/fl-pn-sandy-hook-lucy-richards-arrest-20170403-story.html.
3. Derek Hawkins, "Sandy Hook Hoaxer Gets Prison Time for Threatening 6-Year-Old Victim's Father," *Washington Post*, June 8, 2017, https://www.washingtonpost.com/news/morning-mix/wp/2017/06/08/sandy-hook-hoaxer-gets-prison-time-for-threatening-6-year-old-victims-father/.

CHAPTER FIFTEEN

1. Kathleen McWilliams, "Mother Who Lost Daughter in Newtown Writes Book on Healing," *New Haven Register*, April 8, 2017, https://www.nhregister.com/lifestyle/article/Mother-who-lost-daughter-in-Newtown-writes-book-11313931.php.
2. Kathleen McWilliams, "Five Years After Sandy Hook, Emilie Parker's Family Finds Joy, Solace," Courant.com, December 14, 2017, https://www.courant.com/news/connecticut/hc-emilie-parker-sandy-hook-anniversary-htmlstory.html.
3. Isabel Fattal, "My Life Since the 2012 Sandy Hook Shooting: Alissa Parker's Story," *The Atlantic*, March 24, 2018, https://www.theatlantic.com/family/archive/2018/03/alissa-parker/556409/.
4. Aimee Green, "Aurora Shooting: Portland-Area Man Pleads Guilty to Harassing Relative of Victim," *Oregonian*/OregonLive, June 6, 2013, https://www.oregonlive.com/portland/2013/06/aurora_shooting_portland-area.html.
5. Aimee Green, "Portland Man Pleads Guilty to Harassing Relative of Colorado Theater Shooting Victim," *Oregonian*/OregonLive, June 5, 2013, https://www.oregonlive.com/portland/2013/06/portland_man_pleads_guilty_to_6.html.
6. Associated Press, "Portland Man Arrested in California Charged with Harassing Authorities," *The Bulletin*, October 13, 2013, https://www.bendbulletin.com/localstate/portland-man-arrested-in-california-charged-with-harassing-authorities/article_e15d4712-becc-586b-a5b5-c45f32829826.html.
7. *Oregonian* staff and wire reports, "Man Accused of Bomb Threats Sentenced to Time Served After Medication Returns Competence," OregonLive, March 17, 2015, https://www.oregonlive.com/portland/2015/03/man_accused_of_bomb_threats_at.html.

CHAPTER SIXTEEN

1. Amanda Robb, "Pizzagate: Anatomy of a Fake News Scandal," *Rolling Stone*, November 16, 2017, https://www.rollingstone.com/feature/anatomy-of-a-fake-news-scandal-125877/.
2. Hunt Allcott and Matthew Gentzkow, "Social Media and Fake News in the 2016 Election," National Bureau of Economic Research, Working Paper 23089, 2017, https://doi.org/10.3386/w23089.
3. Amy Chozick, Nicholas Confessore, and Michael Barbaro, "Leaked Speech Excerpts Show a Hillary Clinton at Ease with Wall Street," *New York Times*, October 8, 2016, U.S., https://www.nytimes.com/2016/10/08/us/politics/hillary-clinton-speeches-wikileaks.html.
4. Amy Chozick, Nicholas Confessore, Steve Eder, Yamiche Alcindor, and Farah Stockman, "Highlights from the Clinton Campaign Emails: How to Deal with Sanders and Biden," *New York Times,* October 10, 2016, https://www.nytimes.com/2016/10/10/us/politics/hillary-clinton-emails-wikileaks.html.
5. Gregor Aisch, Jon Huang, and Cecilia Kang, "Dissecting the #PizzaGate Conspiracy Theories," *New York Times*, December 10, 2016, https://www.nytimes.com/interactive/2016/12/10/business/media/pizzagate.html.
6. Benjamin Weiser, "Anthony Weiner Gets 21 Months in Prison for Sexting with Teenager," *New York Times*, September 25, 2017, New York, https://www.nytimes.com/2017/09/25/nyregion/anthony-weiner-sentencing-prison-sexting-teenager.html.
7. Andrew Rossi, dir., review of *After Truth: Disinformation and the Cost of Fake News*," HBO, Abstract Production, https://www.hbo.com/documentaries/after-truth-disinformation-and-the-cost-of-fake-news.

8. Marc Fisher, John Woodrow Cox, and Peter Hermann, "Pizzagate: From Rumor, to Hashtag, to Gunfire in D.C.," *Washington Post*, December 7, 2016, https://www.washingtonpost.com/local/pizzagate-from-rumor-to-hashtag-to-gunfire-in-dc/2016/12/06/4c7def50-bbd4-11e6-94ac-3d324840106c_story.html.
9. Cecilia Kang, "Fake News Onslaught Targets Pizzeria as Nest of Child-Trafficking," *New York Times*, November 11, 2016, https://www.nytimes.com/2016/11/21/technology/fact-check-this-pizzeria-is-not-a-child-trafficking-site.html.
10. Adam Goldman, "The Comet Ping Pong Gunman Answers Our Reporter's Questions," *New York Times,* December 7, 2016, https://www.nytimes.com/2016/12/07/us/edgar-welch-comet-pizza-fake-news.html.
11. Peter Hermann, "Man Who Set Fire at Comet Ping Pong Pizza Shop Sentenced to Four Years in Prison," *Washington Post,* April 23, 2020, https://www.washingtonpost.com/local/public-safety/man-who-set-fire-at-comet-ping-pong-pizza-shop-sentenced-to-four-years-in-prison/2020/04/23/2e107676-8496-11ea-a3eb-e9fc93160703_story.html.
12. Kevin Roose, "Following Falsehoods: A Reporter's Approach on QAnon," *New York Times*, October 3, 2020, Times Insider, https://www.nytimes.com/2020/10/03/insider/qanon-reporter.html.
13. Associated Press, "Fox News, Family of Slain DNC Staffer Seth Rich Settle Suit," April 23, 2021, https://apnews.com/article/election-2020-lawsuits-seth-rich-robbery-52fe8b0d4e8e6256e16fe7ba0251b4f6.

CHAPTER SEVENTEEN

1. "Free Speech Systems LLC. Web Media Kit," Inforwars.com, https://static.infowars.com/ads/mediakit_public.pdf.
2. Alex Seitz-Wald, "Alex Jones: Conspiracy Inc.," *Salon*, May 2, 2013, https://www.salon.com/2013/05/02/alex_jones_conspiracy_inc/.
3. Jim Rutenberg, "In Trump's Volleys, Echoes of Alex Jones's Conspiracy Theories," *New York Times*, February 20, 2017, Business, https://www.nytimes.com/2017/02/19/business/media/alex-jones-conspiracy-theories-donald-trump.html.
4. Manuel Roig-Franzia, "How Alex Jones, Conspiracy Theorist Extraordinaire, Got Donald Trump's Ear," *Washington Post*, November 17, 2016, Style, https://www.washingtonpost.com/lifestyle/style/how-alex-jones-conspiracy-theorist-extraordinaire-got-donald-trumps-ear/2016/11/17/583dc190-ab3e-11e6-8b45-f8e493f06fcd_story.html.
5. Tristan Hallman, "Conspiracy Theorist Alex Jones, Others Rally for Trump Outside GOP Convention," *Dallas Morning News*, July 18, 2016, https://www.dallasnews.com/news/politics/2016/07/18/conspiracy-theorist-alex-jones-others-rally-for-trump-outside-gop-convention/.
6. "Transcript: Donald Trump at the G.O.P. Convention," *New York Times*, July 22, 2016, U.S., https://www.nytimes.com/2016/07/22/us/politics/trump-transcript-rnc-address.html.
7. Glenn Kessler, "Analysis: Stephen Miller's Claim That 'Thousands of Americans Die Year After Year' from Illegal Immigration," *Washington Post*, February 21, 2019, https://www.washingtonpost.com/politics/2019/02/21/stephen-millers-claim-that-thousand-americans-die-year-after-year-illegal-immigration/.
8. Ben Kamisar, "Trump Brings Mothers of Children Killed by Undocumented Immigrants on Stage," *The Hill*, August 23, 2016, https://thehill.com/blogs/ballot-box/292431-trump-brings-mothers-of-children-killed-by-undocumented-immigrants-on-stage.

9. Michael Smith and Lydia Saad, "Economy Top Problem in a Crowded Field," Gallup News, December 19, 2016, https://news.gallup.com/poll/200105/economy-top-problem-crowded-field.aspx.
10. *Frontline*, "United States of Conspiracy," season 2020, episode 2, PBS, https://www.pbs.org/wgbh/frontline/film/united-states-of-conspiracy/.
11. Alan Rappeport, "Hillary Clinton Denounces the 'Alt-Right,' and the Alt-Right Is Thrilled," *New York Times*, August 26, 2016, U.S., https://www.nytimes.com/2016/08/27/us/politics/alt-right-reaction.html.
12. Shaya Tayefe Mohajer, "It Is Time to Stop Using the Term 'Alt Right,'" *Columbia Journalism Review*, August 14, 2017, https://www.cjr.org/criticism/alt-right-trump-charlottesville.php.
13. Rutenberg, "In Trump's Volleys, Echoes of Alex Jones's Conspiracy Theories."
14. Josh Owens, "I Worked for Alex Jones. I Regret It," *New York Times Magazine*, December 5, 2019, https://www.nytimes.com/2019/12/05/magazine/alex-jones-infowars.html.
15. John Oliver, *Last Week Tonight with John Oliver*, "Alex Jones," HBO, July 30, 2017, https://www.youtube.com/watch?v=WyGq6cjcc3Q.
16. Alexander Zaitchik, "Glenn Beck's Shtick? Alex Jones Got There First," *Rolling Stone*, March 4, 2011, https://www.rollingstone.com/culture/culture-news/glenn-becks-shtick-alex-jones-got-there-first-174990/.
17. *Frontline*, "United States of Conspiracy."

CHAPTER EIGHTEEN

1. Connecticut Secretary of State 2016 Presidential Election Official Results database, https://ctemspublic.pcctg.net/#/selectTown.
2. John Voket, "Sandy Hook Parent Frustrated to Learn About Trump Letter," *Newtown Bee*, February 3, 2017, https://www.newtownbee.com/02032017/sandy-hook-parent-frustrated-to-learn-about-trump-letter/.
3. Jim Rutenberg, "In Trump's Volleys, Echoes of Alex Jones's Conspiracy Theories," *New York Times*, February 20, 2017, Business, https://www.nytimes.com/2017/02/19/business/media/alex-jones-conspiracy-theories-donald-trump.html.
4. YouTube Help, "Community Guidelines Strike Basics," n.d., Google.com, accessed August 2, 2021, https://support.google.com/youtube/answer/2802032?hl=en.
5. World Economic Forum, "Davos 2016—Press Conference: The Refugee and Migration Crisis," YouTube, 6:15, https://www.youtube.com/watch?v=dApJnaaNwQY.
6. James Kirchick, "The Disgusting Breitbart Smear Campaign Against the Immigrant Owner of Chobani," *Daily Beast*, September 2, 2016, https://www.thedailybeast.com/the-disgusting-breitbart-smear-campaign-against-the-immigrant-owner-of-chobani.
7. Christine Hauser, "Alex Jones Retracts Chobani Claims to Resolve Lawsuit," *New York Times*, May 17, 2017, https://www.nytimes.com/2017/05/17/us/alex-jones-chobani-lawsuit.html.
8. Lenny Pozner, "Sandy Hook Dad: Expose, Shame Sandy Hook 'Hoaxer' Alex Jones in Public," Courant.com, June 14, 2017, https://www.courant.com/opinion/insight/hc-op-insight-pozner-debunk-alex-jones-0618-20170614-story.html.
9. Charlie Warzel, "Alex Jones Just Released a Father's Day Video to Sandy Hook Parents—but Didn't Apologize," *BuzzFeed News*, June 18, 2017, https://www.buzzfeednews.com/article/charliewarzel/alex-jones-just-released-a-fathers-day-video-to-sandy-hook.

10. Jeremy Fuster, "Alex Jones Sends Condolences to Sandy Hook Families Before Megan Kelly Interview," *The Wrap*, June 18, 2017, https://www.thewrap.com/alex-jones-sends-condolences-sandy-hook-families-megan-kelly-interview.
11. Megyn Kelly, "Alex Jones of 'Infowars,' Conspiracy Theories, and Trump Campaign," NBC News, June 19, 2017, https://www.youtube.com/watch?v=-HzOqZeX3Yk.

CHAPTER NINETEEN

1. John Pirro, "Away from the Spotlight, Sandy Hook Parent Battles Criminal Charges," *News-Times*, May 7, 2013, https://www.newstimes.com/policereports/article/Away-from-the-spotlight-Sandy-Hook-parent-4496775.php.
2. Blythe Bernhard, "Radio Host Protests 'Police State' in Ferguson," STLtoday.com, August 13, 2014, https://www.stltoday.com/news/local/crime-and-courts/radio-host-protests-police-state-in-ferguson/article_3bcf8feb-8e2e-5f9e-abad-36f3de24425e.html.
3. Jonathan Tilove, "In Travis County Custody Case, Jury Will Search for Real Alex Jones," *Austin American-Statesman*, April 18, 2018, https://www.statesman.com/news/20170418/exclusive-in-travis-county-custody-case-jury-will-search-for-real-alex-jones.
4. Ken Herman, "Herman: The Children in Alex Jones' Life—His, Sandy Hook's," *Austin American-Statesman*, April 19, 2017, https://www.statesman.com/news/20170419/herman-the-children-in-alex-jones-life—his-sandy-hooks.
5. Jonathan Tilove, "On the Stand, Alex Jones Testifies He Means What He Says on Infowars," *Austin American-Statesman*, April 20, 2017, https://www.statesman.com/news/20170420/on-the-stand-alex-jones-testifies-he-means-what-he-says-on-infowars.
6. Charlie Warzel, "Alex Jones and the Dark New Media Are on Trial in Texas," *BuzzFeed News*, April 19, 2017, https://www.buzzfeednews.com/article/charliewarzel/alex-jones-and-the-dark-new-media-are-on-trial-in-texas.
7. Jonathan Tilove, "Verdict: Ex-Wife of Alex Jones Wins Joint Custody After Bitter Trial," *Austin American-Statesman*, April 28, 2017, https://www.statesman.com/news/20170428/verdict-ex-wife-of-alex-jones-wins-joint-custody-after-bitter-trial.

CHAPTER TWENTY

1. Reeves Wiedeman, "The Sandy Hook Hoax," *New York*, September 5, 2016, Intelligencer, https://nymag.com/intelligencer/2016/09/the-sandy-hook-hoax.html.
2. Elizabeth Williamson, "A Notorious Sandy Hook Tormentor Is Arrested in Florida," *New York Times*, January 27, 2020, Politics, https://www.nytimes.com/2020/01/27/us/politics/sandy-hook-hoaxer-arrest.html.
3. Liz Wahl, "Truthers: When Conspiracy Meets Reality," Newsy, December 13, 2017, YouTube, https://www.youtube.com/watch?v=NczrIkBGD3M.
4. Mark Hill and Lenny Pozner, "6 Horrifying Realities of Dealing with Sandy Hook 'Truthers,'" *Cracked*, April 26, 2016, https://www.cracked.com/personal-experiences-2232-my-son-died-at-sandy-hook-conspiracy-nuts-think-im-lying.html.
5. Tom Jackman, "Father of Slain Journalist Alison Parker Takes on YouTube over Alleged Refusal to Remove Graphic Videos," *Washington Post*, February 20, 2020, https://www.washingtonpost.com/crime-law/2020/02/20/father-slain-journalist-alison-parker-takes-youtube-over-refusal-remove-graphic-videos/.

6. Troy Griggs, Jasmine C. Lee, Morrigan McCarthy, Brent Murray, Alicia Parlapiano, Joe Ward, and Josh Williams, "Photos from the 'March for Our Lives' Protests Around the World," *New York Times*, March 24, 2018, U.S., https://www.nytimes.com/interactive/2018/03/24/us/photos-march-for-lives.html.
7. Mark D. Wilson, "Man Sues Alex Jones, InfoWars for $1 Million, Says They Defamed Him in Fla. Shooting," *Austin American-Statesman*, April 3, 2018, https://www.statesman.com/news/20180403/man-sues-alex-jones-infowars-for-1-million-says-they-defamed-him-in-fla-shooting.
8. Leonard Pozner and Veronique De La Rosa v. Alex E. Jones, Infowars LLC, and Free Speech Systems LLC, Defendants' Motion to Dismiss Under the Texas Citizens' Participation Act, 345th Judicial District Court in Travis County, Texas, April 16, 2018, p. 16, http://infowarslawsuit.com/wp-content/uploads/2018/10/2018-06-26-Ds-First-Amended-Answer-with-exhibits.final_.compacted.pdf.
9. Neil Heslin v. Alex E. Jones, Infowars, LLC, Free Speech Systems, LLC, and Owen Shroyer, Plantiff's Original Petition and Request for Disclosure, 261st District Court of Travis County, Texas, April 16, 2018, p. 3, https://infowarslawsuit.com/wp-content/uploads/2018/11/2018-04-16-Heslin-Original-Petition-file-stamped.pdf.

CHAPTER TWENTY-ONE

1. Sonia Smith, "When Lemurs Attack: East Texas Edition," *Texas Monthly*, December 11, 2012, https://www.texasmonthly.com/articles/when-lemurs-attack-east-texas-edition/.
2. Christine Hauser, "Alex Jones Retracts Chobani Claims to Resolve Lawsuit," *New York Times*, May 17, 2017, https://www.nytimes.com/2017/05/17/us/alex-jones-chobani-lawsuit.html.
3. Law.com, "Bad News for Alex Jones: Defamation Case Over Charlottesville Theories Gets Green Light," Yahoo.com, March 29, 2019, https://www.yahoo.com/now/bad-news-alex-jones-defamation-120103063.html.
4. Neil Heslin vs. Alex E. Jones, Infowars LLC; Free Speech Systems LLC; and Owen Shroyer, Travis County District Court, Texas, D-1-GN-18-001835, Defendants' First Amended Response to Plaintiff's Motion for Sanctions and Motion for Expedited Discovery and Defendants' Motion for Sanctions, August 27, 2018, https://infowarslawsuit.com/wp-content/uploads/2018/11/2018-08-28-Ds-First-Amended-Response-to-Ps-M-for-Sanctions-and-M-for-Expedited-Discovery-and-ds-M-for-Sanctions.pdf; Declaration of David Jones, p. 4; Bailen letter to Google, Exhibit B-88.
5. Jessica Rosenworcel, "First Things First: Is the Press Still Free?" Remarks at a meeting of the National Association of Broadcasters, Las Vegas, Nevada, April 9, 2018, https://www.fcc.gov/document/commissioner-jessica-rosenworcel-nab-remarks.
6. Elizabeth Williamson, "Biden Nominates Rosenworcel as F.C.C.'s First Female Leader," *New York Times*, October 26, 2021, https://www.nytimes.com/2021/10/26/us/politics/jessica-rosenworcel-fcc.html.

CHAPTER TWENTY-TWO

1. Vice Media, "The Rise of the Crisis Actor Conspiracy Movement," YouTube, 2018, https://www.youtube.com/watch?v=To91BJGKr5I.
2. Serge F. Kovaleski, "The Day the Pastor Was Away and Evil Came Barging into His Church," *New York Times*, November 22, 2017, U.S., https://www.nytimes.com/2017/11/22/us/pastor-frank-pomeroy-sutherland-springs.html.

3. U.S. Attorney's Office for the Western District of Texas, "Lockhart Man Charged with Being a Convicted Felon in Possession of a Firearm," Department of Justice press release, May 23, 2018, https://www.justice.gov/usao-wdtx/pr/lockhart-man-charged-being-convicted-felon-possession-firearm.
4. Associated Press, "Judge Denies Bail to Man Who Said Church Shooting Was Hoax," May 30, 2018, https://apnews.com/article/5f62757020f549b3a58c2ce55ba2111c.
5. Guillermo Contreras, "Judge Sends Sutherland Springs Conspiracy Theorist to Psych Ward," *San Antonio Express-News*, November 16, 2018, https://www.mysanantonio.com/news/local/article/Judge-sends-Sutherland-Springs-conspiracy-13395747.php; Guillermo Contreras, "Held 16 Months, Accused Harasser of Sutherland Springs Survivors Released on Bond," *San Antonio Express-News*, October 17, 2019, https://www.expressnews.com/news/local/article/Held-16-months-accused-harasser-of-Sutherland-14542279.php.
6. Erica Lafferty, et al. v. Alex Emric Jones, et al., case no. UWY-CV18-6046436-S, Connecticut Superior Court, http://civilinquiry.jud.ct.gov/CaseDetail/PublicCaseDetail.aspx?DocketNo=UWYCV186046436S.
7. Democratic National Convention, "FULL: Erica Smegielski—Mom Died at Sandy Hook—Democratic National Convention," ABC 15, July 27, 2016, https://www.youtube.com/watch?v=3MNvp9Tf8_k.
8. Rebecca Webber, "Our Mother Was the Sandy Hook Principal," *Glamour*, February 22, 2013, https://www.glamour.com/story/our-mother-was-the-sandy-hook-principal-glamour-march-2013.
9. Erica Lafferty, "Mr. Trump, Denounce Alex Jones: Sandy Hook Principal's Daughter," *USA Today*, November 25, 2016, https://www.usatoday.com/story/opinion/2016/11/25/donald-trump-sandy-hook-alex-jones-column/94335420/.
10. Francine Wheeler, "Weekly Address: Sandy Hook Victim's Mother Calls for Commonsense Gun Responsibility Reforms," April 13, 2013, https://obamawhitehouse.archives.gov/the-press-office/2013/04/13/weekly-address-sandy-hook-victim-s-mother-calls-commonsense-gun-responsi.
11. Elizabeth Williamson, "Truth in a Post-Truth Era: Sandy Hook Families Sue Alex Jones, Conspiracy Theorist," *New York Times*, May 23, 2018, U.S., https://www.nytimes.com/2018/05/23/us/politics/alex-jones-trump-sandy-hook.html.
12. David Edward McCraw, *Truth in Our Times: Inside the Fight for Press Freedom in the Age of Alternative Facts* (New York: All Points Books, 2019), 259–62.

CHAPTER TWENTY-THREE

1. Stuart Tomlinson, "Judge Increases Bail to $100,000 for Kevin Purfield, the Man Accused of Harassing Families of Victims of Colorado Mass Theater Shooting," *Oregonian*/OregonLive, April 11, 2013, https://www.oregonlive.com/portland/2013/04/judge_increases_bail_to_100000.html.
2. Stuart Tomlinson, "Man Accused of Bomb Threats Sentenced to Time Served After Medication Returns Competence," *Oregonian*/OregonLive, March 16, 2015, https://www.oregonlive.com/portland/2015/03/man_accused_of_bomb_threats_at.html.
3. Scott Travis, "For Parkland Shooting Victim Alaina Petty, the AR-15 Was Her Favorite Gun," *South Florida Sun-Sentinel*, July 28, 2018, https://www.sun-sentinel.com/local/broward/parkland/florida-school-shooting/fl-ryan-petty-ar15-alaina-20180728-story.html.

4. Nicholas Confessore, "Cambridge Analytica and Facebook: The Scandal and the Fallout So Far," *New York Times*, April 4, 2018, https://www.nytimes.com/2018/04/04/us/politics/cambridge-analytica-scandal-fallout.html.
5. "Mark Zuckerberg Testimony: Senators Question Facebook's Commitment to Privacy," *New York Times*, April 10, 2018, https://www.nytimes.com/2018/04/10/us/politics/mark-zuckerberg-testimony.html.
6. "Data Protection: Rules for the Protection of Personal Data Inside and Outside the EU," European Commission, https://ec.europa.eu/info/law/law-topic/data-protection_en.
7. Karen Hao, "How Facebook Got Addicted to Spreading Misinformation," *MIT Technology Review*, March 11, 2021, https://www.technologyreview.com/2021/03/11/1020600/facebook-responsible-ai-misinformation/.
8. Andrew Perrin, "Americans Are Changing Their Relationship with Facebook," Pew Research Center, September 5, 2018, https://www.pewresearch.org/fact-tank/2018/09/05/americans-are-changing-their-relationship-with-facebook/.
9. Kara Swisher, "Zuckerberg: The Recode Interview," *Vox*, October 8, 2018, https://www.vox.com/2018/7/18/17575156/mark-zuckerberg-interview-facebook-recode-kara-swisher.
10. Monika Bickert, "Removing Holocaust Denial Content," Meta-Facebook, October 12, 2020, https://about.fb.com/news/2020/10/removing-holocaust-denial-content.
11. Leonard Pozner and Veronique De La Rosa, "An Open Letter to Mark Zuckerberg: Our Child Died at Sandy Hook—Why Let Facebook Lies Hurt Us Even More?," *Guardian*, July 25, 2018, https://www.theguardian.com/commentisfree/2018/jul/25/mark-zuckerberg-facebook-sandy-hook-parents-open-letter.
12. Kurt Wagner and Rani Molla, "Facebook's User Growth Has Hit a Wall," *Vox*, July 25, 2018, https://www.vox.com/2018/7/25/17614426/facebook-fb-earnings-q2-2018-user-growth-troubles.
13. Janet Burns, "Sandy Hook Parents Decry Facebook's Weak Moderation in Open Letter to Mark Zuckerberg," *Forbes*, July 27, 2018, https://www.forbes.com/sites/janetwburns/2018/07/27/sandy-hook-parents-decry-facebooks-weak-moderation-in-open-letter-to-mark-zuckerberg.
14. Casey Newton, "Facebook's Forecast for the Future Looks Suddenly Bleak," *The Verge*, July 26, 2018, https://www.theverge.com/2018/7/26/17615330/facebook-earnings-forecast-user-growth-revenue.
15. Raz Robinson, "Sandy Hook Parents Call Out Zuckerberg and Facebook in a Powerful Open Letter," Yahoo Entertainment, July 26, 2018, https://www.yahoo.com/entertainment/sandy-hook-parents-call-zuckerberg-174832904.html.
16. Rupert Neate, "Over $119bn Wiped Off Facebook's Market Cap After Growth Shock," *Guardian*, July 26, 2018, https://www.theguardian.com/technology/2018/jul/26/facebook-market-cap-falls-109bn-dollars-after-growth-shock.
17. Kevin Roose, "Facebook and YouTube Give Alex Jones a Wrist Slap," *New York Times*, July 27, 2018, https://www.nytimes.com/2018/07/27/technology/alex-jones-facebook-youtube.html.
18. Elizabeth Williamson, "In Alex Jones Lawsuit, Lawyers Spar over an Online Broadcast on Sandy Hook," *New York Times*, August 1, 2018, https://www.nytimes.com/2018/08/01/us/politics/infowars-sandy-hook-alex-jones.html.
19. Elizabeth Williamson, "Lawyers for Neo-Nazi to Defend Alex Jones in Sandy Hook Case," *New York Times*, July 2, 2018, https://www.nytimes.com/2018/07/02/us/politics/sandy-hook-alex-jones-lawyers.html.

20. Luke O'Brien, "Alex Jones' Lawyer Violated Legal Ethics by Soliciting Porn Bribes. Just How Dirty Is Marc Randazza?," *HuffPost*, December 27, 2018, https://www.huffpost.com/entry/alex-jones-lawyer-marc-randazza_n_5c1c283ae4b08aaf7a86b9e4.
21. Williamson, "Lawyers for Neo-Nazi to Defend Alex Jones."
22. Michael Kunzelman, "Target of Online Trolls Suing Neo-Nazi Website's Publisher," *Los Angeles Times*, April 18, 2017, https://www.latimes.com/nation/ct-daily-stormer-intimidation-lawsuit-20170418-story.html.
23. Matt Pearce, "Daily Stormer Asks to Dismiss Trolling Lawsuit, Says Neo-Nazi Memes Posed 'No True Threat' to Jewish Woman," *Los Angeles Times*, December 1, 2017, https://www.latimes.com/nation/la-na-daily-stormer-lawsuit-20171201-story.html.
24. Elizabeth Williamson, "Alex Jones, Pursued over Infowars Falsehoods, Faces a Legal Crossroads," *New York Times*, July 31, 2018, https://www.nytimes.com/2018/07/31/us/politics/alex-jones-defamation-suit-sandy-hook.html.
25. "Mark C. Enoch," Glast, Phillips & Murray law firm, http://gpm-law.com/enoch.html.
26. Leonard Pozner and Veronique De La Rosa v. Alex E. Jones, Infowars LLC, and Free Speech Systems LLC, Plaintiffs' Motion to Dismiss Under the Texas Citizens Participation Act, District Court, Travis County, Texas, July 25, 2018, Exhibit A, Affidavit of Fred Zipp, p. 23.
27. Fairleigh Dickinson University's Public Mind Poll of 809 registered voters was conducted nationally by landline and cell phone from April 13 through April 17, 2016, and has a margin of error of +/–4.1 percentage points. Fairleigh Dickinson University, "Fairleigh Dickinson University's PublicMind Poll Finds Trump Supporters More Conspiracy-Minded than Other Republicans," May 4, 2016, http://publicmind.fdu.edu/2016/160504/final.pdf.
28. Williamson, "In Alex Jones Lawsuit, Lawyers Spar over an Online Broadcast."

CHAPTER TWENTY-FOUR

1. David Leffler, "Alex Jones Is Mad as Hell," *Austin Monthly*, October 2020, https://www.austinmonthly.com/alex-jones-is-mad-as-hell/.
2. Harmon Leon, "The Alex Jones Origin Story: On Austin Public Access TV, His Act Was Never an Act," *Observer*, April 17, 2019, https://observer.com/2019/04/alex-jones-austin-public-access-tv-origin-story.
3. Charlie Warzel, "Alex Jones Just Can't Help Himself," *BuzzFeed News*, May 3, 2017, https://www.buzzfeednews.com/article/charliewarzel/alex-jones-will-never-stop-being-alex-jones.
4. Nate Blakeslee, "Alex Jones Is About to Explode," *Texas Monthly*, March 2010, https://www.texasmonthly.com/news-politics/alex-jones-is-about-to-explode.
5. Leffler, "Alex Jones Is Mad as Hell."
6. Leon, "The Alex Jones Origin Story."
7. *Inside Edition*, "Ex-Wife of 'Infowars' Host Alex Jones: 'He's a Really Unhappy, Disturbed Person,'" YouTube, May 30, 2017, https://www.youtube.com/watch?v=FV8OWkqDQjU.
8. *Inside Edition*, "Ex-Wife of 'Infowars' Host Alex Jones: 'He's a Really Unhappy, Disturbed Person.'"
9. Alexander Zaitchik, "Meet Alex Jones," *Rolling Stone*, March 2, 2011, https://www.rollingstone.com/culture/culture-news/meet-alex-jones-175845.
10. Elizabeth Williamson and Emily Steel, "Conspiracy Theories Made Alex Jones Very Rich. They May Bring Him Down," *New York Times*, September 7, 2018, https://www.nytimes.com/2018/09/07/us/politics/alex-jones-business-infowars-conspiracy.html.
11. Williamson and Steel, "Conspiracy Theories Made Alex Jones Very Rich."

12. Elizabeth Williamson, "Three Hours Up Close with Alex Jones of Infowars," *New York Times*, September 9, 2018, https://www.nytimes.com/2018/09/09/insider/alex-jones-infowars-media-interview.html.
13. Callum Borchers, "Alex Jones Is a Narcissist, a Witness Testifies. And He's Undermining His Own Attorneys," *Washington Post,* April 20, 2017, https://www.washingtonpost.com/news/the-fix/wp/2017/04/20/alex-jones-is-a-narcissist-a-witness-says-and-possibly-the-worst-client-ever.
14. Jonathan Tilove, "Wednesday Wrap-Up: Alex Jones Child Custody Case Awaits Closing Arguments," *Austin American-Statesman*, April 27, 2017, https://www.statesman.com/news/20170427/wednesday-wrap-up-alex-jones-child-custody-case-awaits-closing-arguments.
15. Allan Smith, "'It's Performance Art': Lawyer for Alex Jones Says InfoWars Founder Is 'Playing a Character,'" *Business Insider*, April 17, 2017, https://www.businessinsider.com/lawyer-alex-jones-infowars-playing-character-acting-2017-4.
16. Jonathan Tilove, "On the Stand, Alex Jones Testifies He Means What He Says on Infowars," *Austin American-Statesman*, April 20, 2017, https://www.statesman.com/news/20170420/on-the-stand-alex-jones-testifies-he-means-what-he-says-on-infowars.
17. Williamson and Steel, "Conspiracy Theories Made Alex Jones Very Rich."
18. Rex Jones, "Alex Jones' Son Challenges David Hogg to a Gun Debate," posted July 31, 2018, video, 2:10, https://www.youtube.com/watch?v=FSB0rnIlfL4.
19. Elizabeth Williamson, "Alex Jones, Pursued over Infowars Falsehoods, Faces a Legal Crossroads," *New York Times*, July 31, 2018, https://www.nytimes.com/2018/07/31/us/politics/alex-jones-defamation-suit-sandy-hook.html.
20. Benjamin Fearnow, "Alex Jones Threatens to Shoot 'Pedophile' Robert Mueller, Accuses Zuckerberg of Facebook 'Shadow Ban,'" *Newsweek*, July 24, 2018, https://www.newsweek.com/alex-jones-threatens-shoot-pedophile-robert-mueller-accuses-zuckerberg-1038500.
21. Alex Dobuzinskis, "YouTube Removes Videos from Conspiracy Theorist Alex Jones: Infowars Website," Reuters, July 25, 2018, https://www.reuters.com/article/us-youtube-infowars/youtube-removes-videos-from-conspiracy-theorist-alex-jones-infowars-website-idUSKBN1KF36Z; Williamson, "Alex Jones, Pursued over Infowars Falsehoods."
22. Charlie Warzel, "Alex Jones Just Released a Father's Day Video to Sandy Hook Parents—But Didn't Apologize," *BuzzFeed News*, June 18, 2017, https://www.buzzfeednews.com/article/charliewarzel/alex-jones-just-released-a-fathers-day-video-to-sandy-hook.
23. Williamson, "Three Hours Up Close with Alex Jones."
24. Williamson, "Three Hours Up Close with Alex Jones."
25. KC Baker and Adam Carlson, "It Took Years for Sandy Hook Dad to Take Down Christmas Tree Son Helped Put Up Before Massacre," *People*, December 15, 2017, https://people.com/crime/sandy-hook-victim-jesse-lewis-dad-christmas-5-years-later.
26. Elizabeth Williamson, "Truth in a Post-Truth Era: Sandy Hook Families Sue Alex Jones, Conspiracy Theorist," *New York Times*, May 23, 2018, U.S., https://www.nytimes.com/2018/05/23/us/politics/alex-jones-trump-sandy-hook.html.

CHAPTER TWENTY-FIVE

1. Jack Nicas, "Alex Jones Said Bans Would Strengthen Him. He Was Wrong," *New York Times*, September 4, 2018, https://www.nytimes.com/2018/09/04/technology/alex-jones-infowars-bans-traffic.html.

2. Jane Coaston, "YouTube, Facebook, and Apple's Ban on Alex Jones, Explained," *Vox*, August 6, 2018, https://www.vox.com/2018/8/6/17655658/alex-jones-facebook-youtube-conspiracy-theories.
3. Sarah Wells, "Here Are the Platforms That Have Banned Infowars So Far," *TechCrunch*, August 8, 2018, https://techcrunch.com/2018/08/08/all-the-platforms-that-have-banned-infowars.
4. Avie Schneider, "Twitter Bans Alex Jones and InfoWars; Cites Abusive Behavior," NPR, September 6, 2018, https://www.npr.org/2018/09/06/645352618/twitter-bans-alex-jones-and-infowars-cites-abusive-behavior.
5. Coaston, "YouTube, Facebook, and Apple's Ban."
6. Casey Newton, "How Alex Jones Lost His Info War," *The Verge*, August 7, 2018, https://www.theverge.com/2018/8/7/17659026/alex-jones-deplatformed-misinformation-hate-speech-apple-facebook-youtube.
7. Brian Stelter, "Reliable Sources: Alex Jones Has Been 'Deplatformed.' Now What?," CNN, August 7, 2018, https://money.cnn.com/2018/08/07/media/reliable-sources-08-06-18/index.html.
8. Jack Nicas, "Alex Jones and Infowars Content Is Removed from Apple, Facebook and YouTube," *New York Times*, August 6, 2018, https://www.nytimes.com/2018/08/06/technology/infowars-alex-jones-apple-facebook-spotify.html.
9. Ryan Mac and Craig Silverman, "'Mark Changed the Rules': How Facebook Went Easy on Alex Jones and Other Right-Wing Figures," *BuzzFeed News*, February 21, 2021, https://www.buzzfeednews.com/article/ryanmac/mark-zuckerberg-joel-kaplan-facebook-alex-jones.
10. Sapna Maheshwari and John Herrman, "This Company Keeps Lies About Sandy Hook on the Web," *New York Times*, August 13, 2018, https://www.nytimes.com/2018/08/13/business/media/sandy-hook-conspiracies-leonard-pozner.html.
11. Eric Johnson, "The Tech Industry Needs to Reckon with the Dark Side of Advertising, WordPress CEO Matt Mullenweg Says," *Vox*, August 2, 2018, https://www.vox.com/2018/8/2/17641412/matt-mullenweg-wordpress-media-privacy-advertising-data-reckoning-kara-swisher-decode-podcast.
12. Nicholas Fandos, "Alex Jones Takes His Show to the Capitol, Even Tussling with a Senator," *New York Times*, September 5, 2018, https://www.nytimes.com/2018/09/05/us/politics/alex-jones-marco-rubio-inforwars.html.
13. Elizabeth Williamson and Emily Steel, "Conspiracy Theories Made Alex Jones Very Rich. They May Bring Him Down," *New York Times*, September 7, 2018, https://www.nytimes.com/2018/09/07/us/politics/alex-jones-business-infowars-conspiracy.html.
14. Elizabeth Williamson, "Judge Rules Against Alex Jones and Infowars in Sandy Hook Lawsuit," *New York Times*, August 30, 2018, https://www.nytimes.com/2018/08/30/us/politics/alex-jones-infowars-sandy-hook-lawsuit.html.
15. Nicas, "Alex Jones Said Bans Would Strengthen Him."
16. State of Connecticut, "Final Report of the Sandy Hook Advisory Commission," Dedication, March 6, 2015, https://portal.ct.gov/Malloy-Archive/Working-Groups/Sandy-Hook-Advisory-Commission.
17. Nicholas Rondinone, "Autopsy Confirms Sandy Hook Father Jeremy Richman's Death as Suicide," *Hartford Courant*, March 27, 2019, https://www.courant.com/breaking-news/hc-br-jeremy-richman-death-suicide-20190327-jwbzqmkjjfh7bjfdheii34wpkq-story.html.

18. Choose Love Movement, http://chooselovemovement.org.
19. Nicholas Rondinone, "Wife of Sandy Hook Father Jeremy Richman Speaks Out for the First Time Since His Suicide, Saying He 'Succumbed to the Grief That He Could Not Escape,'" *Hartford Courant*, March 28, 2019, https://www.courant.com/breaking-news/hc-br-jennifer-hensel-jeremy-richman-fundraiser-note-20190328-lfdw4wsttjfs5igybbdgyjrlnu-story.html.
20. Elizabeth Williamson, "Two Sides of Ted Cruz: Tort Reformer and Personal Injury Lawyer," *New York Times*, January 20, 2016, https://www.nytimes.com/2016/01/21/opinion/two-sides-of-ted-cruz-tort-reformer-and-personal-injury-lawyer.html.
21. Mike McIntyre, "Ted Cruz Didn't Report Goldman Sachs Loan in a Senate Race," *New York Times*, January 13, 2016, https://www.nytimes.com/2016/01/14/us/politics/ted-cruz-wall-street-loan-senate-bid-2012.html.
22. Maggie Haberman, "Donald Trump Accuses Ted Cruz's Father of Associating with Kennedy Assassin," *New York Times*, May 3, 2016, https://www.nytimes.com/politics/first-draft/2016/05/03/donald-trump-ted-cruz-father-jfk.
23. BBC News, "All the Things Ted Cruz Has Said About Trump," September 23, 2016, https://www.bbc.com/news/election-us-2016-37457578.
24. Alexandra Hutzler, "Ted Cruz Defends Conspiracy Theorist Alex Jones, Asks Who Made Facebook the 'Arbiter of Political Speech,'" *Newsweek*, July 28, 2018, https://www.newsweek.com/ted-cruz-defends-alex-jones-over-social-media-suspensions-1046604.
25. Ted Cruz, Twitter post, July 28, 2018, 10:05 a.m., https://mobile.twitter.com/tedcruz/status/1023207746454384642.
26. Anne Allred, Dori Olmos, and Sam Clancy, "Photo Shows Hawley Giving Fist Pump to Trump Supporters Before Capitol Violence," KSDK-TV, January 7, 2021, https://www.ksdk.com/article/news/politics/national-politics/senator-josh-hawley-photo-capitol-trump-supporters/63-3b5d7611-9d07-41bd-8113-a9ccf74ebb68.

CHAPTER TWENTY-SIX

1. Leonard Pozner v. James Fetzer et al., Dane County Circuit Court, Wisconsin, 2018, CV003122.
2. Alex Jones, "Theories About September 11th: American Scholars Symposium," filmed June 25, 2006, at 9/11 and the Neocon Agenda conference, https://www.c-span.org/video/?193155-1/september-11th-terrorist-attacks.
3. Mike Mosedale, "The Man Who Thought He Knew Too Much," *City Pages*, June 28, 2006, https://web.archive.org/web/20130927054545/http://www.citypages.com/2006-06-28/news/the-man-who-thought-he-knew-too-much.
4. Richard Grenier, "On the Trail of America's Paranoid Class: Oliver Stone's *JFK*," *National Interest* 27 (1992): 76–84, https://www.jstor.org/stable/42896811.
5. Mosedale, "The Man Who Thought He Knew Too Much."
6. John Gravois, "A Theory That Just Won't Die: Across America a Small but Fanatical Cadre of Conspiracy-Minded Academics Believe the U.S. Government Engineered 9/11. This Is Their Story," *National Post* (formerly *Financial Post*, Canada), July 28, 2006.
7. In an April 12, 2004, letter from the university's vice-chancellor for academic administration, obtained through a public records request, the university determined that Fetzer's "continued pursuit of a romantic relationship" with a coworker, including repeated appearances in the person's office and "insistence upon a 'closed door relationship,' constitutes verbal and

physical conduct of a sexual nature that has created an intimidating, hostile, and offensive work environment." Fetzer signed a termination agreement after he returned to campus in fall 2005 and left for good in late May 2006.

8. Public Justice, "2020 Trial Lawyer of the Year Finalist: Pozner v. Fetzer," August 10, 2020, video, 2:30, https://www.youtube.com/watch?v=Al85Rb9kbmY.
9. Chris Rickert, "Judge Finds Sandy Hook Conspiracy Theorist from Oregon in Contempt of Court," *Wisconsin State Journal*, September 14, 2019, https://madison.com/wsj/news/local/crime-and-courts/judge-finds-sandy-hook-conspiracy-theorist-from-oregon-in-contempt-of-court/article_d20808de-65d0-5b81-b3cc-fe20d31bbf28.html.

CHAPTER TWENTY-SEVEN

1. Kaster Lynch Farrar & Ball LLP, YouTube channel, https://www.youtube.com/channel/UCeeCy2sW9BRXjlfsIOA5DgA/videos.
2. Kaster Lynch Farrar & Ball LLP, "Alex Jones / Sandy Hook Video Deposition, Part I," videotaped deposition, March 2019, https://www.youtube.com/watch?v=I7siWJ86g40. Bankston's full exchange with Jones referenced here begins at the 1:25 mark and continues through 12:26.
3. Kaster Lynch Farrar & Ball LLP, "Alex Jones / Sandy Hook Video Deposition, Part II," videotaped deposition, March 2019, https://www.youtube.com/watch?v=XES-AydpIoc.
4. *New York Times*, "Coronavirus in the U.S.: Latest Map and Case Count," accessed September 30, 2021, https://www.nytimes.com/interactive/2021/us/covid-cases.html.
5. Howard Koplowitz, "Trump Booed over Vaccines in Alabama: White House Offers Support, Alex Jones Calls Him 'Dumbass,'" AL.com, August 23, 2021, https://www.al.com/politics/2021/08/trump-booed-over-vaccines-in-alabama-white-house-offers-support-alex-jones-calls-him-dumbass.html.
6. Vanessa Romo, "Poison Control Centers Are Fielding a Surge of Ivermectin Overdose Calls," NPR, September 4, 2021, https://www.npr.org/sections/coronavirus-live-updates/2021/09/04/1034217306/ivermectin-overdose-exposure-cases-poison-control-centers.
7. Brett Bachman, "Alex Jones Gobbles Ivermectin On-Air During Bizarre Rant: 'You Think I'm Easy to Kill?,'" *Salon*, September 4, 2021, https://www.salon.com/2021/09/04/alex-jones-gobbles-ivermectin-on-air-during-bizarre-rant-you-think-im-easy-to-kill.
8. Kevin Roose, "How a Viral Video Bent Reality," *New York Times*, September 8, 2021, https://www.nytimes.com/2021/09/08/technology/loose-change-9-11-video.html.
9. Michael Wines and Nick Corasaniti, "Arizona Vote Review 'Made Up the Numbers,' Election Experts Say," *New York Times*, October 1, 2021, https://www.nytimes.com/2021/10/01/us/arizona-election-review.html.
10. Center for Countering Digital Hate, "The Disinformation Dozen," March 2021, https://www.counterhate.com/disinformationdozen.
11. Davey Alba, "YouTube Bans All Anti-Vaccine Misinformation," *New York Times*, September 29, 2021, https://www.nytimes.com/2021/09/29/technology/youtube-anti-vaxx-ban.html.
12. Elizabeth Williamson, "Alex Jones's Podcasting Hecklers Face Their Foil's Downward Slide," *New York Times*, April 18, 2021, https://www.nytimes.com/2021/04/18/us/politics/alex-jones.html.
13. Elizabeth Williamson, "How a Small Town Silenced a Neo-Nazi Hate Campaign," *New York Times*, September 5, 2021, https://www.nytimes.com/2021/09/05/us/politics/nazi-whitefish-charlottesville.html.

14. Elizabeth Williamson, "How a Small Town Silenced a Neo-Nazi Hate Campaign."
15. Jeff Schultz, "How Much Data Is Created on the Internet Each Day?" *Micro Focus Blog*, August 6, 2019, https://blog.microfocus.com/how-much-data-is-created-on-the-internet-each-day/.
16. Paul Farhi, "Win Points for McCain!," *Washington Post*, August 7, 2008, https://www.washingtonpost.com/wp-dyn/content/article/2008/08/06/AR2008080603589.html.
17. Nicholas Confessore, Gabriel J. X. Dance, Richard Harris, and Mark Hansen, "The Follower Factory," *New York Times*, January 27, 2018, https://www.nytimes.com/interactive/2018/01/27/technology/social-media-bots.html.
18. Brian Niemietz, "'Hoax' Kills COVID-Denying Anti-Vaxxer Who Worked for the C.I.A.: Report," New York *Daily News*, August 30, 2021, https://www.nydailynews.com/news/national/ny-hoax-covid-19-denier-cia-anti-vax-20210831-leumibd7tfhnferzxn3jm3so7m-story.html.
19. Roger Berkowitz, "Why Arendt Matters: Revisiting *The Origins of Totalitarianism," Los Angeles Review of Books*, March 18, 2017, https://lareviewofbooks.org/article/arendt-matters-revisiting-origins-totalitarianism.
20. J. M. Berger, "Our Consensus Reality Has Shattered," *The Atlantic*, October 9, 2020, https://www.theatlantic.com/ideas/archive/2020/10/year-living-uncertainly/616648.
21. Liz Wahl, "Truthers: When Conspiracy Meets Reality," Newsy, December 13, 2017, YouTube, https://www.youtube.com/watch?v=NczrIkBGD3M.
22. Austin Vining and Sarah Matthews, "Overview of Anti-SLAPP Laws," Reporters Committee for Freedom of the Press, accessed October 3, 2021, https://www.rcfp.org/introduction-anti-slapp-guide.
23. Lafferty v. Jones, hearing transcript, Judicial District of Fairfield at Bridgeport, Connecticut, June 18, 2019, p. 50.
24. Graig Graziosi, "Alex Jones Says He Paid $500,000 for Rally That Led to Capitol Riot," *Independent*, January 11, 2021, https://www.independent.co.uk/news/world/americas/us-election-2020/alex-jones-paid-rally-capitol-riot-b1784603.html.
25. United States of America v. Samuel Christopher Montoya, U.S. District Court for the District of Columbia, Affidavit in Support of Criminal Complaint and Arrest Warrant, April 8, 2021, https://www.justice.gov/usao-dc/case-multi-defendant/file/1386671/download.
26. Michael S. Schmidt and Maggie Haberman, "The Lawyer Behind the Memo on How Trump Could Stay in Office," *New York Times*, October 2, 2021, https://www.nytimes.com/2021/10/02/us/politics/john-eastman-trump-memo.html.
27. Ben Sasse, "QAnon Is Destroying the GOP from Within," *The Atlantic*, January 16, 2021, https://www.theatlantic.com/ideas/archive/2021/01/conspiracy-theories-will-doom-republican-party/617707.

EPILOGUE

1. "Portland Man Charged with Stalking the Oregonian/OregonLive Editor," *Oregonian*/OregonLive, January 29, 2019, https://www.oregonlive.com/news/2019/01/man-charged-with-stalking-the-oregonianoregonlive-editor.html.

INDEX